ELECTRONICS for MOTOR MECHANICS

LES STACKPOOLE

MAL MORRISON

ALAN GREGORY

Longman Cheshire

Longman Cheshire Pty Limited
Longman House
Kings Gardens
95 Coventry Street
Melbourne 3205 Australia

Offices in Sydney, Brisbane, Adelaide and Perth. Associated companies,
branches and representatives throughout the world.

Copyright © Longman Cheshire Pty Ltd 1992
First published 1992

Edited by Louis de Vries
Designed by Judith Summerfeldt
Set in 10/12 Garamond
Produced by Longman Cheshire Pty Ltd
Printed in Malaysia — CLP

National Library of Australia
Cataloguing-in-Publication data

Stackpoole, Les
    Electronics for motor mechanics.

    ISBN 0 582 86821 1.

    1. Automobiles—Electronic equipment.
    2. Automobiles—Electronic equipment—
    maintenance and repair. I. Morrison, Mal. II.
    Gregory, Alan. III. Title.

629.2549

CONTENTS

7   ELECTRONIC COMPONENTS   72
—CONSTRUCTION AND
OPERATION

8   BRAKE SYSTEMS   97

9   ELECTRONICALLY   133
CONTROLLED VEHICLE
STEERING SYSTEMS

1

INTRODUCTION

The merging of electronics with automotive systems has, for the most part, been successful. The driver of the vehicle is only aware of the intrusion of electronics when the vehicle is fitted with digital instrument displays. If the driver could 'see' into the vehicle systems he would undoubtedly be surprised by the level of technology and the extent of applied electronics. Most of the advanced vehicle systems are now controlled by electronics. The success of these control systems is evident by the lack of knowledge that drivers have of their existence. Many electronic systems have proven to be so reliable that repairs are almost unknown.

The servicing technician, unlike the driver, is very aware of the presence of electronics within the vehicles. Servicing procedures have undergone revolutionary changes. The idea of 'try this' or 'tinkering' are concepts from the past. The technician servicing the modern vehicle must understand the system before attempting repairs. Lack of knowledge and poor servicing procedures will result in expensive mistakes and the service outlet may quickly lose business to competitors. One of the

mainstays from the past was to substitute known 'good' parts as a solution to a difficult diagnostic problem. This approach, if used with electronic systems, can quickly reduce a stock of good parts to unusable scrap. Substitution of parts is no longer a suitable alternative to knowledge and sound diagnostic procedures.

The servicing outlet, when faced with staff retraining problems, may decide to 'retreat' and specialise in vehicle areas or components not influenced by electronics. An examination of the chapters within this book should convince servicing personnel that, any course other than retraining will result in their employment being of limited tenure. Electronics have spread through almost every vehicle system, therefore avoiding electronics means restricting repair work to a rapidly decreasing population of older vehicles. The chapters within this book have been written as a confidence builder for the trained mechanic wanting to understand how system operation has changed. The control methods for several popular systems are comprehensively explained. Each system description is designed to bridge the gap between routine servicing procedures and the wider knowledge required for understanding of system operation. This book is not a repair manual but is intended to complement manufacturers' service and overhaul manuals, thereby providing the technician with improved comprehension of the repair task.

Before specialising in any specific chapter, a résumé of where electronics and electronic control systems have emerged will provide the reader with an overview of the material to be covered.

CHARGING SYSTEM

The charging system was one of the first vehicle systems to accept electronics. The early DC generator was replaced by an alternator in the early 1960s. The early alternators were 6 diode types with field current control provided by electro-mechanical relays and regulators. A

number of these units, with design improvements, continued to be fitted as original equipment for twenty years. Many manufacturers replaced the field relay by using a 9 diode type alternator. These units still tended to use electro-mechanical voltage control for a number of years but the electro-mechanical control units were superseded by solid state voltage regulators. These solid state units were originally mounted external to the alternator but as the components were reduced in size it was possible to locate the voltage regulator within the alternator casing. The modern alternator is a self-contained charging unit capable of sustained high amperage output and excellent reliability.

FUEL SYSTEMS

The carburettor and fuel pump underwent many changes over a period of years. The need

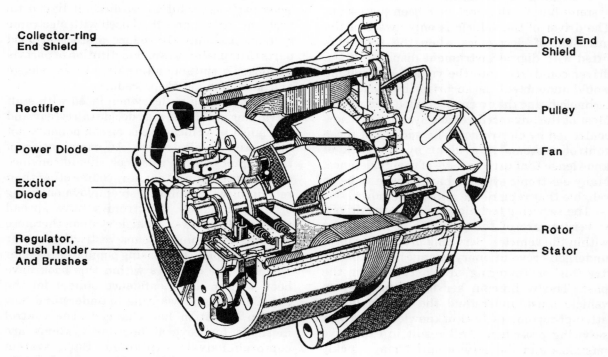

Collector-ring End Shield
Rectifier
Power Diode
Excitor Diode
Regulator, Brush Holder And Brushes
Drive End Shield
Pulley
Fan
Rotor
Stator

Figure 1-1 Typical alternator design.

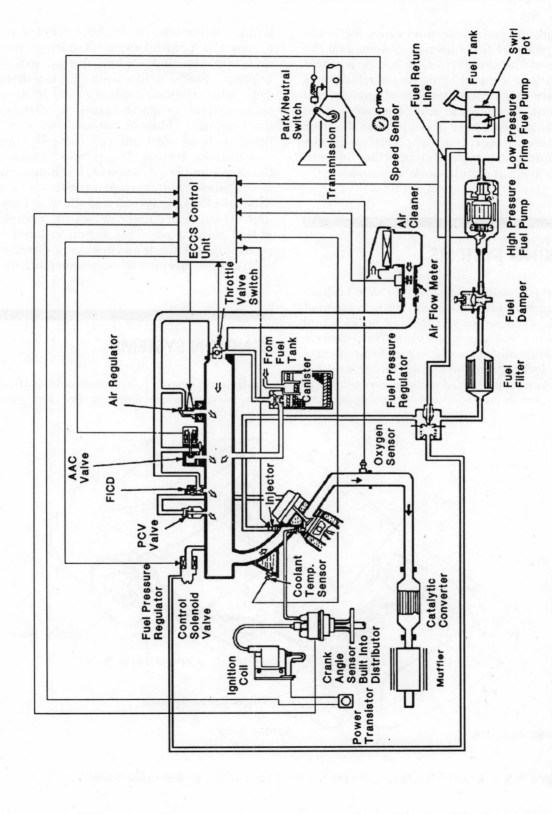

Figure 1-2 Electronic Fuel Injection.

to accurately monitor air/fuel ratios led to the development of fuel injection systems and the demise of the carburettor. A few carburettor applications are still being manufactured but these are very limited in sales volume. Several different mechanical and electronic fuel injection systems were developed but the clear leader for widespread application is the electronic pulse type fuel injection in both throttle body and multi-point variations.

BRAKING SYSTEMS

Most of the visible changes to vehicle braking systems were associated with the development of 'split' systems, power assisted braking and the use of disc brake wheel units. None of these worthwhile advances addressed the problem of brake locking and skidding under braking conditions. Experienced rally drivers tended to use a particular technique of 'pumping' the brakes under adverse braking conditions to lessen the possibility of skidding and to maintain steering control during extreme braking situations. Realising that few drivers could attain the skill levels of a rally driver, the brake system designers turned to electronics for a solution. Modern vehicles can now be fitted with an electronically controlled anti-lock braking system. The vehicle 'thinks' for the driver and can detect wheel locking conditions. The electronic system interfaces with the hydraulic braking system and the wheel brake unit, which is in danger of locking, is rapidly pulsed on and off. This action reduces the possibility of skidding and maintains directional control when braking in adverse conditions.

IGNITION SYSTEMS

The ignition system has advanced dramatically over recent years. Contact breaker points

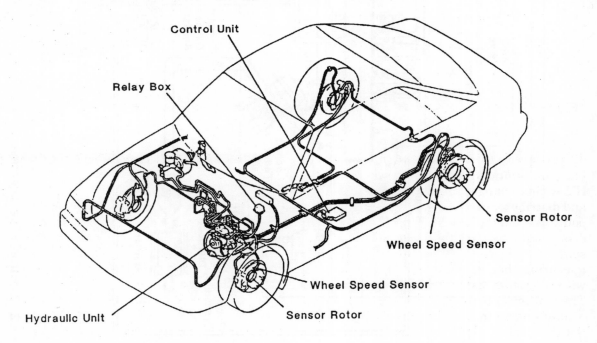

Figure 1-3   Anti-lock braking system fitted to a front wheel drive vehicle.

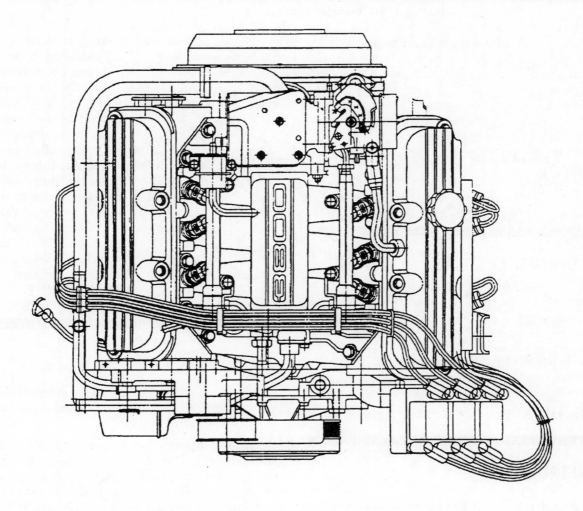

Figure 1-4 Distributorless engine fitted with multiple ignition coils and an EST system.

disappeared, to be replaced by electronic triggers. Ignition coil energy levels increased. HT wiring changed in insulation capabilities and distributor caps increased in size. Ignition point timing mechanisms were improved then disappeared totally to be replaced by electronic spark timing (EST). New systems are now appearing that have replaced the distributor with a system using multiple ignition coils. The ignition is becoming an integrated part of a total engine management package and the ignition 'firing' point can be varied by other system inputs.

STEERING SYSTEMS

Most drivers are aware of the advantages of power assisted steering but many are dissatisfied by the lack of 'road feel' experienced at high road speeds. Electronics and computer control have now combined to provide power steering that, while still basically a pump driven hydraulic unit, is related to vehicle road speed. A few systems are also available that provide rear wheel steering capability which is also responsive to changes in vehicle speed.

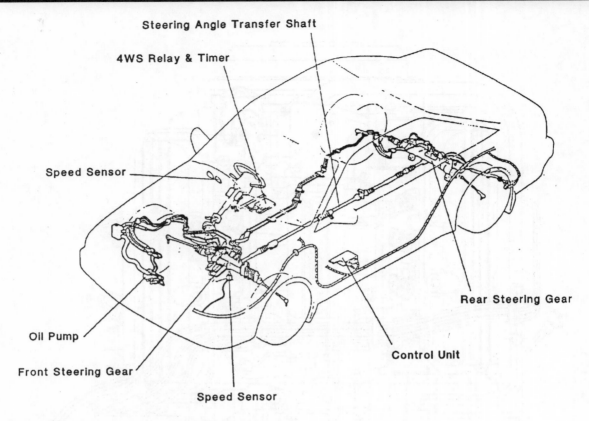

Steering Angle Transfer Shaft

4WS Relay & Timer

Speed Sensor

Oil Pump

Front Steering Gear

Speed Sensor

Rear Steering Gear

Control Unit

Figure 1-5 Four wheel steering system.

INSTRUMENTS

Digital displays tell the driver that the car is fitted with electronics, but the extent of the electronics application is perhaps not fully understood. It is no longer necessary to drive the speedometer head by cable; this task can be performed by a transducer at the transmission and fed by insulated wire to the speedometer display. Trip computers are linked

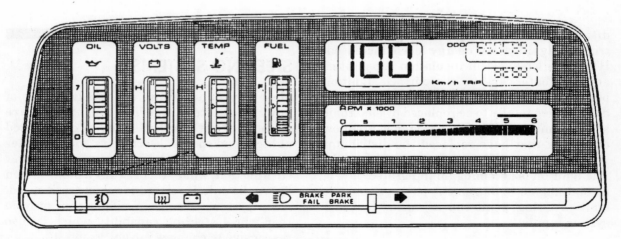

Figure 1-6 Vacuum fluorescent instrument display.

into the fuel injection system, fuel supply quantity and the odometer. The range of system check lights conceal a sophisticated network of sensors continually monitoring the wellbeing of many engine, transmission, braking and lighting components. The area of visual display is one of significant sales appeal and changes will continue to appear at regular intervals for many years into the future.

HEATING/COOLING

Older drivers may recall when the addition of a heater to a vehicle was considered an optional extra. Once drivers accepted the idea of the comfort offered by heating it was only a question of time before consumer demand led to the introduction of vehicle air conditioning. This trend continued with the development of integrated heating/cooling systems until the present fully integrated climate control systems arrived. Some recreational vehicles even have inbuilt refrigerators.

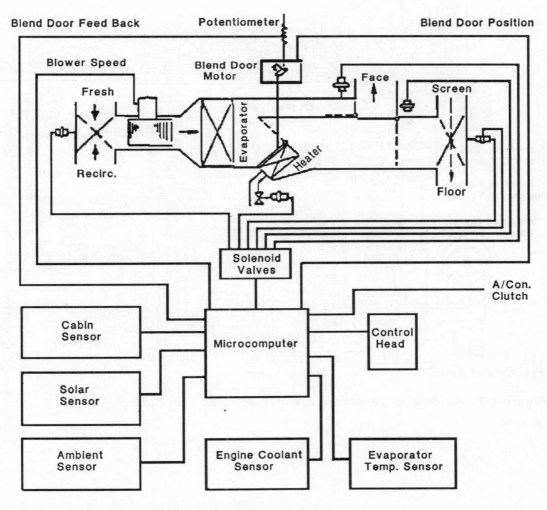

Figure 1-7 Integrated heating and cooling system.

AUTOMATIC TRANSMISSIONS

Repairs to automatic transmissions have tended to be the function of specialist workshops. The introduction of electronic controls onto what is still basically a hydraulic unit is unlikely to change this repair trend for the small workshop. The servicing technician in the larger workshop cannot ignore this area because diagnostic skill will now become more important. The transmission is no longer a self-contained unit.

Faults can occur which may be related to external sensors or computers. If the transmission is removed and sent to a specialist rebuilder for every malfunction a number of expensive mistakes are sure to occur. A $2000 overhaul which could have been avoided by replacement of a $30 throttle sensor is difficult to explain to a customer. It is essential to verify the electronic function of the transmission before dismantling commences. This diagnostic skill will become more important as the new fully integrated engine and transmission computers are fitted to a wider range of models.

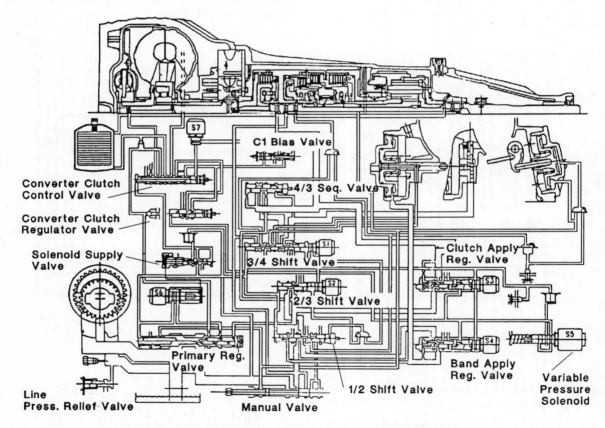

Figure 1-8 Automatic transmission with electronic shift control.

ACCESSORIES

Many components, which were regarded as 'extras' on earlier model cars, are now fitted as standard on base line models. There are many electrical and electronic accessories, most of which are relatively simple in operation and can be easily replaced if a malfunction occurs. The range of accessories is vast and would require a book written specifically for these units. One item chosen for inclusion within this book is the cruise control. This unit has interaction with vehicle speed and throttle positions and therefore considered to be of sufficient interest to the reader to warrant a system explanation.

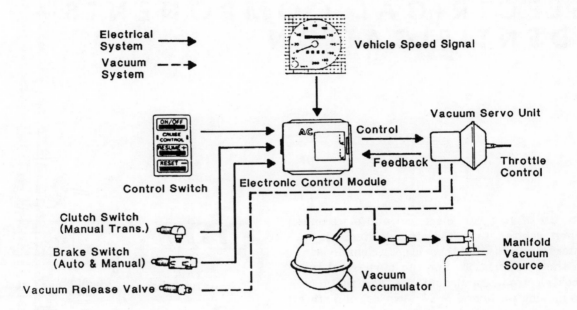

Figure 1-9 Typical electronic cruise control.

2

ELECTRICAL COMPONENTS— IDENTIFICATION

The use of the terms 'electrical' and 'electronic' when related to automotive systems is so interchangeable that there is not a clear division in their application. For the purpose of this book, electronic components are those that have no moving parts and are known as solid state devices, e.g. transistors and diodes.

RESISTORS

A resistor is used in an electronic circuit to control current flow. Resistors are made of many different materials, such as ceramic, carbon, metal oxides, metallic salts and wire wound types. They are formed in many shapes and sizes.

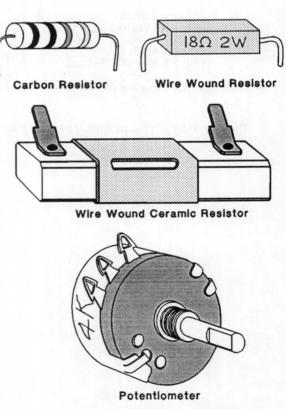

Carbon Resistor   **Wire Wound Resistor**

Wire Wound Ceramic Resistor

Potentiometer

Figure 2-1   Various types of resistors.

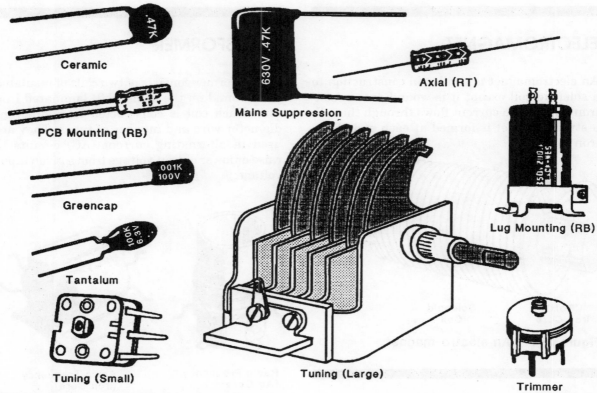

Ceramic

PCB Mounting (RB)

Greencap

Tantalum

Tuning (Small)

Mains Suppression

Axial (RT)

Lug Mounting (RB)

Tuning (Large)

Trimmer

Figure 2-2 Various types of capacitors.

CAPACITORS

A capacitor is used to reduce the magnitude of varying or transient voltages by 'trapping' or temporarily storing electrical energy. The shape and size of the capacitor is determined by the application and location in a circuit. They are constructed of metal, polyester, ceramic and tantalum.

SOLENOID COIL

A solenoid coil consists of many turns of insulated (enamel coated) wire closely wound in layers onto a cylindrical former. When current flows through the coil, a strong magnetic field is formed through its centre.

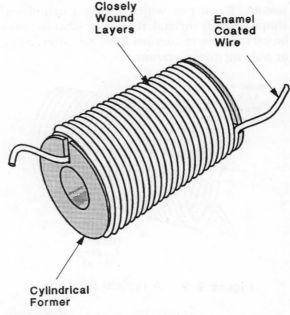

Closely Wound Layers

Enamel Coated Wire

Cylindrical Former

Figure 2-3 A solenoid coil.

ELECTROMAGNET

An electromagnet is similar in construction to a solenoid coil except it is wound onto a 'soft' iron core. When current flows through the coil, a strong magnet is formed at each end of the iron core.

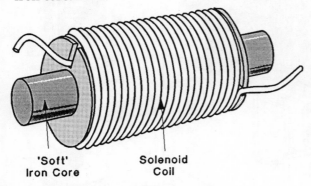

'Soft'
Iron Core

Solenoid
Coil

Figure 2-4   An electro-magnet.

CHOKES

A choke is an insulated (enamel coated) wire wound into a coil which forms a cylindrical shape with air through its centre. Chokes resist quick changes in current flow, e.g. alternating or pulsing direct current.

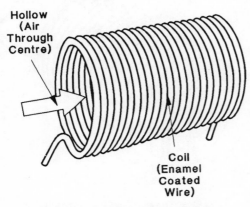

Hollow
(Air
Through
Centre)

Coil
(Enamel
Coated
Wire)

Figure 2-5   A typical choke.

TRANSFORMER

A transformer consists of two coils of insulated wire wound on the same 'soft' laminated iron core. Each coil is constructed from different diameter wire and number of turns. They are used in alternating current (AC) circuits to raise or lower output voltage from a given input voltage.

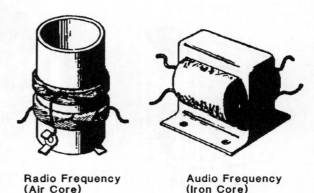

Radio Frequency
(Air Core)

Audio Frequency
(Iron Core)

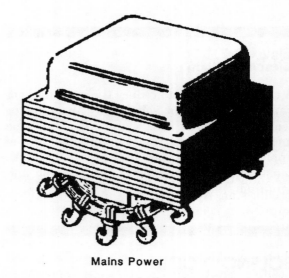

Mains Power

Figure 2-6 Various types of transformers.

IGNITION COIL

An ignition coil is a pulse transformer that steps up voltage from the battery and alternator to a voltage high enough to 'jump' the gap at a spark plug.

Various types of ignition coils are used in modern motor vehicles. They are divided into two groups; those for point ignition systems and those for electronic ignition systems.

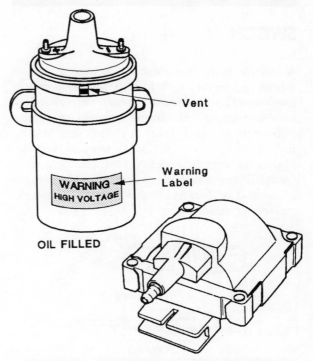

OIL FILLED

Figure 2-8 Coils used with electronic ignition systems.

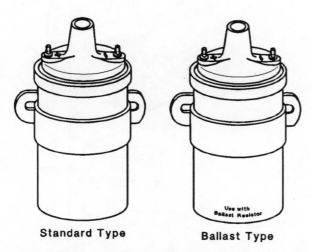

Standard Type

Ballast Type

Figure 2-7 Coils used with various types of point ignition systems.

PULSE GENERATOR

An inductive type pulse generator is used to produce an AC signal. Its application is to supply an electronic control unit with an accurately timed signal voltage. This device has many applications and can provide information related to wheel speed, crankshaft position, ignition firing point and other types of input signals. For further details on the operation of various types of pulse generators refer to Chapter 7.

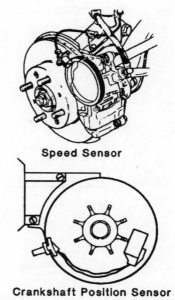

Speed Sensor

Crankshaft Position Sensor

Figure 2-9 Examples of inductive type pulse generators.

SWITCH

A switch, in its simplest form, will 'make' or 'break' a circuit. In addition to the common mechanically operated contact set, switches may be thermal, electromechanical, magnetic or electronic (solid state). A switch may also be used to direct current flow in several circuits, e.g. single pole, rotary switch. Complex switches are available which provide control of many different circuits with a single switching action.

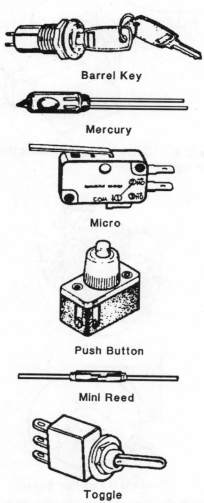

Barrel Key

Mercury

Micro

Push Button

Mini Reed

Toggle

Figure 2-10 Examples of different switches that are currently being used in electronics.

RELAY

A relay is an electromechanical device which allows a small signal current to switch a much larger current, e.g. fuel pump relay.

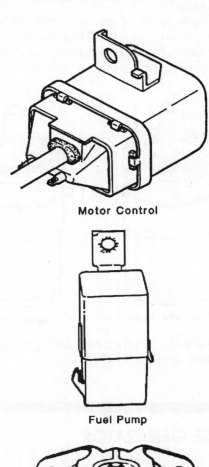

Motor Control

Fuel Pump

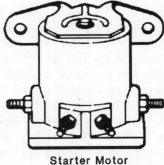

Starter Motor

Figure 2-11 Types of relays.

BATTERY

The electrical storage and base power supply for a motor vehicle is the battery. Passenger vehicles operate from a 12 volt (nominal) battery supply; however the battery construction can vary considerably from a standard battery to the more advanced types which contain a switchable battery within the main battery case.

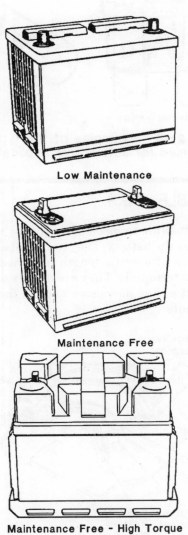

Low Maintenance

Maintenance Free

Maintenance Free - High Torque

Figure 2-12 Various types of passenger vehicle batteries.

VOLTAGE STABILISERS

Voltage stabilisers (Often referred to as voltage regulators) are usually electromechanical thermal devices which can be a separate unit attached to the rear of the instrument panel. A number of manufacturers prefer to fit the voltage stabiliser as an integral part of a gauge. They are used to provide a constant voltage level to the instrument gauges.

Modern electronic components may contain solid state voltage regulators.

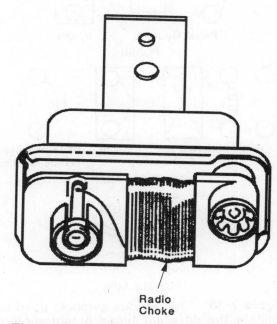

Radio
Choke

Figure 2-13 An instrument voltage regulator fitted with a radio choke.

Figure 2-14 A solid state regulator.

PROTECTION DEVICES

Failure of an electrical component or wiring harness can cause a massive increase in current flow, generating excessive heat and presenting a potential fire hazard. To protect the vehicle, a number of devices are installed in the wiring. The common protection devices used are:

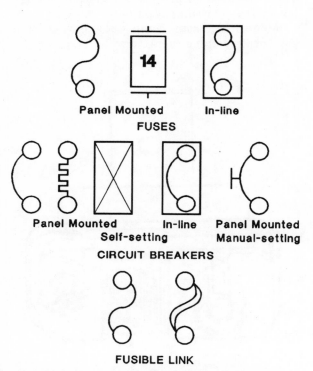

Panel Mounted   In-line

FUSES

Panel Mounted   In-line   Panel Mounted
Self-setting   Manual-setting

CIRCUIT BREAKERS

FUSIBLE LINK

Figure 2-15 The various symbols used to indicate the different types of automotive protection devices.

1 FUSES

Fuses are connected into a circuit to protect one or a limited number of components. They are located in a fusebox, or an individual holder when used with non-standard accessories. A fuse assembly consists of a plastic holder containing a current rated wire fuse, attached to two terminals. The ampere rating of the fuse is selected to suit the current flow within the circuit.

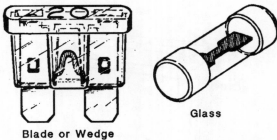

Blade or Wedge   Glass

Figure 2-16 Examples of the types of fuses used for automotive applications.

2 FUSIBLE LINKS

Fusible links are connected into the main supply wires near the battery. They consist of a current rated wire of specific length attached to two insulated terminals. They are mainly designed to protect wiring harnesses and not individual components.

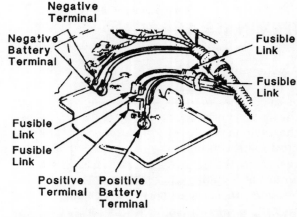

Figure 2-17 Location of fusible links.

3 CIRCUIT BREAKERS

Circuit breakers are connected to individual circuits that are subjected to high current flows. They are generally located between the switch and the component. Circuit breakers can be automatic or manual reset devices.

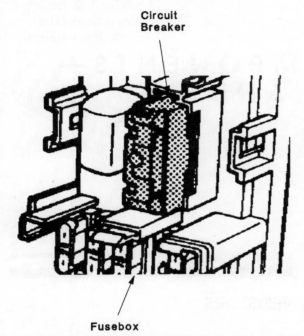

Circuit Breaker

Fusebox

Figure 2-18 A manual reset circuit breaker installed in a fusebox.

TERMINALS AND CONNECTORS

A wide range of terminals and connectors are required to connect wires to components and/ or to connect wiring looms together. Although the appearance will vary considerably between the various types, all connectors and terminal blocks share a common requirement. Electrical resistance must not be introduced into the circuit at the join; the terminal materials must be selected to prevent corrosion. The metal current carrying parts must be insulated from adjacent wires and the vehicle body.

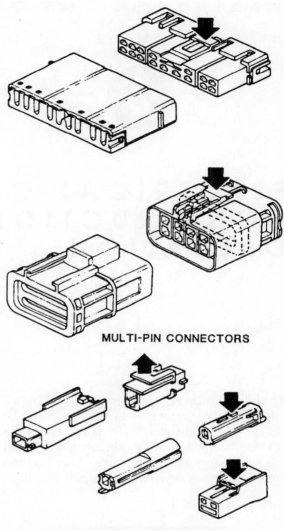

MULTI-PIN CONNECTORS

SINGLE PIN CONNECTORS

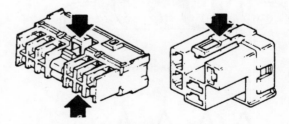

COMPONENT TERMINAL BLOCKS

Figure 2-19 Examples of connectors and terminal blocks. The arrow indicates the direction that each locking device must be moved to release the section.

3

ELECTRICAL COMPONENTS— CONSTRUCTION AND OPERATION

The vast number of components in an automotive electrical system, make it impractical to describe the construction and operation of every component. The components selected are representative of the types found in the various electronic systems in the motor vehicle.

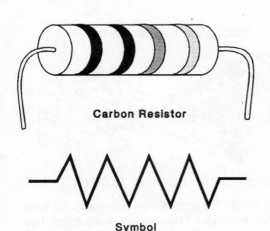

Carbon Resistor

Symbol

Figure 3-1 Carbon resistor and its symbol.

RESISTORS

The two types of resistors are:
- Fixed;
- Variable.

FIXED RESISTORS

The common type is the carbon composition resistor. The value of these resistors can be changed by altering the amount of carbon particles in their construction. More carbon gives less resistance and different coloured rings are used to indicate the resistance value in ohms.

Automotive systems may also require the use of a wire wound resistor when high wattage applications requiring accurate current control are used. This type of resistor consists of a

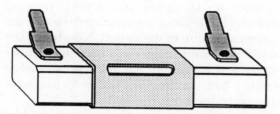

Figure 3-2 Wire wound ceramic resistor.

resistance wire wound onto a ceramic core. The resistance wire is protected and insulated by fitting the wound coil into a ceramic casing.

VARIABLE RESISTOR

It is sometimes necessary to vary the value of a resistor within a circuit, e.g. controlling the volume of a radio, dimming a light or changing the voltage signal from a throttle position sensor. When this action is required, a variable resistor (potentiometer) is fitted to the circuit. The two common types are the carbon strip and the wire wound element; the resistance value within the stated range is determined by the position of a sliding or rotary 'wiper arm'. The maximum resistance value, in ohms, is stamped on the body of the potentiometer.

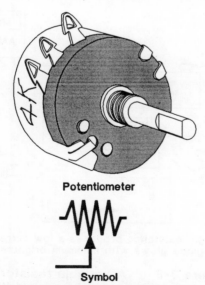

Potentiometer

Symbol

Figure 3-3 Potentiometer and symbol.

When the wiper arm is positioned at the end of the resistive element, maximum resistance is introduced into the circuit. This reduces the current to a minimum in this section of the circuit.

Repositioning the wiper arm to the other end of the resistive element reduces the resistance in the circuit to a minimum value. Current flow in this section of the circuit will increase to a maximum.

The way in which resistors affect circuit operation can be illustrated by a simple experiment.

Construct a basic circuit (see Figure 3-5) containing a battery, switch, globe and wires. Turn the switch to the on position and note how brightly the globe glows.

Add a resistor into the circuit (see Figure 3-6) and repeat the experiment. When the switch

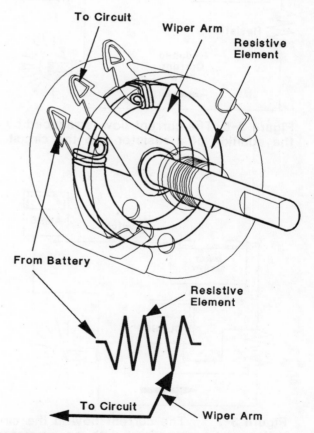

Figure 3-4 Potentiometer in its maximum resistance position.

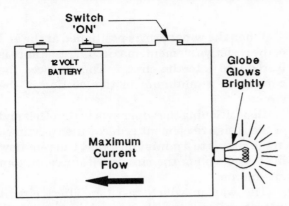

Figure 3-5 The maximum current flow is controlled by the resistance of the globe.

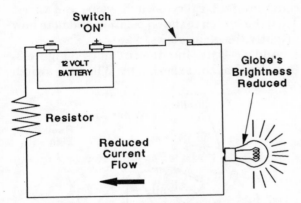

Figure 3-6 The current flow is reduced by the addition of a resistor into the circuit.

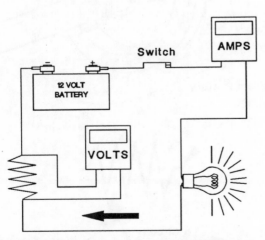

Figure 3-7 The current flow in the circuit can be measured with an ammeter and the voltage drop across the resistor can be measured with a voltmeter.

is turned on the globe glows with reduced brilliance. The resistor has restricted the amount of current flowing in the circuit. Now touch the resistor with your fingers and note the resistor is quite warm. Energy has been consumed to cause this heating effect and the power available to the globe has been reduced.

It is possible to prove what has occurred by connecting an ammeter and a voltmeter into the circuit (see Figure 3-7). When the resistor is in the circuit and the globe is illuminated, a reduction in current flow and a voltage drop across the resistor can be measured.

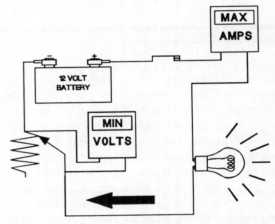

A low resistance produces a high current flow and the globe glows with maximum brightness.

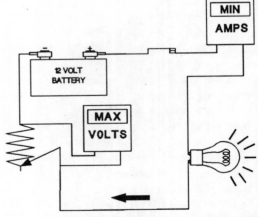

A high resistance produces a low current flow and the globe glows with minimum brightness.

Figure 3-8 A variable resistor (potentiometer) can be used to control the current flow in a circuit.

In this circuit the fixed resistor has limited the globe to a constant level of brightness.

By replacing the fixed resistor (see Figure 3-8) with a variable resistor the brightness of the globe can be varied. Leave the voltmeter and ammeter in the circuit and turn the knob on the resistor. When the globe glows brightly the voltmeter reading will be low and the current will be high. As the knob is turned in the opposite direction the globe will reduce in brightness. The voltmeter reading will increase and the current flow will reduce.

Resistors limit current flow in a circuit.

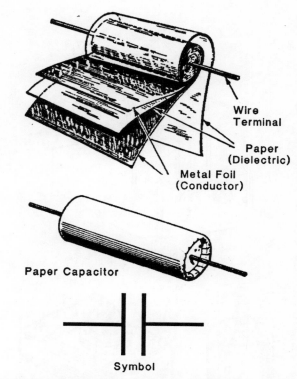

Figure 3-9 **Fixed capacitor and its symbol.**

CAPACITORS

Capacitors store an electrical charge. Two types of capacitors are:

- Fixed;
- Variable.

FIXED CAPACITORS

The simplest type of capacitor has two 'conductor' plates separated by an insulating material (dielectric) such as paper, mica and polystyrene. Both plates will be connected to external terminals on the capacitor. One terminal will be connected to the negative and the second terminal connected to the positive part of a circuit. When the circuit is switched 'on', an excess of electrons will appear on the negative plate and at the same time electrons will leave the positive plate. The potential difference between the two plates will continue to rise until the voltage across the capacitor is equal to the supply voltage. The capacitor is said to be in a 'charged state'.

The conductors are usually in the form of flat strips which, for large value capacitors, may be rolled into a tube to reduce their overall size.

The capacitance value of a capacitor is expressed in farads, however this unit of measurement is extremely large and the general

unit encountered is the microfarad (uF, one-millionth of a Farad).

The capacitance value is stamped on the body of the capacitor and is accompanied by a voltage value. If the capacitor is placed in a circuit where the working voltage exceeds the stated value it will be damaged.

A special group of fixed capacitors, known as electrolytics, have a further marking on the casing indicating the polarity of the connecting leads.

Note: Within the motor trade capacitors are commonly called condensers.

VARIABLE CAPACITORS

These capacitors are commonly used to tune radios. Smaller (trimmer) variable capacitors are used to 'fine tune' circuits and are preset in production.

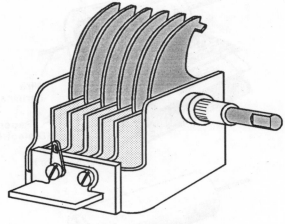

Variable Capacitor

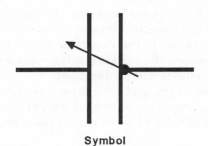

Mica

Trimmer

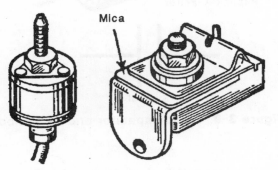

Symbol

Figure 3-10 Variable capacitors and their symbol.

The operation of a variable capacitor is similar to a 'fixed' capacitor; however, large variable capacitors consist of one or more fixed plates separated from a similar number of moving plates by an air gap. The position of the movable plates is adjusted by a shaft and knob. The maximum capacitance is achieved when

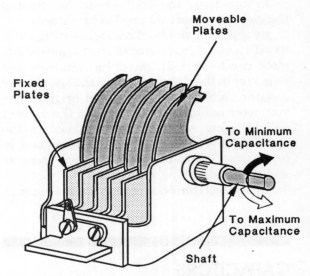

Figure 3-11 Operation of a variable capacitor.

the plates are fully overlapped. When the fixed plates are not overlapping the movable plates, a minimum capacitance is introduced to the circuit.

ELECTROMAGNETISM

Electromagnetic devices have many applications in the automobile. These devices can be grouped under two broad headings:

- Constant current;
- Variable current.

CONSTANT CURRENT APPLICATIONS

When a current is passed through a conductor (wire), a weak magnetic field forms around the conductor (see Figure 3-12).

By forming the current-carrying wire into a tight loop (see Figure 3-13) the magnetic field is concentrated inside the loop.

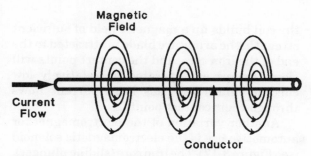

Figure 3-12 Conductor, current flow and magnetic field.

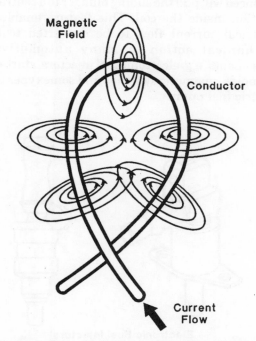

Figure 3-13 All magnetic fields travel in the same direction through the centre of the loop.

To form a **solenoid coil**, continue to wind the wire into a series of successive loops forming a coil (see Figure 3-14). The magnetic fields surrounding each loop combine to produce a single magnetic field. The strength of the resultant magnetic field will be the sum of the individual coils. The field formed around the coil will have a north and south pole similar to a bar magnet.

By inserting an iron core through the centre of the solenoid coil (see Figure 3-15) an

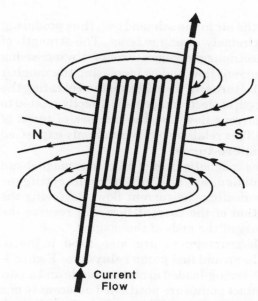

Figure 3-14 By combining the fields of individual coils, a strong magnetic field is produced.

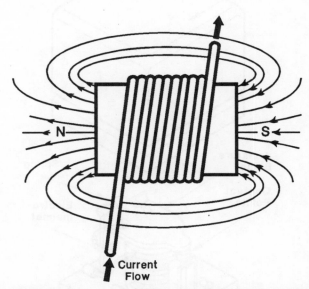

Figure 3-15 Inserting an iron core through a coil strengthens the magnetic field.

electromagnet is formed. The poles remain at the same ends provided the current flow continues in the same direction. The iron core concentrates the magnet field more effectively than the air in the solenoid coil, thus producing an extremely strong magnet. The strength of the electromagnet can be increased by increasing the current flow and/or increasing the number of coils (turns) of wire. The method of rating the strength of electromagnets is directly related to the current flow and the number of turns of wire. This relationship is commonly expressed as **ampere-turns.**

The magnetic field of an electromagnet can be turned on and off by connecting or disconnecting the current flow. Reversing the direction of the current flow will reverse the polarity at the ends of the magnet.

Electromagnets are also used in horn, headlamp and fuel pump **relays** (see Figure 3-16). A spring-loaded armature blade and a pair of contact points are positioned adjacent to one end of the iron core. When current flow through

the coil builds up a magnetic field of sufficient strength, the armature blade is attracted to the end of the iron core, and the contact points will close. Relays are operated by relatively low current values yet pass large current flows through their contact points.

Another variation of the electromagnet for automotive use is the electromagnetic solenoid (see Figure 3-17). The iron core (sliding plunger), which is free to move, is positioned so one end protrudes from the coil. By passing current through the coil the electromagnetic field produced will pull the sliding plunger to a central position inside the coil. This feature enables electrical current flow to be converted to a mechanical action in many automotive components, e.g. electronic fuel injectors, starter solenoids, magnetic switches and some types of electric fuel pumps.

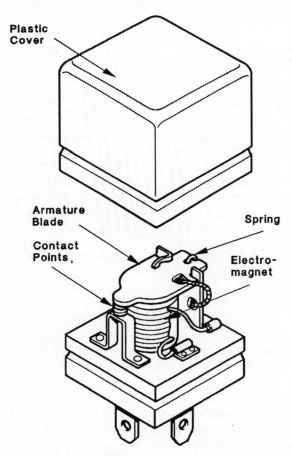

Figure 3-16 Typical relay.

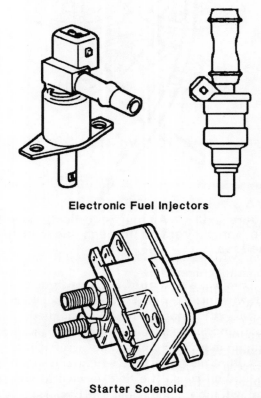

Electronic Fuel Injectors

Starter Solenoid

Figure 3-17 Typical electro-magnetic solenoid devices.

magnetic field around a single current carrying wire is very small. However, when the wire is wound into a solenoid coil, the induced voltage is greatly increased. This concept can be applied in several of the devices used in automobiles.

SELF INDUCTION COILS

To limit unwanted noise and interference from a circuit (e.g. radio hum or static), it is necessary to filter current flowing in a wire so that the alternating current (AC), causing the problem, is smoothed.

The filtering action occurs when an inductance of suitable characteristics is created within the coil which opposes the noise or interference. By changing the physical features of the **choke coil** such as the windings, core and wire spacing, it is possible to smooth different frequency ranges of the AC.

MUTUAL INDUCTION COILS

When wire (B) is placed close to a wire (A) in which there is a pulsating primary current, an induced pulsating current will be created in wire (B). The action is called **mutual induction** (see Figure 3-19).

When a wire that is formed into a coil (secondary) is placed in close proximity to a coil (primary) in which there is a pulsating primary current, an induced pulsating current will be created in the secondary coil (see Figure 3-20).

It is not necessary to have an alternating current (AC) input to achieve this pulsating action. A similar result will occur if the input is obtained by switching direct current (DC) on and off.

In the illustration (see Figure 3-21), the switch is open, therefore no current is flowing in the primary coil. A magnetic field is not formed so there is no movement of the field across the secondary coil, resulting in no induced current in the secondary coil windings.

The moment the switch is closed (see Figure 3-22), current starts to flow in the primary coil and the magnetic field forms. The increasing current flow causes the magnetic field to expand

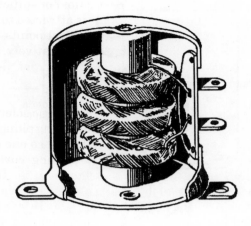

Figure 3-18 Different types of chokes.

VARIABLE CURRENT APPLICATIONS

You will recall, from the section describing constant current flow in a wire, that a magnetic field formed around the wire when current flow was present. By increasing and decreasing the rate of current flow the magnetic field will expand and contract around the wire. This action **induces** a voltage in the wire. The process is known as **self-inductance.** It is important to note that the induced voltage is called **counter-electromotive force** because it tends to oppose the current flow that produced the electromagnetic field.

The induced voltage developed by the moving

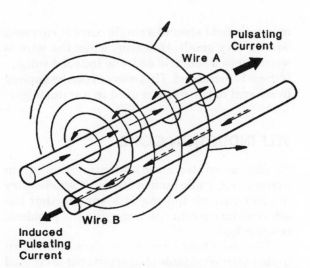

Figure 3-19 Mutual induction between two wires.

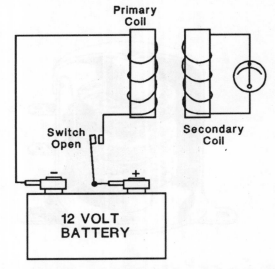

Figure 3-21 No current flow, no magnetic field, no current in secondary coil.

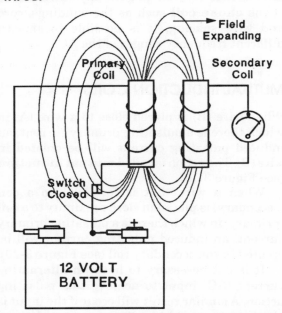

Figure 3-20 The magnetic field builds up, the instant the switch is closed.

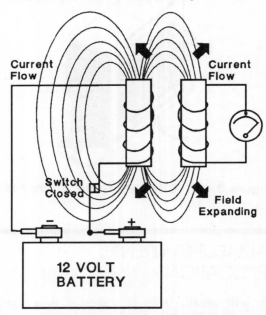

Figure 3-22 The magnetic field cuts across the secondary coil.

across the secondary coil. The moving magnetic field causes an induced current to flow in the secondary coil windings.

When the primary coil current attains a constant flow rate (see Figure 3-23), the magnetic field is still present but is neither increasing nor decreasing in strength. The lack of movement of the field across the secondary coil means there

will be no current induced in the secondary coil windings.

By opening the switch, current flow in the primary coil ceases (see Figure 3-24). The magnetic field collapses across the secondary coil towards the primary coil. The moving magnetic field causes an induced current to flow in the secondary coil windings. This current flow is in the opposite direction to the flow caused during magnetic field build-up.

The purpose of the iron core placed in the centre of each coil is to increase the magnetic field strength which will raise the induced current to a usable level.

Mutual induction is the principle used in automotive ignition coils, alternators and generators.

IGNITION COIL/TRANSFORMER

An ignition coil for automotive use is a form of pulse transformer, therefore an understanding of transformer principles will be of assistance when considering coil operation.

TRANSFORMER OPERATION

A transformer which has the same number of windings (see Figure 3-25) on both primary and secondary coils has a voltage/amperage

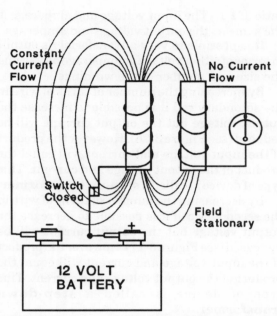

Figure 3-23 No current flow in the secondary coil due to a stationary magnetic field.

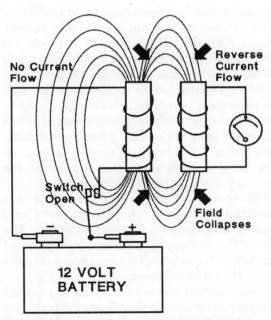

Figure 3-24 The magnetic field cuts across the secondary coil in the opposite direction.

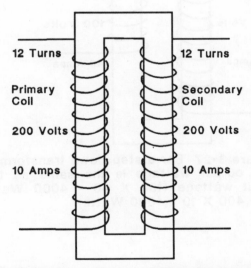

Figure 3-25 For a 1:1 transformer, the output is the same as the input (200 X 10 = 2 000 Watts for both coils).

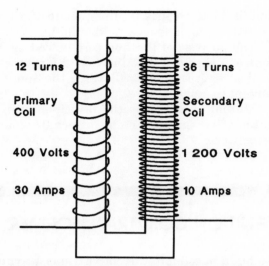

Figure 3-26 For a step-up transformer, the output wattage is the same as the input wattage (1200 X 10 = 12 000 Watts and 400 X 30 = 12 000 Watts).

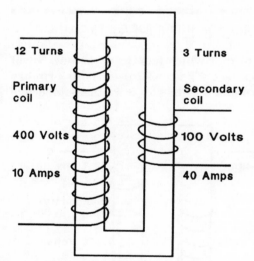

Figure 3-27 For a step-down transformer, the output wattage is the same as the input wattage (100 X 40 = 4000 Watts and 400 X 10 = 4000 Watts).

ratio of 1:1. (The input voltage and amperage is the same as the output voltage and amperage.)

It is possible to vary a transformer's output in terms of either voltage or amperage by varying the size and number of the windings.

By increasing the number of turns of wire in the secondary coil it is possible to increase the output voltage but the output current will be reduced (see Figure 3-26). However the product of the input voltage and current will equal the product of the output voltage and current. This type of device is called a **step-up transformer**.

By decreasing the number of turns of wire in the secondary coil it is possible to decrease the output voltage but the output current will be increased (see Figure 3-27). However the product of the input voltage and current will equal the product of the output voltage and current. This type of device is called a **step-down transformer**.

IGNITION COIL OPERATION

An ignition coil is a form of pulse transformer which is used to step up the vehicle battery voltage to a voltage high enough to produce a spark at the spark plug electrode. The spark produced must be of sufficient intensity to ignite the air–fuel mixture in the combustion chamber. The secondary winding, consisting of many thousands of turns of fine wire, is wound onto a laminated iron core. The primary winding, consisting of a few hundred turns of relatively thick wire, is wound on the outside of the secondary winding (see Figure 3-28).

The ends of the primary winding are connected externally via the positive and negative terminals on the insulator cap. The ends of the secondary winding are connected internally to the positive terminal and externally via the HT terminal tower.

When current starts to flow in the primary winding a magnetic field is formed which cuts across the secondary windings (see Figure 3-29). The magnetic field increases in strength until the primary current flow reaches a constant value. When the primary current is interrupted

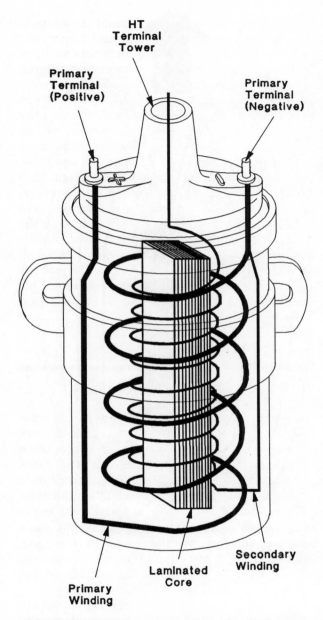

Figure 3-28 The construction of an ignition coil.

by the switching of 'points' or by an electronic control module, the magnetic field collapses across both windings inducing a very high voltage in the secondary winding. A relatively lower voltage is also induced in the primary winding. The output voltage available at the HT terminal will be affected by several factors:

1 SELF-INDUCTION

The magnetic field around the primary winding does not reach its full strength at the instant the current starts to flow. A very short time is needed to build up the field strength. You will recall, during your study of self-inductance, a counter-voltage is induced in each loop of the coil. The counter-voltage opposes primary circuit voltage therefore maximum current flow will not occur until the magnetic field reaches maximum strength. At this point the counter-voltage disappears and primary current flow is at its maximum value.

Manufacturers have minimised the time delay in magnetic field growth (rise time) by redesigning the coil and ignition supply circuits. (See Chapter 2 for identification of the different ignition coil types.)

2 AMPERE-TURNS RATIO

The ratio of the number of turns of the secondary windings to that of the primary windings determines the output voltage of the coil. For a ratio of 100:1 and an induced primary voltage of 300 volts, the output voltage would be 30 000 volts. The output voltage has increased by 100 times the value of the input voltage, however the output amperage has decreased to one-hundredth of the input amperage.

The current flowing in the primary windings has a substantial effect on the output voltage. This is noticeable in an ignition circuit when the ballast resistor is bypassed during engine cranking.

3 FIRING VOLTAGE

An ignition coil may have the capacity to develop a peak voltage in the range of 20 000 to 50 000 volts. When coupled into an ignition system this peak level voltage will not be attained under normal engine operating conditions. The voltage produced by the ignition coil will be determined by engine operating conditions but generally, never exceeds two-thirds of peak level voltage.

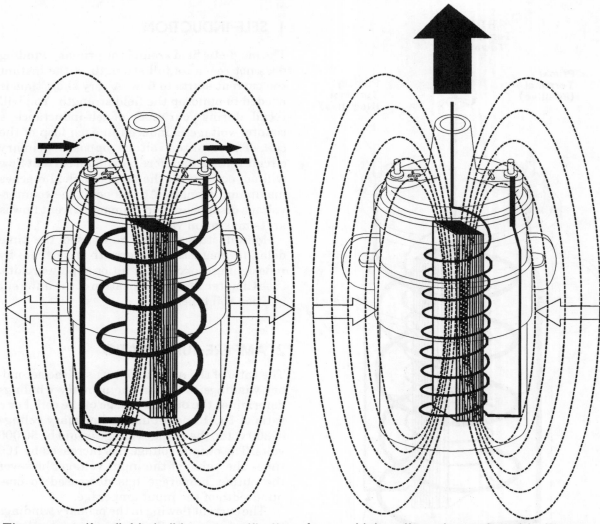

The magnetic field builds up until the primary current reaches a constant value.

A very high voltage is produced in the secondary winding by the collapsing magnetic field.

Figure 3-29 The operation of an ignition coil.

SWITCHES

MECHANICAL

Current flow in a vehicle wiring harness is normally controlled by mechanical switches. When a switch controls the current flow in a single wire the switch is referred to as a single pole, single throw (SPST) device. The 'pole' refers to the input contact and the 'throw' refers to the output contact of the switch. If a switch has a single input contact and two output contacts supplying alternate circuits, it would be described as a single pole, double throw (SPDT) device.

The simplest form of mechanical switch is the SPST (see Figure 3-30). These switches can be either continuous or momentary in operation.

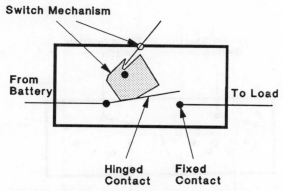

Figure 3-30 Single pole single throw switch (continuous type).

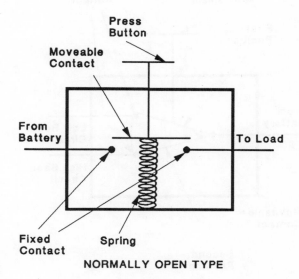

NORMALLY OPEN TYPE

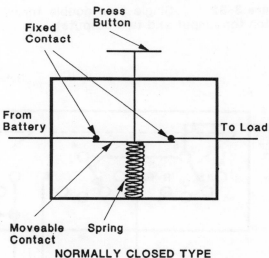

NORMALLY CLOSED TYPE

Figure 3-31 Single pole single throw switch (momentary type).

The continuous type switch contains a hinged contact arm which is always in one of two positions. When current is flowing the hinged contact is touching the fixed contact. In the open position (no current) the hinged contact is held away from the fixed contact by the switch mechanism.

The momentary type switch consists of a movable spring-loaded contact bar and two fixed contacts (see Figure 3-31 upper). A circuit is formed through the switch by pressing the operating button which moves the contact bar to a position that bridges (joins) the two contacts. Current will flow through the switch until the spring-loaded button is released. This design of switch would be known as a normally open type. Applications for this switch include controlling horn operation or the stop lights.

This type of switch can be redesigned to allow current flow, in the circuit, until the operating button is pressed (see Figure 3-31 lower). While pressure is applied to the operating button, the spring-loaded contact bar will be held away from the fixed contacts and the circuit will be opened. This design of switch would be known as a normally closed type. Courtesy light switches operated by the vehicle doors are one application of this switch.

To control a headlight circuit requires a slightly more complex switch (see Figure 3-32). The dip switch must be capable of switching between high and low beam circuits. The switch does not have an off position. Current from the main headlight switch passes through the single input pole to one of two output throws (high and low beam).

When the driver operates the dip switch (e.g. when the headlights are on low beam), a latching mechanism disconnects the low beam circuit and 'switches on' the high beam circuit. This design of switch can be described as a single-pole, double-throw (SPDT).

Many circuits within the vehicle require switches capable of controlling multiple inputs and outputs (see Figure 3-33). A switch

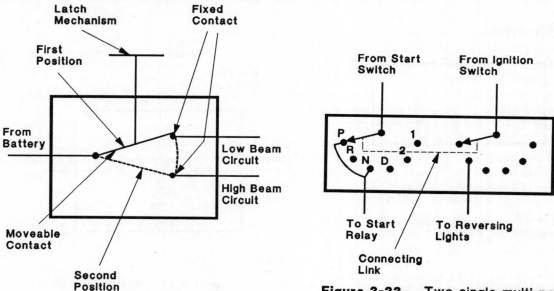

Figure 3-32 Single pole double throw switch (one input and two outputs).

Figure 3-33 Two single multi-pole multi-throw switches combined into one case.

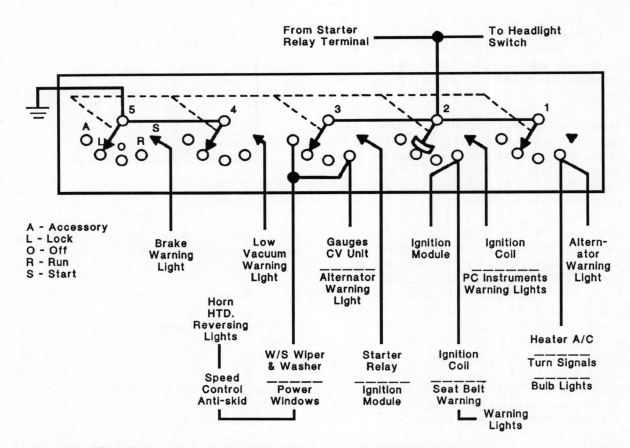

A - Accessory
L - Lock
O - Off
R - Run
S - Start

Figure 3-34 Five switches combined into one multi-ganged switch.

combining the circuits of reversing lights and a transmission neutral start switch is fitted to many automatic transmissions. Two single pole multiple throw switches are linked together (ganged) to form a single multiple-pole, multiple-throw switch (MPMT). The term 'ganged' refers to the operation of multiple switches within a single casing. Each switch is mechanically linked together so operating the switch control causes each individual input wiper contact to sweep across a set of fixed output throws. It should be noted that the input poles and the output throws are electrically insulated from each other.

A modern ignition switch is an example of a multi-ganged electrical switch.

THERMAL

These switches are activated by temperature, e.g. electrical current flow, coolant temperature.

One application of a thermal switch is a flasher unit for use in trafficator or circuit monitoring devices (alerts driver to a system malfunction). See Figure 3-35 for a simplified drawing of the construction of this type of device. A heating element is connected in series with a bi-metallic strip and contact points. The contact points are normally closed and when current flows in the circuit the heating element will cause the bi-metallic strip to deflect. The contact

points will open, current will stop flowing and the bi-metallic strip will start to cool. When the strip returns to its original temperature and position, the contact points will close. Current will start to flow and the cycle will be repeated.

Figure 3-36 is an example of a normally open bi-metallic switch used in engine cooling systems. The switch is screwed into the engine water jacket and connected to a warning light mounted in the instrument panel. Voltage is applied to the warning light when the ignition switch is in the 'on' position. Under normal engine operating conditions the warning light will remain 'off'. When the engine overheats, the coolant temperature causes the bi-metallic strip to deflect, the contacts close completing the circuit to earth. The warning light is switched 'on'.

The thermo-time switch has a set of contacts that are normally closed by a bi-metallic strip (see Figure 3-37). The strip is subjected to heating by an electrical winding and engine coolant. This design feature allows the contacts to respond to changes in engine coolant temperature. Below a specified engine coolant

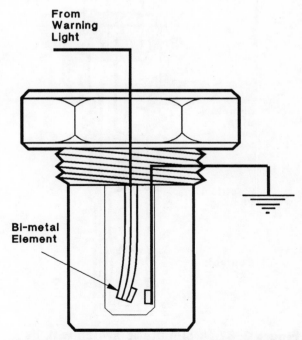

Figure 3-36 A bi-metallic switch with its contact points in a normally open position.

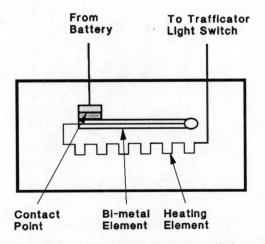

Figure 3-35 A thermal type switch used in a trafficator circuit.

temperature the contacts remain closed until a voltage is applied to the electrical winding. The current flow heats the bi-metallic strip to open the contacts. The time taken for the contacts to open rates the switch for different applications.

ELECTROMECHANICAL

The most common application of this type of switch is the relay (see Figure 3-38).

A relay contains two electrical circuits. The control circuit, which has low current flow, consists of an electromagnet, supply and earth wires. The power circuit contains a spring-loaded armature, contact set, supply wire and a wire that connects to the load (e.g. headlight).

When a switch in the control circuit is closed, current will flow through the solenoid to form a strong magnetic field. The armature is attracted to the iron core of the solenoid by the magnetic

field. This movement closes the contact points and current will flow to the load (e.g. the headlight will be turned 'on').

In heavy current applications (starter motors) a different type of relay is used. Generally these units are known as starter solenoids or electromagnetic switches (see Figure 3-39).

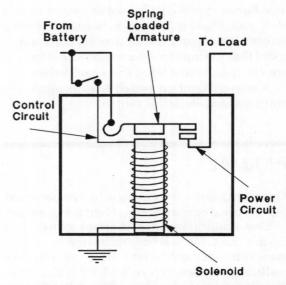

Figure 3-38 This type of electro-magnetic switch has two circuits - a control circuit and a power circuit.

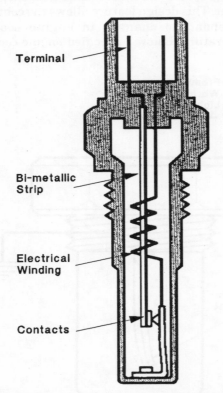

Figure 3-37 A bi-metallic switch with its contact points in a normally closed position.

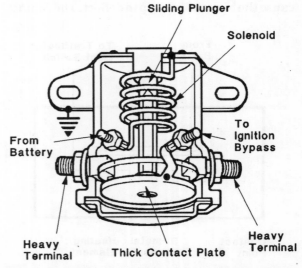

Figure 3-39 A heavy current relay used as a starter solenoid.

The control circuit contains a solenoid, supply and earth wires. The sliding plunger, which is fitted with a thick contact plate, is pulled into the solenoid when current flows in the control circuit. The power circuit is completed when the contact plate bridges two heavy terminals. This action allows current to flow from the battery to the starter motor.

Note: The ignition bypass terminal is used in conjunction with a ballast resistor ignition circuit which has no effect in the starter operation.

Modern starter solenoids reduce the current required in the control circuit after current starts to flow in the power circuit. This is achieved by fitting two windings (pull in and hold) in the solenoid.

POWER SUPPLY

A modern vehicle power supply can be compared to the state's industrial and domestic power system. Each have a base power unit and substations to alter the voltage at given points to the required level.

The base power supply of the modern vehicle is the lead acid battery (12 volts); however, not all the components operate from battery voltage. Many electronic systems operate on a lower voltage (e.g. 5 volts) while some display modules require a much higher voltage (e.g. 38 volts).

The common lead acid battery is a device that delivers electrical energy when a chemical reaction takes place between the lead plates and the electrolyte (acid).

During normal operating conditions the system voltage can vary from as low as 9 volts (at cranking), to 14.7 volts (normal load). As this variance in voltage is not acceptable to electronic circuits, a method of regulating the supply voltage must be used to keep the voltage within an acceptable range. This range may be above or below battery voltage. The devices used may be electromechanical, solid state electronics and a combination of solid state electronics coupled with a step-up transformer.

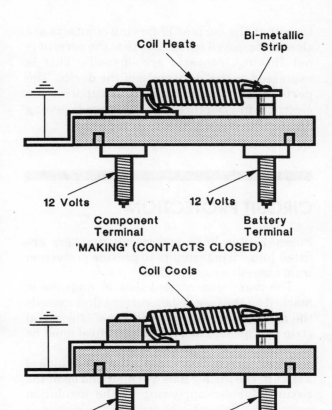

Figure 3-40 The operation of an instrument panel voltage regulator.

A simple form of instrument panel voltage regulator (IVR) is used to supply a reference voltage of approximately 5 volts from a 12 volt input. This device consists of a thermal bi-metallic switch and a radio choke (see Figure 3-40).

The 5 volt reference output is obtained by 'making' (closing) and breaking (opening) a 12 volt supply. The output voltage is determined by the design of the bi-metallic and/or the heating element. The internal contacts of the regulator open and close rapidly as the bi-metallic strip heats and cools. This pulsating action will cause the output voltage to vary between 0 and 12 volts but will have an average of 5 volts. The average output voltage is a direct result of the time that the current is flowing (contacts are closed) compared to the time that the current is not flowing (contacts are opened). This is expressed as the 'duty cycle' of the device. The purpose of the radio choke is to suppress the electrical interference caused by rapid opening and closing of the contacts in the IVR.

CIRCUIT PROTECTION

Fuses, fusible links and circuit breakers are fitted into wiring circuits to provide protection from excessive current flow.

The maximum current flow in amperes is marked on the fuse. When current flow exceeds the rated value, the fuse will 'blow' (the metal strip will melt). The replacement fuse must be the same rated value and type.

When current flow exceeds the rated value of a circuit, the fusible link will melt and open the circuit. Bubbles appearing in the insulation cover of the link indicate the link has melted or has severely over-heated (see Figure 3-42). If either condition occurs, the fault must be located and rectified prior to fitting a replacement link of the same size and type.

Fuses and fusible links are 'go' (circuit complete) or 'no go' (circuit permanently opened). When a vehicle is wired so that all its lights are connected through a single fuse and the fuse blows, a dangerous situation will be created. To avoid this risk, a number of manufacturers have replaced the lighting fuse with an automatic resetting circuit breaker.

These circuit breakers are of the thermal bi-metallic type consisting of a shaped strip and two contacts (see Figure 3-43). When excessive current flows, the strip will heat, deflect and open the lighting circuit. The unit is designed so rapid cooling will take place and the contacts will close to restore the circuit. The headlights will flash on and off rapidly. However, sufficient illumination is provided to bring the vehicle to a stop safely while the current flow in the circuit is reduced preventing permanent damage to the wiring. This action will continue until the

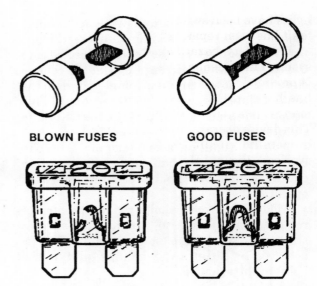

BLOWN FUSES　　**GOOD FUSES**

Figure 3-41　The same rated value and type of fuse must be fitted.

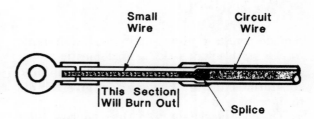

Figure 3-42　A fusible link is located near the battery to protect the complete circuit.

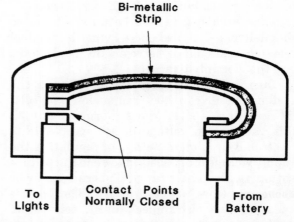

Figure 3-43　A resetting circuit breaker is used to protect the headlight circuits.

lights have been switched off and/or the circuit fault has been repaired.

If a safety hazard does not occur when a system fails, adequate protection for the circuit is provided by a fuse. However a circuit which has high current loads can be fitted with a manual reset circuit breaker (see Figure 3-44). This device operates in a similar manner to the bi-metallic circuit breaker when a current overload occurs. However the circuit is not reconnected when the bi-metallic strip cools but must be manually reset by pressing a button located on the circuit breaker casing.

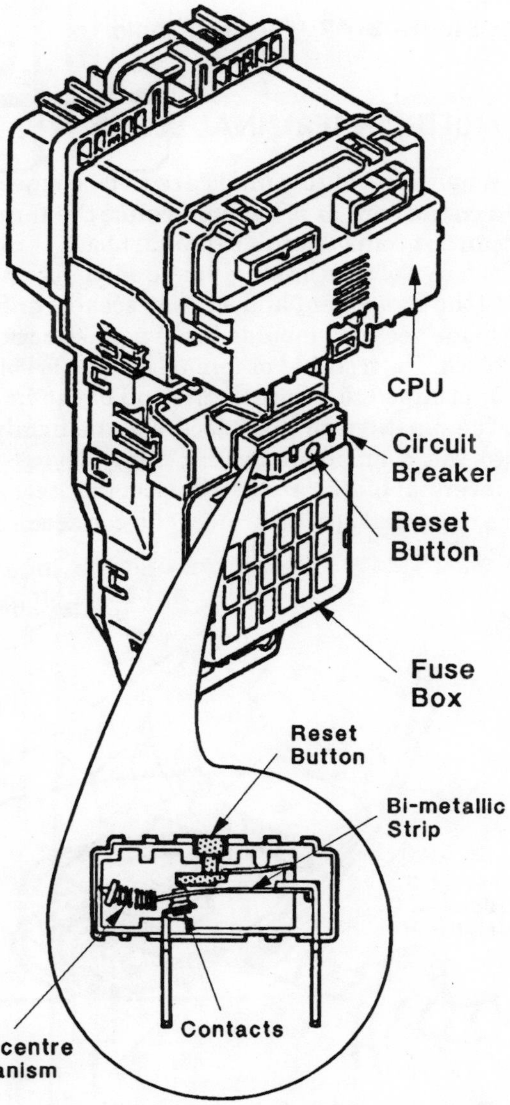

TERMINALS AND CONNECTORS

Terminals and connectors are an important feature of automotive vehicle electrical wiring systems. Terminals allow the wires to be attached to a component. Connectors are placed at various points in a wiring harness to allow sections of the harness to be installed into the body-work of the vehicle.

TERMINALS

Terminals are fitted to the ends of the wires in a harness and different methods are used to attach them to the component. The terminal type selected allows easy removal or testing of the component. There is a vast number of terminal types available but most requirements will be satisfied by the types illustrated in the following text.

BATTERY TERMINAL

A battery terminal is soldered to the cable (wire) and a clamping bolt causes the terminal to grip the battery post.

Figure 3-44 A manual reset circuit breaker fitted into the fusebox and CPU housing.

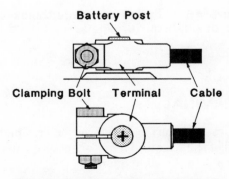

Figure 3-45 Battery terminal.

SPADE TERMINAL

Spade terminals consist of a 'male' section which is attached to, or formed on, the component and a 'female' section that is soldered or crimped to the end of a wire. A sliding fit between the sections allows the wire to be easily connected and disconnected.

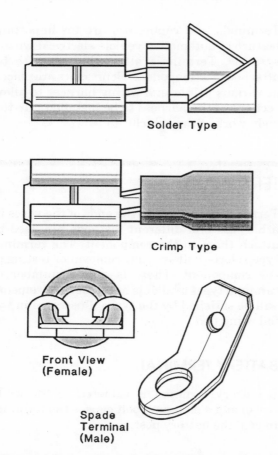

Figure 3-46 Types of spade, quick connector or push-on terminals.

EYE TERMINAL

This terminal may be soldered or crimped to the end of the wire and a nut or setscrew firmly attaches it to the component.

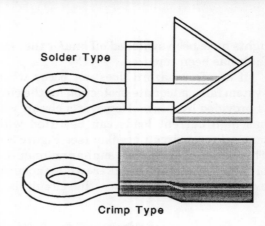

Figure 3-47 Eye terminals.

MULTI-PIN TERMINAL BLOCK

When several terminals are to be connected to a component at a common point, the terminals can be grouped into a terminal block. It consists of a row of pins or male spade sections attached to the component and a row of sockets or female spade sections moulded into a plastic terminal block. Each socket or female section is soldered or crimped to a wire at the end of the harness. The plastic terminal block may be firmly held to the component by a locking device. Some terminal blocks are fitted with an internal and an external seal to protect the connectors from moisture.

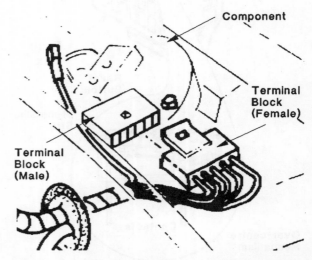

Figure 3-48 A typical multi-pin terminal block.

CONNECTORS

Connectors are used to join two or more sections of the electrical wiring harness together. They consist of a plastic moulded plug and socket with one or more wires attached. The internal pins can be of round or flat cross-section. The connector may have a locking device to ensure a good attachment and a seal to protect the connectors from moisture.

SINGLE PIN CONNECTOR

Connectors of this type are used for single wire connection of either a supply or earth wire.

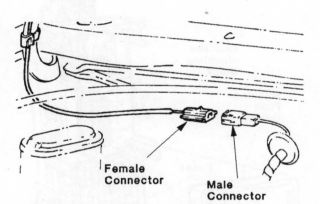

Figure 3-49 A single pin connector.

TWO PIN CONNECTOR

The main applications for this type of connector are:

- after-market accessories.
- supply and earth return for components which are insulated from the body of the vehicle. (e.g. lights in plastic holders).

To prevent the supply and earth wires being accidentally cross-connected, the pins within the connector may be offset. If the pins are equally spaced, the connector body can be shaped so connection is only possible when the correct pins are aligned.

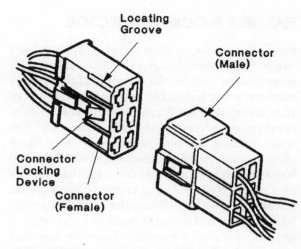

Figure 3-51 A typical multi-pin connector.

MULTI-PIN CONNECTOR

Wiring harness complexity, in a modern vehicle, requires the use of many of these connectors. They are available in a wide range of shapes, sizes and pin numbers. The shape is usually determined by the location of the connector in the harness, e.g. crescent shaped when mounted adjacent to the steering column. Various methods are used to prevent cross-connection between wires; these may take the form of pin spacing, key and groove or connector body shape.

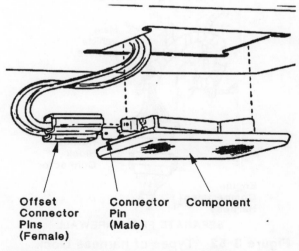

Figure 3-50 A two pin connector.

HARNESS BLOCK CONNECTOR

For convenience, a number of manufacturers use wiring system designs which include rigidly mounted connectors. An example of this type is commonly located on the engine firewall where the main vehicle harness and the engine compartment harness are linked together. The main vehicle harness terminates in a block (similar in appearance to a component terminal block) which is attached to the firewall by screws or a rubberised seal. The engine compartment harnesses are plugged into the rigidly mounted harness block. Cross-connection is prevented by pin spacing, connector shape or key and groove.

FUSE BOX

A major requirement, as electrical systems have become more complex, is to design into the vehicle a 'helping hand' for the servicing mechanic. Where possible, the wiring harnesses are manufactured so the fuses and relays are concentrated into a single mounting block. These blocks are protected by a cover which has, moulded into the surface, the identification details of each fuse and relay. When the cover is removed, the mechanic has access to major sections of the electrical system for testing and diagnosis.

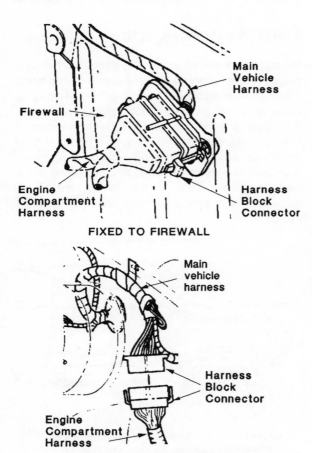

Figure 3-52 Types of harness block connectors.

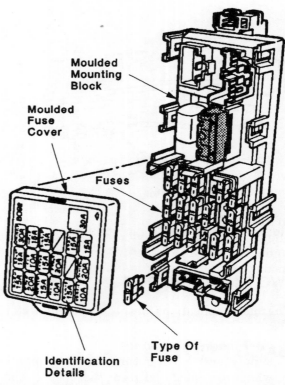

Figure 3-53 An example of an integrated fuse box.

4

CIRCUITS

A basic circuit consists of a power supply, a switch and a work unit connected together with wires. However, automotive circuits are more complex. As more components are added to the basic circuit they have an effect on those already in the circuit. This chapter describes these effects so that the reader can gain a basic understanding of circuits for purposes of modifying, servicing and testing.

BASIC CIRCUIT

A basic circuit (see Figure 4-1) consists of:

- a power supply—in a vehicle the prime source of electrical energy is derived from a 12 volt battery;
- supply and return paths—the supply path or conductor is an insulated wire. The return path is a combination of a wire and the vehicle body;

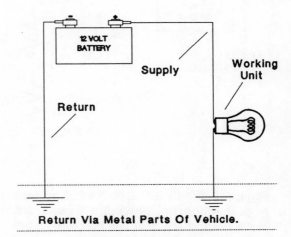

Return Via Metal Parts Of Vehicle.

Figure 4-1 Basic automotive circuit.

- the working unit—often called the load, the working unit is the globe, motor, resistor, coil or any other device within the circuit which converts electricity to another form of energy. (All illustrations in this section will have resistor symbols in substitution for the working units.)

Obviously the circuit illustrated has limited application within a motor vehicle. Most circuits require a battery, circuit protection, switches, connectors, working units and conductors. The arrangement of these parts within a circuit will vary depending on the type of circuit.

The three forms of electrical circuits are the:

- series;
- parallel; and
- a combination of both known as series/parallel.

SERIES CIRCUIT

A series circuit is a single path where all the current that flows through one working unit also flows through all other working units in the circuit.

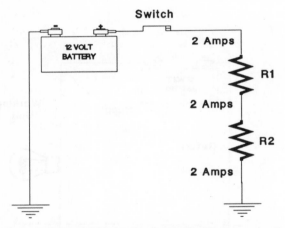

Figure 4-2 SERIES CIRCUIT: Any current (e.g. 2 Amperes) flowing will pass through both R1 and R2 and will have the same value at any test point.

PARALLEL CIRCUIT

A parallel circuit allows the current to follow alternative paths to complete the circuit. The amount of current flow through each path will

depend on the resistance value in each path. As the resistance value is increased the current flow will be decreased.

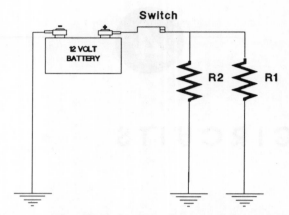

Figure 4-3 PARALLEL CIRCUIT: R1 and R2 operate independently of each other and the current flow in the alternate paths to earth will be determined by the resistance value in each path.

SERIES/PARALLEL CIRCUIT

As the name suggests, this circuit is a combination of a series circuit and a parallel

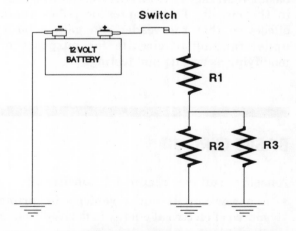

Figure 4-4 SERIES PARALLEL CIRCUIT: All current in the circuit flows through R1 but is then split through R2 and R3. The combined current flow through R2 and R3 will be equal to the current flow at R1.

circuit. In the example (see Figure 4-4) shown *all* current must pass through resistor 1 but alternative paths are available through resistor 2 or resistor 3.

CURRENT IN A CIRCUIT

The flow of electrons through a conductor is called current, and is measured in amperes. Current in a circuit can be described by the 'conventional' or the 'electron' theory.

- Conventional theory—current is considered to flow from the positive terminal of the voltage supply through the circuit and returns to the negative terminal.
- Electron theory—current is considered to flow from the negative terminal of the voltage supply through the circuit and returns to the positive terminal.

Note: The conventional theory of current flow is widely accepted by industry. All references to current flow in this book will be based on the conventional theory.

Regardless of the type of circuit used, all the electrons (current) leaving one of the voltage supply terminals must return to the other terminal.

SERIES CIRCUIT CURRENT

The current in a series circuit will pass through all the components. The circuit illustrated (Figure 4-5) has four resistors connected end to end to form the path for the current. Total current is determined by dividing the supply voltage by the total resistance of the circuit, i.e. 12 volts divided by 12 ohms equals 1 ampere. The 1 ampere of current must pass through each of the resistors.

PARALLEL CIRCUIT CURRENT

The current in a parallel circuit will be the same at the beginning of the circuit (leaving the supply) as it will be at the end of the circuit (returning to the supply). However, all the current does not have to pass through all the components in the circuit. The circuit illustrated (Figure 4-6) has four resistors connected in parallel to form the paths for the current. It can be seen that part of the current is diverted through each of the parallel resistors. The current through each path is inversely proportional to the value of the resistor in that

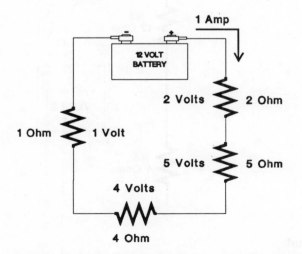

Figure 4-5 Current flow in a series circuit.

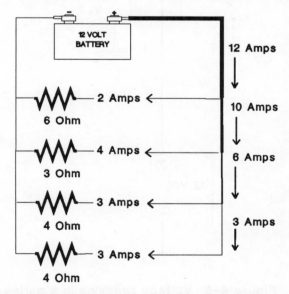

Figure 4-6 Current flow in a parallel circuit.

path. An increase in resistor value causes a decrease in current. Conversely, a decrease in resistor value will cause an increase in current.

SERIES-PARALLEL CIRCUIT CURRENT

As depicted in Figure 4-7, the section of the circuit containing the 2-ohm resistor is a series circuit (single current path). The remaining resistors form a parallel circuit (two current paths available). All current in the circuit must pass through the 2-ohm resistor and is then shared between the two parallel resistors. The current through each path is inversely proportional to the value of the resistor in that path. An increase in resistor value causes a decrease in current. Conversely, a decrease in resistor value will result in an increase in current.

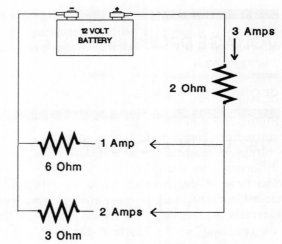

Figure 4-7   **Current flow in a series/ parallel circuit.**

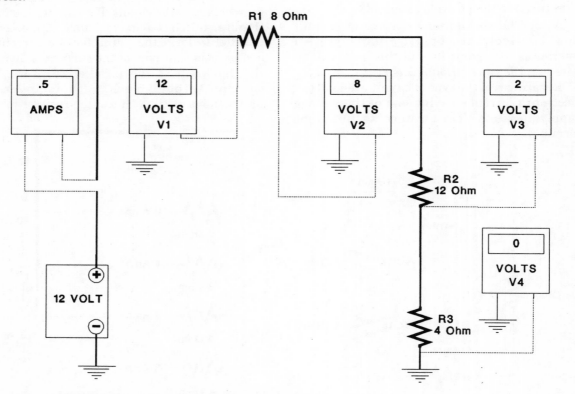

Figure 4-8 Voltage readings in a series circuit.

VOLTAGE DROP IN A CIRCUIT

SERIES CIRCUIT VOLTAGE DROP

Figure 4-8 shows a series circuit with four voltmeters connected at various points. Each voltmeter reading has a different value. The difference in reading between each of the voltmeters is the voltage drop between the points at which they are connected into the circuit.

- V1 is connected at a short distance from the battery along the main cable. The reading is the same as the voltage of the battery because the resistance between the two points is too small to record a different value.
- V2 is connected on the negative side of resistor R1. The reading is 8 volts. When

this voltage reading is compared to the reading of V1, it will be noted that a difference of 4 volts has occurred. For current to pass through a resistor, work must be done, and energy expended (in this case, converted to heat). The voltage was 'lost' from the circuit as the current (electrons) was pushed through the resistor. This loss can be expressed as 4 volts **voltage drop** across the resistor.

- V3 is connected to the negative side of resistor R2. The reading is 2 volts. A voltage drop of 6 volts has occurred across resistor R2. The energy loss from the circuit resulted in a heating of the resistor.
- V4 is connected to the negative side of resistor R3. The reading is zero volts. This indicates that there is no potential difference between the connection point of the voltmeter and the negative terminal of the battery.

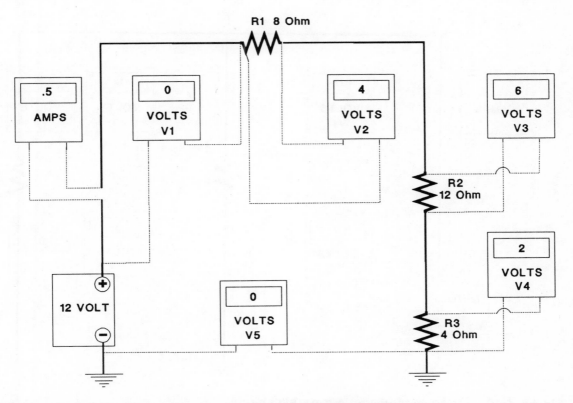

Figure 4-9 Voltage drop across resistors in a series circuit.

The voltmeters have been reconnected (see Figure 4-9) across each of the resistors. The voltage readings will be the potential difference (voltage drop) between the ends of the resistors. When these readings are added together, the result must equal the supply voltage, which in this case is 12 volts.

SUMMARY OF VOLTAGE DROP IN THE SERIES CIRCUIT

- Voltage drop is caused by current flowing through resistance.
- The voltage drop across each resistance is proportional to the value of the resistance.
- If resistances are equal, the voltage drop across them will also be equal.
- The sum of the voltage drops in the circuit must equal the supply voltage.

PARALLEL CIRCUIT VOLTAGE DROP

Figure 4-10 shows a parallel circuit with four voltmeters connected at various points. The difference in reading between each of the voltmeters is the voltage drop between the points at which they are connected into the circuit.

- V1 and V2 are connected at a short distance from the battery along the main cable. The readings are the same as the voltage at the battery. These voltmeters are connected across the resistors to earth.
- V3 and V4 are connected on the negative side of resistor R1 and R2. The readings are zero volts. When these voltage readings are compared to the readings of V1 and V2, it will be noted that a difference of 12 volts has occurred.

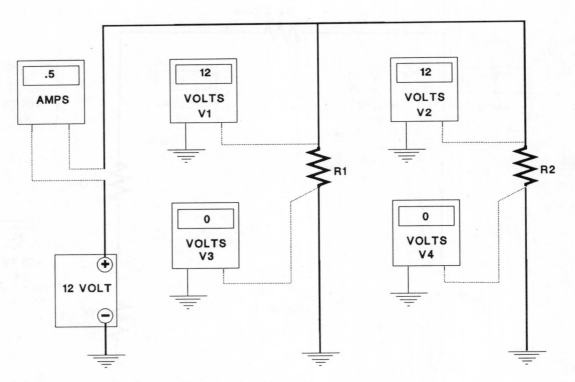

Figure 4-10   Voltage readings in a parallel circuit.

SUMMARY OF VOLTAGE DROP IN THE PARALLEL CIRCUIT

• All resistors have the same supply voltage.
• Each resistor has a direct connection to the battery therefore the voltage drop across each resistor will be the supply voltage.

SERIES PARALLEL CIRCUIT VOLTAGE DROP

Figure 4-11 shows a series-parallel circuit with three voltmeters connected at various points. The difference in reading between each of the voltmeters is the voltage drop between the points at which they are connected into the circuit.

• V1 is connected at a short distance from the battery along the main cable. The reading is the same as the voltage at the battery.

• V2 and V3 are connected on the negative side of resistor R1. The readings are both 8 volts. When these voltage readings are compared to the reading of V1, it will be noted that a difference of 4 volts has occurred. This loss can be expressed as 4 Volts voltage drop across the resistor R1.

SUMMARY OF VOLTAGE DROP IN THE SERIES-PARALLEL CIRCUIT

• The voltage drop across the series resistor will decrease the voltage available to the remainder of the circuit.
• Each parallel resistor has a direct connection to the reduced voltage therefore the voltage drop across each resistor will be equal.
• The sum of the voltage drops across the series resistor and one of the parallel resistors will equal the battery voltage.

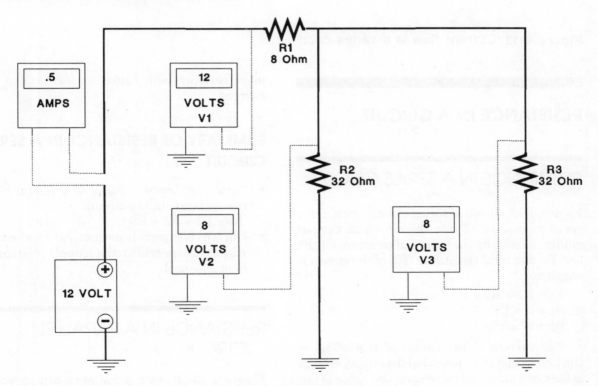

Figure 4-11   Voltage readings in a series / parallel circuit.

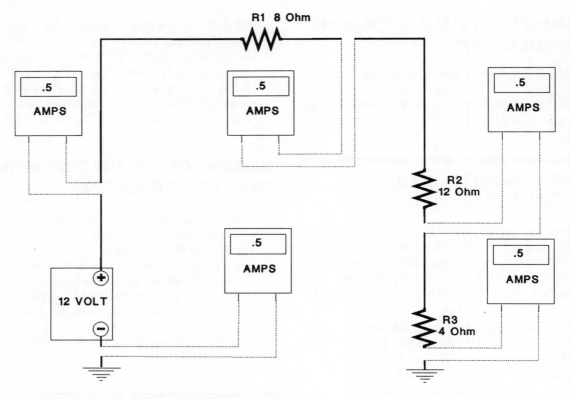

Figure 4-12 Current flow in a series circuit.

RESISTANCE IN A CIRCUIT

RESISTANCE IN A SERIES CIRCUIT

Figure 4-12 shows a series circuit with three resistors connected into the circuit at various points. By adding the values of resistors R1, R2 and R3, the total resistance (Rt) of the circuit is obtained.

Rt = R1 + R2 + R3
Rt = 8 + 12 + 4
Rt = 24 ohms

The current in the circuit is dependent on the total resistance, provided the supply voltage is held constant. By increasing the value of one or more resistors, the current will decrease. Alternatively, a decrease in the value of one or

more resistors will cause an increase in the current.

SUMMARY OF RESISTANCE IN A SERIES CIRCUIT

- Total resistance is equal to the sum of all the resistors in the circuit.
    Rt = R1 + R2 + R3
- Current is inversely proportional to the total resistance provided the supply voltage is held constant.

RESISTANCE IN A PARALLEL CIRCUIT

Figure 4-13 shows a parallel circuit in which the current has two flow paths, each path containing a resistance. Because the current

---

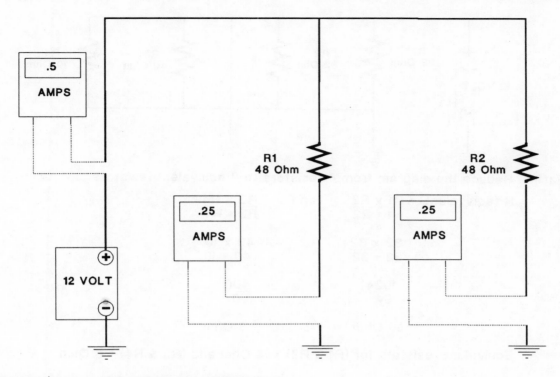

Figure 4-13 Current flow in a parallel circuit.

has alternate paths, the total resistance cannot be calculated by adding the value of the resistors. To calculate the value of an equivalent resistor (Rt) that would replace the two resistors in the circuit, it is necessary to use the following formula.

$$Rt = \frac{R1 \times R2}{R1 + R2}$$

$$Rt = \frac{48 \times 48}{48 + 48}$$

$$Rt = \frac{2304}{96}$$

$$Rt = 24 \text{ ohms}$$

A more complicated formula is available for circuits containing more than two parallel resistances. Generally the automotive service persons' needs are met by dividing the circuit's resistors into pairs and applying the simple formula shown. The formula has to be applied several times until a single equivalent resistance value is obtained (see Figure 4-14).

SUMMARY OF RESISTANCE IN A PARALLEL CIRCUIT

- The value of the equivalent resistance Rt will be less than the smallest resistor in the circuit.

- Rt is calculated by the formula
$$Rt = \frac{R1 \times R2}{R1 + R2}$$

RESISTANCE IN A SERIES-PARALLEL CIRCUIT

Figure 4-15 shows a circuit with three resistors. R1 is connected as a series resistor, R2 and R3 are parallel resistors. All current in the circuit

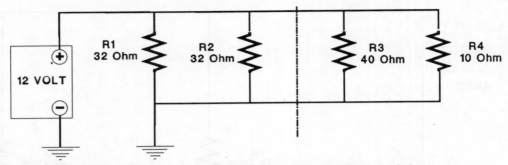

Part 1: Reduce the diagram from 4 resistors to 2 equivalent resistors.

$$R \text{ (equivalent)} = \frac{R1 \times R2}{R1 + R2} \quad \text{and} \quad \frac{R3 \times R4}{R3 + R4}$$

$$\frac{32 \times 32}{32 + 32} \qquad \frac{40 \times 10}{40 + 10}$$

$$\frac{1024}{64} \qquad \frac{400}{50}$$

$$16 \text{ Ohm} \qquad 8 \text{ Ohm}$$

Equivalent resistors for (R1 & R2) = 16 Ohm and (R3 & R4) = 8 Ohm.

The previous diagram can be redrawn as shown below.

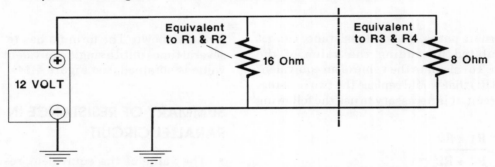

Part 2: Reduce the diagram from 2 resistors to 1 equivalent resistor.

$$Rt \text{ (equivalent)} = \frac{R1 \times R2}{R1 + R2}$$

$$\frac{16 \times 8}{16 + 8}$$

$$\frac{128}{24}$$

$$Rt = 5.33 \text{ Ohm}$$

The total circuit resistance (R1 & R2 & R3 & R4) is equal to 5.33 Ohm.

Figure 4-14 Changing parallel resistances to a single equivalent resistance value.

will flow through R1, then divide and flow through R2 and R3. To calculate the total resistance of this circuit the parallel circuit equivalent resistance must be calculated first. The equivalent resistance value is substituted for the parallel resistors in the circuit, then the total resistance can be calculated by adding R1 and the equivalent resistance value (see Figure 4-16).

SUMMARY OF RESISTANCE IN A SERIES-PARALLEL CIRCUIT

- All resistors in the parallel section of the circuit must be reduced to an equivalent resistance.
- All the resistors in the series section of the circuit must be added to the parallel equivalent resistance to give the **total resistance** of the circuit.

ENERGY LOSSES IN A CIRCUIT

The main cause of energy loss in a circuit is heat. As current passes through a resistor (resistance), heat is emitted to the surroundings. **Unwanted resistance** (see Figure 4-17) in a circuit is a common reason for energy loss. It is generally formed in connectors

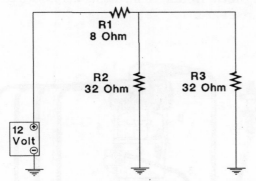

Figure 4-15   Resistances in a series/parallel circuit.

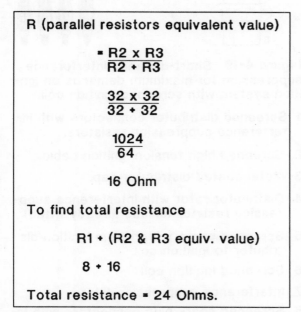

Figure 4-16 Calculation of total resistance in a series/parallel circuit.

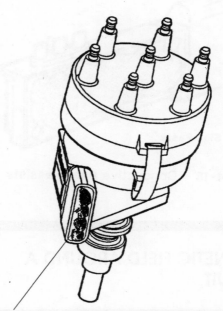

APPLY 0.9 GRAM OF DOW CORNING HEAT SINK COMPOUND 340 OR EQUIVALENT TO INSIDE OF TFI SOCKET ON DISTRIBUTOR.

Figure 4-17   To prevent water ingress, which could cause corrosion (unwanted resistance), a sealing compound is applied to the terminal socket.

or terminals when they are loose, dirty or corroded. This unwanted resistance causes excessive heat and a voltage drop in the circuit. It is best detected by connecting a voltmeter to each side of the connector or terminal while the circuit has current passing through it.

Two important aspects of an electronic circuit are the design and the layout. These two factors must be considered in order to ensure that heat (temperature) sensitive devices are protected from excessive heat.

Some electronic devices must be mounted on **heat sinks** to prevent them from overheating. A common type of heat sink used for automotive applications is an aluminium plate placed in an air stream or attached to a large thick piece of metal (see Figure 4-18). A heat conductive paste is smeared on the mating surfaces of the electronic device and the heat sink before they are screwed together.

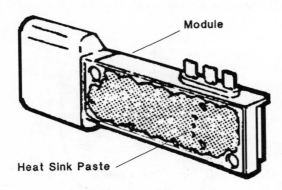

Figure 4-18 Conductive paste assists cooling.

MAGNETIC FIELD AROUND A CIRCUIT

DIRECT CURRENT (DC) CIRCUITS

Wires (conductors) that have current passing through them have weak magnetic fields formed around them. Generally, these weak magnetic fields have little effect on wires of other circuits that are bundled in the same harness. However, when the voltage or the current is very high, a magnetic field can develop to a strength that will induce unwanted current into an adjacent wire. The unwanted current can cause serious damage to electronic components ('burn them out'). This may occur, for example, when the high tension leads (spark plug leads) are placed alongside low tension leads.

Two methods used to prevent **unwanted magnetic fields** from cutting across wires in a circuit are:

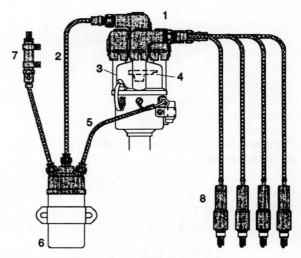

Figure 4-19 Short-distance interference suppression for maximum demands on ignition system with screened ignition coil:

1 Screened distributor connectors with interference suppression resistors.

2 Screened high tension ignition cable.

3 Metal coated distributor cap.

4 Distributor rotor with interference suppression resistor (unless already fitted).

5 Screened primary lead from ignition distributor to ignition coil.

6 Screened ignition coil.

7 Interference suppression filter.

8 Screened spark plug connectors with interference suppression resistors.

- to ensure wires that are subjected to high voltages or current are isolated (kept away) from other wires;
- to place a metal shield around the wires or install a metal box around the circuit.

ALTERNATING CURRENT (AC) CIRCUITS

Wires (conductors) that have AC passing through them have weak pulsating magnetic fields formed around them. Generally, these weak pulsating magnetic fields have little effect on wires of other circuits that are bundled together in the same loom. However, when the voltage or the current is very high, a strong pulsating magnetic field is formed around the wire. The strength of this pulsating magnetic field can be increased by winding the wire around a metal core or by forming a coil (solenoid).

To prevent excessive heat being generated, thin sheets of metal (laminations) are used to form the core (see Figure 4-20) instead of a solid piece of metal. Transformers have laminated metal or ceramic cores.

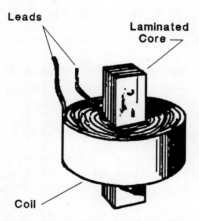

Figure 4-20 Laminated core is fitted within a coil of wire.

INDUCTANCE AND CAPACITANCE IN A CIRCUIT

Ohm's Law is ideal when steady flow conditions exist however, when current varies with time Ohm's Law does not fully describe the events that are taking place in a circuit. To explain the conditions that exist when current and voltage are changing it is necessary to introduce two new terms: inductance and capacitance.

CAPACITANCE IN A SERIES CIRCUIT

Figure 4-21 shows the capacitors connected end to end and then connected through a switch to the power supply. The total capacitance of the circuit is determined by calculating an equivalent value of the two capacitors. To calculate the value of an equivalent capacitor (Ct), it is necessary to use the following formula:

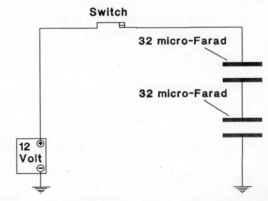

Figure 4-21 Capacitors connected in series within a circuit.

$$Ct = \frac{C1 \times C2}{C1 + C2}$$

$$Ct = \frac{32 \times 32}{32 + 32}$$

$$Ct = \frac{1024}{64}$$

Ct = 16 microfarads.

It is unlikely that the automotive service person will be required to solve circuits containing more than two capacitors. Should it be necessary to evaluate the capacitance of a complex circuit, the formula can be applied in a similar manner to that developed when solving parallel resistors.

CAPACITANCE IN A PARALLEL CIRCUIT

Figure 4-22 shows the capacitors connected in parallel and then connected through a switch to the power supply. The total capacitance (Ct) of the circuit is determined by adding the values of the individual capacitors.

Ct = C1 + C2
Ct = 32 + 32
Ct = 64 microfarads

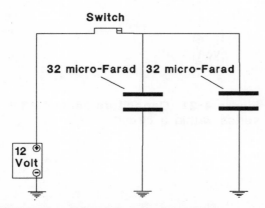

Figure 4-22 Capacitors connected in parallel within a circuit.

INDUCTANCE IN A SERIES CIRCUIT

Figure 4-23 shows a series circuit with two inductors connected into the circuit. By adding the values of the inductors L1 and L2 the total Inductance (Lt) of the circuit is obtained.

Lt = L1 + L2
Lt = 16 + 16
Lt = 32 milli-Henry

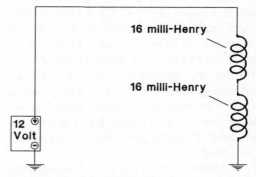

Figure 4-23 Inductors connected in series within a circuit.

INDUCTANCE IN A PARALLEL CIRCUIT

Figure 4-24 shows a parallel circuit in which the current has two flow paths, each path containing an inductor. Because the current

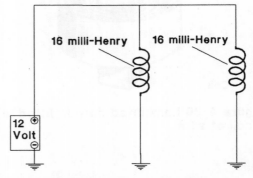

Figure 4-24 Inductors connected in parallel within a circuit.

has alternate paths, the total inductance (Lt) cannot be calculated by adding the value of the inductors. To calculate the value of an equivalent inductor (Lt) that would replace the two inductors in the circuit, it is necessary to use the following formula.

$$Lt = \frac{L1 \times L2}{L1 + L2}$$

$$Lt = \frac{16 \times 16}{16 + 16}$$

$$Lt = \frac{256}{32}$$

$$Lt = 8 \text{ milli-Henry}$$

5

AMPS, VOLTS AND OHMS

Many of the component operations on a vehicle are physical—the method of operation is visible and can be readily understood.

When considering electrical and electronic systems the problem of visualising the operation becomes more difficult. The effect of an electrical current is observable—a starter motor turns or a lamp is illuminated—but electricity cannot be 'seen' in the same way as it is possible to observe the flow of oil or fuel.

To test electrical circuits and components a basic knowledge of the flow rate (current), the flow pressure (voltage) and the resistance to flow (resistance) of electricity is required.

AMPERE

Current is the flow rate of electrons per second in a conductor. An electron is very small and the number passing a point in the conductor is huge (6.8 billion billion electrons per second), therefore to describe current flow in electrons per second is not practical for every day use. For convenience the term ampere is used (1 ampere = 6.8 billion billion electrons per second).

A general statement to describe current would be: *The flow of electrons through a wire (conductor) is called current and is measured in amperes.*

VOLTAGE

A force must be applied to cause electrons to move along a wire. To supply this force it is necessary to have a positive charge at one end and a negative charge at the other end of the wire. While a charge difference exists between the ends of the wire, electricity will flow through the wire. By increasing the difference in charge values at each end of the wire a greater force can be produced. This force can be described as a potential difference or by the more commonly used term, 'voltage'. The unit of measurement is the volt.

Voltage is produced between two points when a positive charge exists at one point and a negative charge exists at the other point.

RESISTANCE

All materials have resistance to current flow. By careful selection of materials it is possible to have virtually no flow of electrons or extremely high electron flow rate. Materials with a low electron flow are glass, ceramics and plastics; these materials can be used as insulators. To allow a high electron flow, materials such as copper, aluminium and gold are used. These materials are classed as conductors.

When a short section of wire, smaller in diameter than the main wire, is placed in the circuit, it acts as a restrictor. This feature, coupled with selection of materials, can be used to create a heating effect, as seen in a headlamp globe. The special wire in the globe restricts the flow of electrons and glows brightly producing light for the headlamp.

Unfortunately resistance can be a problem as well as a benefit. Wiring which is too small in cross-section, or a corroded connection, can also act as a resistance, slowing the flow rate of electrons and preventing the component at the end of the wire from operating correctly. Resistance values are measured in ohms.

The basic unit of resistance is the ohm and is defined as the resistance that will allow 1 ampere to flow when the potential difference is 1 volt.

OHM'S LAW

Ohm's law, which governs the relationship between current flow, voltage and resistance, can be expressed as an equation. Electrical

convention uses 'I' for current, 'E' or 'V' for volts and 'R' for resistance when explaining or calculating problem solutions using Ohm's law. To simplify this process we will use the S.I. convention and of V for volts (electrical pressure), A for ampere (current flow) and Ω for ohms (resistance).

An extension to Ohm's law is wattage. Wattage is the power consumed by an electrical or electronic component when a current flows through it to produce work (movement, light etc.).

The application of Ohm's law can be best related to a component connected to a battery. Generally, this component will have a watts rating which involves volts, amps and resistance.

Take the situation of fitting driving lights to the front of a vehicle (see Figure 5-2). If each light is rated at 100 watts therefore the power consumption of the two lights will be 200 watts. When these are connected to a 12 volt system, the current draw will be 17 amps.

Calculations are:

100 watt x 2 lights =
200 watts total power

$$\frac{200 \text{ watts}}{12 \text{ volts}} = 17 \text{ amps}.$$

From the current flow calculations, the mechanic can see the need for a relay to bypass the headlight switch (see Figure 5-3). If a relay was not used, the wiring to the headlight switch and the switch contacts would be incapable of carrying the combined current flow for both the headlights and the driving lights. Obviously, the relay selected will require an ampere rating greater than 17 amps.

This example is one of many applications of Ohm's law. You can use this law to assist you in problem solving or the designing of an electrical circuit by understanding a few basic concepts.

CURRENT FLOW

To calculate current flow (amp) when voltage and resistance are known:

Given that the combined resistance of an

OHM'S LAW

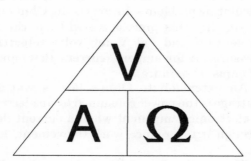

A simple approach to Ohm's Law is to consider the statement, 'Current flowing in a circuit is proportional to the voltage and inversely proportional to the resistance' ; as three values contained in a triangle. To change the symbols into 3 useful formulae it is only necessary to cover the unknown value as shown.

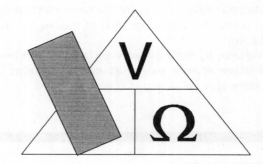

To find A with V and Ω known, cover A.

The formula becomes:

$$A = V / \Omega$$

AMPERES = VOLTS divided by OHMS

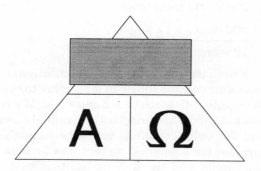

To find V with A and Ω known, cover V.

The formula becomes:

$$V = A \times \Omega$$

VOLTS = AMPERES x OHMS

Figure 5-1 Ohm's Law in simplified form.

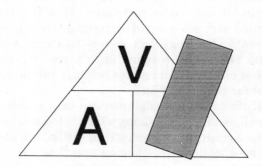

To find Ω with V and A known, cover Ω.

The formula becomes:

$$\Omega = V / A$$

OHMS = VOLTS divided by AMPERES

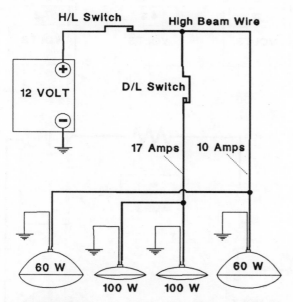

Figure 5-2 Two 100 Watt driving lights, connected to the battery through the high beam wire and the headlight switch, more than double the current through the switch.

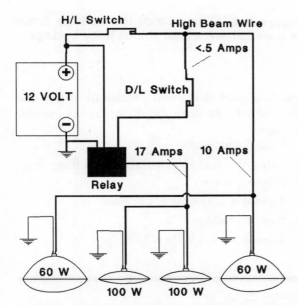

Figure 5-3 The addition of a relay enables 17 Amperes to be supplied to the driving lights, but less than .5 Ampere is added to the current through the switch.

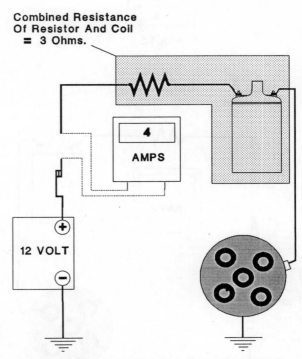

Figure 5-4 Current flow in the circuit depends on the voltage available and the total circuit resistance.

ignition coil and ballast resistor is 3 ohms and the vehicle system is 12 volts, the current flow through the primary circuit is 4 amps.

Calculation using $A = V/\Omega$:

$$\frac{12 \text{ volts}}{3 \text{ ohms}} = 4 \text{ amps}$$

VOLTAGE

To calculate the voltage drop across a component when the current flow and the resistance are known:

Given that an ignition coil primary winding has a resistance of 1.8 ohms and a ballast resistor has a resistance of 1.2 ohms with a current flow of 4 amps through the circuit. Calculate the voltage applied to the ignition coil primary winding when the vehicle is fitted with a 12 volt battery.

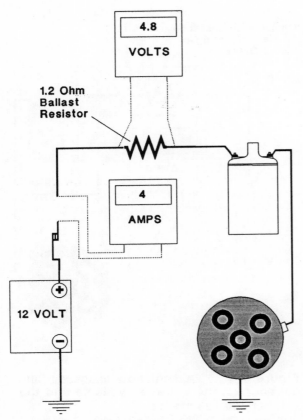

Figure 5-5 The voltage at the coil terminal will be less than 12 Volts when a ballast resistor is in the circuit.

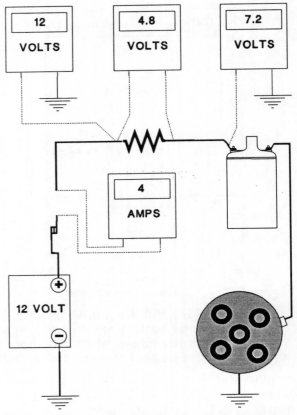

Figure 5-6 The sum of the voltage drops in a circuit will equal the battery voltage.

Before commencing the calculation, it is necessary to consider the problem.

1 The voltage available at the coil will be 12 volts minus the voltage drop across the resistor.
2 Calculate the voltage drop across the resistor.
3 Subtract this voltage (step 2) from 12 volts.

 a Calculation using $V = A \times \Omega$:

 Ballast resistor voltage drop (see Figure 5-5),

 4 amps x 1.2 ohms = 4.8 Volts

 b Voltage at coil,

 12 volts—4.8 volts = 7.2 volts

The result can be checked by using the same method to calculate the voltage drop across the coil. The sum of the coil voltage drop and the resistor voltage drop should equal the system voltage.

 Calculations are:

a using $V = A/\Omega$ to calculate the voltage drop across the coil,

 4 amps x 1.8 ohms = 7.2 volts

b System voltage,

 7.2 volts + 4.8 volts = 12 volts

RESISTANCE

To calculate the resistance (ohms) of a component, circuit or a connector, the current flow and voltage drop needs to be measured.

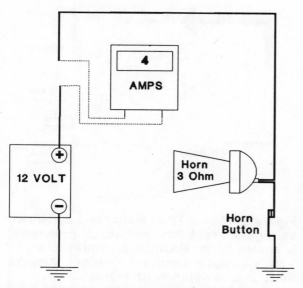

Figure 5-7   Total circuit resistance can be calculated if the system voltage and the current flow are known.

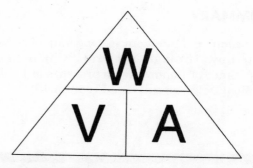

Figure 5-8 Three formulae can be developed from the 'WATTAGE' triangle. Cover the unknown value and the formula will be shown (refer Figure 5-1).
To find:
a) Watts when Volts and Amps are known
    W = V multipled by A
b) Volts when Watts and Amps are known
    V = W divided by A
c) Amps when Watts and Amps are known
    A = W divided by V

Given a horn circuit (see Figure 5-7) that has a current flow of 4 amps and a voltage drop across the horn of 12 volt, calculate the resistance of the circuit.
Calculation of resistance:

Using Ω = V/A,

$$\frac{12 \text{ volts}}{4 \text{ amps}} = 3 \text{ ohms}$$

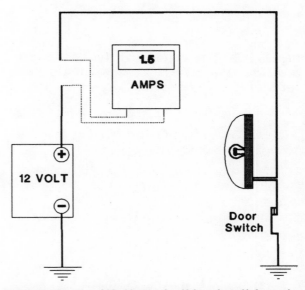

Figure 5-9   Wattage in this circuit is calculated from the battery voltage multiplied by the current flow.

WATTAGE

To calculate the wattage consumed by an electrical or electronic component, the current flow and voltage drop needs to be measured.

Given a dome light circuit (see Figure 5-9) which has a current flow of 1.5 amps and a voltage drop of 12 volts, calculate the wattage consumed by the globe.
Calculation of wattage:

Using W = A x V,

1.5 amps X 12 volts = 18 watts

SUMMARY

The common statement is: One amp of current will flow through a circuit which has a resistance of 1 ohm when a pressure of 1 volt is applied to the circuit (see Figure 5-10).

HEAT

As a component or circuit consumes power (wattage), heat is given off to its surroundings. This heat can be used to an advantage by fitting an element (demister) in a rear window, however, it can be a problem when too much current is passed through a circuit causing the wires or the component to 'burn out'.

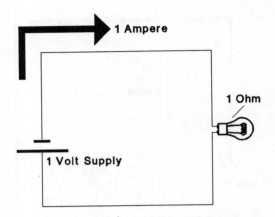

Figure 5-10   The relationship between voltage, current flow and circuit resistance is shown in the illustration. A pressure of 1 Volt will cause a current flow of 1 Ampere through a resistance of 1 Ohm.

6

ELECTRONIC COMPONENTS— IDENTIFICATION

The science of electronics is not new but advances in miniaturisation have provided devices that can be easily adapted for automotive applications. These electronic devices operate on much lower voltages and current flows than the electrical devices in the motor vehicle. Although they may look similar to electrical devices, it is important that they be recognised as electronic devices because of the different methods required for their servicing and testing.

integrated circuits and other solid state devices. Instead of wires , a series of very thin copper conductors are used to link the components. The copper conductors resemble a road map of a small town and are normally on one side only of an insulated board. The components are mounted on one side of the board, opposite the conductors. The mounting terminals of each component pass through holes in the board and are soldered to the copper conductors.

CIRCUIT BOARDS

The modern solid state devices are relatively small in size and do not require the sturdy mounting chassis used with older valve and individual component circuits. A mounting board known as a printed circuit (PC board) is used to cross-connect and mount transistors,

SEMI-CONDUCTORS

A conductor allows current to flow to the point where it is required. Materials such as silver and copper are classed as good conductors because they allow current to flow freely. When a material does not allow current to flow, it may be classified as an insulator or a non-conductor, e.g. rubber or PVC.

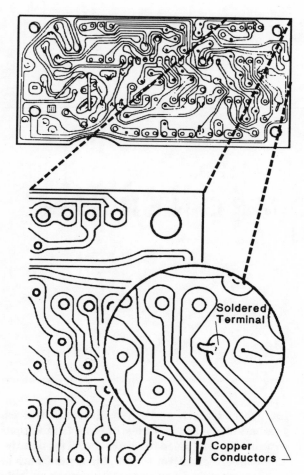

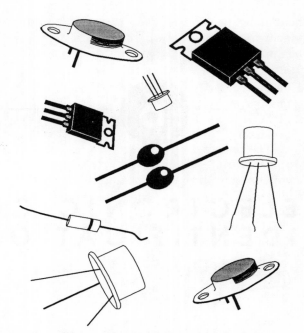

Figure 6-2   Common semi-conductors.

Figure 6-1   Leads from components are passed through the PC board and soldered to the copper tracks (conductors).

A semi-conductor is an electronic component which acts as a conductor under certain conditions and as a non-conductor under other conditions. There are many different types of semi-conductors (see Figure 6-2) but the thermistor, diode and transistor are the most widely used in automotive application.

THERMISTORS

A thermistor is a semi-conductor device in which the resistance value will vary considerably with temperature change. The thermistor can be

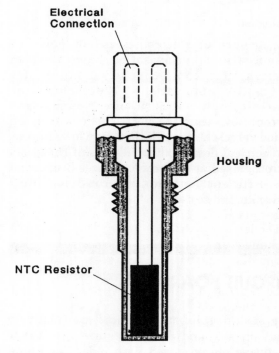

Figure 6-3   NTC resistor used as an engine coolant temperature sensor.

heated by passing current through it or by placing it in a position where external heat can be applied to it. The two groups of thermistors are:

1 NTC (negative temperature coefficient) resistors;
2 PTC (positive temperature coefficient) resistors.

NTC resistors are used in a variety of applications, including temperature compensation in electronic amplifier circuits. Figure 6-4 shows a wide range of shapes and sizes of these resistors.

PTC resistors are used for current limiters and time switches.

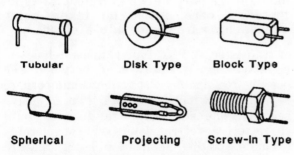

Tubular   **Disk Type**   **Block Type**

Spherical   **Projecting**   **Screw-in Type**

Figure 6-4   Shapes of NTC resistors.

DIODE

A diode is a semi-conductor device; this means that under certain conditions the device will behave as an insulator and under a different set of conditions it becomes a conductor. This remarkable behaviour is caused by the materials and construction of the diode.

Diodes are available in a large range of packages and sizes. The devices used in automobiles range in size from a few millimetres to 25 mm in diameter. The current, in the circuit to be controlled, will determine the type and size of the diode required.

Zener diodes are similar in appearance to a small signal diode, however their identification markings are different. To determine their application, a suitable reference chart is essential. They are used in circuits as voltage conscious switches.

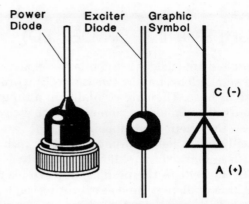

Power Diode   Exciter Diode   Graphic Symbol

C (-)

A (+)

Figure 6-5   Typical alternator diodes.

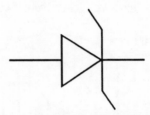

Figure 6-6   Zener diode symbol.

PHOTO DIODES

The casing of a photo diode is cylindrical with a plastic or glass window in one end. The connection terminals are in the opposite end. They are designed to detect light and can be used in conjunction with a light source and a slotted plate to provide a switching action.

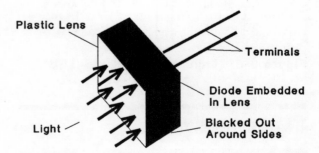

Plastic Lens

Terminals

Diode Embedded In Lens

Light

Blacked Out Around Sides

Figure 6-7   Photo diode: Light passing through the transparent lens falls on the PN junction of the diode, allowing a small current to flow. The diode may also be packaged with a 'window', at one end of the case, acting as the lens.

LIGHT EMITTING DIODE (LED)

Light emitting diodes are similar in appearance to a small globe, having two connecting prongs at their base. They are available in a range of colours. They must be connected into a circuit observing their correct polarity. The 'anode' usually has a longer prong than the 'cathode' which is marked by a 'flat' on the body of the LED adjacent to the prong. They are used in instrument displays and as operating condition lights (CPU air/fuel ratio indicator).

SIGNAL LIGHT 'PACKAGE'

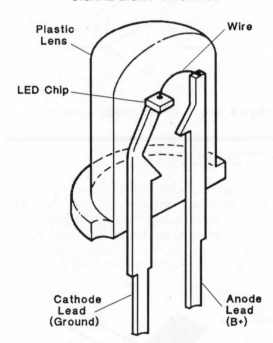

Figure 6-8 Light emitting diode (LED).

TRANSISTORS

The transistor is a semi-conductor device containing three semi-conductor regions, each of which has an external connection point. These points or terminals are called base, collector and emitter. The type of material used in the semi-conductor regions will determine the type

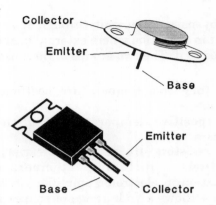

Figure 6-9 The illustration shows common arrangements of the collector, base and emitter connection pins. The pin identification will vary between transistor types, therefore care should be taken before connecting a transistor into a circuit.

of transistor. A PNP type transistor has one negative region sandwiched between two positive regions. A NPN type has one positive region sandwiched between two negative regions. The size, shape and operational characteristics of transistors vary considerably depending on their application. Transistors are used for many switching and amplifying functions.

THYRISTORS

Thyristor is the general name applied to the group of semi-conductor devices known as SCRs (silicon controlled rectifiers) and Triacs.

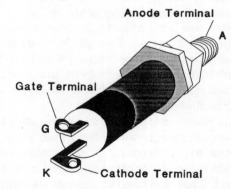

Figure 6-10 Large 'bolt-in' thyristor.

Generally the type used in automotive applications is the SCR.

The function of a thyristor can be compared to a relay that is turned on by a control current but turned off by removal of the main flow current. The internal components can be represented as two PN or NP diodes in series. In addition to the conventional two leads for anode and cathode connections, the thyristor has a third connection known as the 'gate'. The thyristor is available in a wide range of sizes and current handling capabilities. Thyristors may be packaged as shown in Figure 6-10, but are also available in packages which resemble small transistors. The connection terminals are identified as K, A, and G (cathode, anode and gate). This semi-conductor device is used in a range of units as diverse as CDI ignitions and alternator (24V) over voltage protection controls.

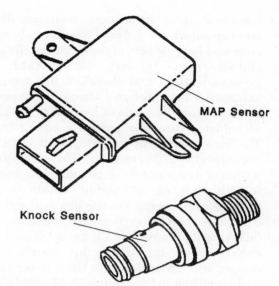

Figure 6-12 Examples of piezo-resistive manifold absolute pressure (MAP) and piezoelectric detonation (knock) sensors.

OXYGEN SENSORS

The oxygen sensor (Lambda Sensor) consists of a ceramic body contained in a metal housing. The sensor is mounted into the wall of the exhaust manifold and is used to monitor the oxygen content of the exhaust gas. The computer is linked to the sensor by either a shielded wire or, on some types, three wires.

A piezoresistive sensor is housed in a plastic or metal container. Several electrical wires permit the sensor to be connected to the vehicle's electrical circuits. A rubberised hose connects the unit to the inlet manifold of the engine. Piezoresistive devices are commonly used as Manifold Absolute Pressure (MAP) sensors and are generally located on the vehicle's firewall near the engine's inlet manifold.

A piezoelectric sensor is housed in a metal body which is threaded at one end and has a hexagonal head at its other end. Several wires permit the sensor to be connected into the vehicle's electrical circuits. The main application for a piezoelectric device is a knock (engine detonation) sensor. These units may be screwed into the engine's cylinder head or block.

Figure 6-11 Oxygen (Lambda) sensor.

CRYSTAL SENSORS

Two types of crystal sensors, used for automotive applications, are piezoelectric or piezoresistive devices.

INTEGRATED CIRCUITS

An integrated microcircuit, popularly known as a 'computer chip', is a miniaturisation of many of the components previously described.

Devices such as transistors, resistors, diodes and capacitors are reduced to extremely small sizes and become part of the circuit of the chip. Inductors (coils) are not suitable for miniaturisation, and therefore are mounted externally. The mechanic should not expect to see miniature versions of transistor packages all soldered into a copper track circuit board. The construction of an integrated circuit is completely different to the concept of individually packaged components. It is possible to build an integrated circuit containing 2000 or more components yet the physical size will be smaller than one 'normal' transistor. The following diagrams (see Figure 6-14) show the construction technique by which these components are included in the microcircuit.

In addition to the readily recognisable 'dual in-line' package it is possible to have integrated circuits available in packages resembling a power transistor or a small metal cylinder from which the mounting pins protrude.

Integrated circuits have many applications within automotive computers and electronic units. They perform functions such as counting pulses, pulse shaping, pulse generator, amplifying signals, controlling voltage levels, triggers, switches, signal converters, display drivers, voltage comparators, current limiting, decision making and data storage.

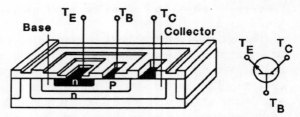

Diagram Of Integrated NPN Transistor.

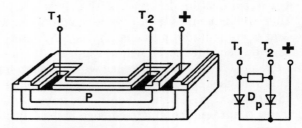

Integrated Resistor.

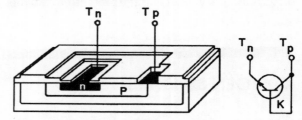

Integrated Diode.

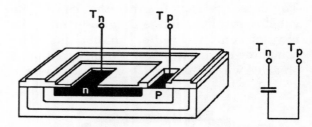

Integrated Capacitor.

Figure 6-14 Micro-components formed on n-type islands supported on a substrate of p type material.

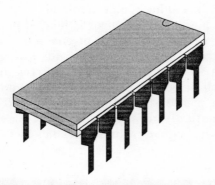

Figure 6-13 Typical integrated circuit 'computer chip'.

ELECTRONIC CONTROL UNIT

Electronic control unit (ECU) is a name that has been given to many devices that are housed in a metal or plastic box. These devices range from a simple integrated circuit (IC) to a microcomputer. Other names that have been used are the 'black box', the ECA (electronic control assembly), the electronic module, the MPU (main processing unit), the CPU (central processing unit), the 'microprocessor', and 'the computer'. Some of these units are connected to the electrical system by three or four wires and others may have more than twenty wires.

The main difference between these devices is the method used to process the input information before an output is sent to a work unit. This process varies considerably between an electronic module and a microcomputer.

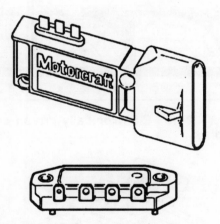

Figure 6-15 Two types of electronic ignition modules.

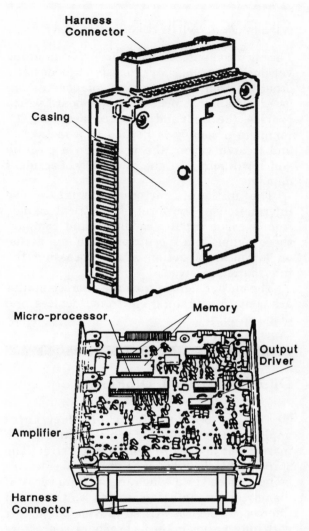

Figure 6-16 Micro-computer and casing.

ELECTRONIC MODULE

An ignition module or ignitor is a common type of electronic module. Electronic modules are located on or near the distributor. Several wires connect the module into the ignition system. The function of an ignition module is to control the primary circuit current of the ignition system.

MICROCOMPUTER

The microcomputer is a compact unit containing various ICs, transistors, resistors, diodes and other electronic devices. The components are assembled onto a printed circuit board which provides the input and output connections. The complete assembly is housed in a metal box and located within the vehicle in a position which will minimise the possibility of accident damage.

The function of a microcomputer is to receive information in the form of electrical signals from various inputs, condition and compare these signals to a specific program and decide on a course of action after processing the information received.

The output signals from the microcomputer are sent to actuators, electronic devices and display units.

DIGITAL VISUAL DISPLAYS

The most obvious form of electronics application, in the motor vehicle, must be the instrument display in front of the driver. The standard black and white dials and needles are being replaced by a wide range of liquid crystal display, light emitting diodes and vacuum fluorescent devices. These provide the driver with information in the form of bars, characters or graphic symbols. Many of these displays will not function effectively at the standard 12 volts normally available in the vehicle wiring circuit. Depending on the type of display it may be necessary to increase or decrease the supply voltage.

Clocks, speedometers, tachometers, fuel, voltage, temperature, oil pressure, charge rate and system monitoring are typical of the functions which are now displayed electronically.

ANALOGUE VISUAL DISPLAYS

Although not as apparent to the driver, many of the familiar needle (analogue) type instruments are driven by electronic devices. The instrument panel in Figure 6-18 is a fully electronic unit with an analogue tachometer and speedometer combined with digital clock and instruments.

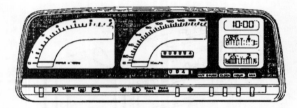

Figure 6-18   Electronically driven analogue display.

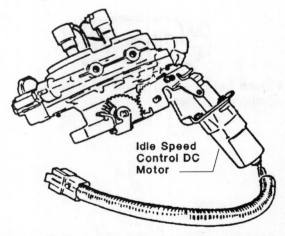

Idle Speed Control DC Motor

Figure 6-19   Idle speed can be varied by a computer controlled DC motor.

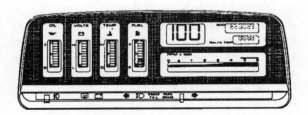

Figure 6-17   Fluorescent display panel.

STEPPING MOTORS

A stepping motor is a specialised type of electric motor. The internal construction of the motor allows accurate positioning of a pin or a metering device. The motor is normally computer (ECU) controlled and is suitable for use when accurate control of air, vacuum or linkage position justifies the expense of the motor and control system.

ELECTRONIC COMPONENTS— CONSTRUCTION AND OPERATION

To fully describe the construction and operation of every electronic component, that could be encountered in the systems of a motor vehicle, is not within the scope of this book. The components that have been selected are widely used in automotive electronics. The description will cover the inputs, process and outputs of the devices without delving into their characteristics and behaviour, as they relate to physics and chemistry.

SEMI-CONDUCTORS

THERMISTORS

NTC resistors are formed from sintered compounds of metal oxides and metallic salts. They are used in a variety of applications, including temperature compensation in electronic amplifier circuits and to measure temperature of the engine coolant (see Figure 7-1) and transmission fluid.

As the temperature rises, the resistance of an NTC thermistor decreases.

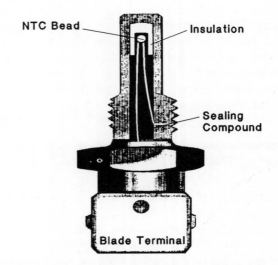

Figure 7-1   NTC resistor used as an engine coolant temperature sensor.

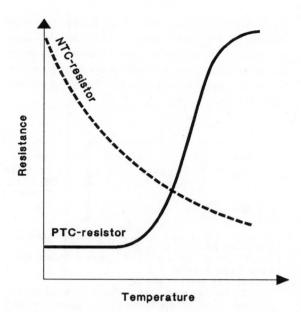

Figure 7-2 PTC-resistors have greater resistance to current flow as their operating temperature increases. NTC-resistors have less resistance to current flow with temperature increase.

PTC resistors are formed from sintered ceramic material containing additives which control the limits of their resistance.

They are usually disc shaped and special solder is used to attach the two connecting leads. PTC resistors are used for current limiters and time switches. They are also used as protection devices for electric motors, such as those used in electric window winders. As the temperature rises, the resistance of a PTC thermistor increases.

DIODES

Most diodes are manufactured from germanium or silicon crystals. These materials are modified by 'doping' the base material with phosphorus (to produce 'N' type material) or boron (to produce 'P' type material). By joining a layer of the N type crystal to a layer of P type crystal a diode is formed and the point at which they are

DIODE

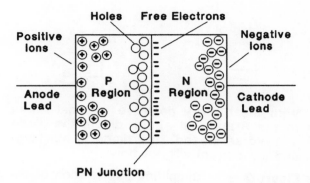

A thin wafer of silicon coated on opposite faces with phosphorus and Boric compound is treated in a furnace. The result is a semi-conductor material with an N and a P region separated by a PN junction. The N region has an excess of electrons and the P region has a lack of electrons (holes). The electrical properties of this material are the foundation of solid state electronics.

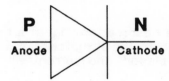

A diode will allow current flow if the P (anode) terminal is positive. No current will flow if the N (cathode) terminal is positive.

Figure 7-3 Diode construction and symbol.

joined is called the PN junction. Each layer has an external connection point. These points will be marked anode (positive) or cathode (negative).

When the N material is connected to the positive terminal of a battery and the P material is connected to the negative terminal of the battery (reverse direction), a very high resistance is formed at the PN junction. This state prevents a current flow through the circuit (see Figure 7-4).

When the connections are reversed (forward direction)—battery negative to the N material and battery positive to the P material—a low

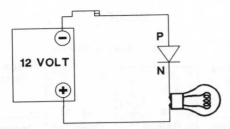

With the battery negative terminal linked to the P material of the diode: the diode is 'reverse biased' (the diode is blocking) and the current will not flow.

Figure 7-4 Simplified explanation of the diode action when reverse biased.

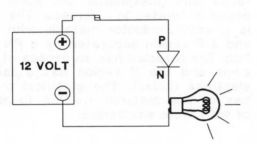

When the positive terminal of a battery is connected to the P material, the diode is 'forward biased' and current will flow.

Figure 7-5 Simplified explanation of the diode action when forward biased.

resistance is formed at the PN junction. This state allows current to flow and the diode is said to be in a conductive state (see Figure 7-5).

The PN junction can be compared to a one-way valve in a fuel pump. The inlet valve allows fuel to enter the fuel pump but prevents it from returning to the tank. In a similar manner current in a circuit will pass through a diode in one direction but will be prevented from flowing in the opposite direction.

ZENER DIODE

A zener diode is similar in construction to a 'normal' diode, however the semi-conductor material selected and the degree of doping can alter its operational characteristics. The zener

diode will block current flow below a preset voltage. When this voltage is exceeded the diode becomes conductive in the reverse direction. Because this action simulates a voltage switch, the zener diode is suitable for charging voltage regulation.

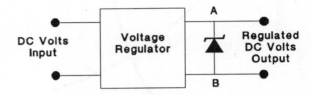

Controlled voltage output:
Zener diode 'blocks' and current does not flow from A to B.
Momentary over-voltage:
The Zener reverse bias voltage is exceeded. Current is diverted through the Zener diode clamping the output voltage to a voltage slightly above the normal value set by the regulator.

Figure 7-6 A Zener diode may be fitted as an over-voltage protection device.

PHOTO DIODE

A photo diode may be used in conjunction with a light source and a transistor to form a light-sensitive switch. When the light is directed

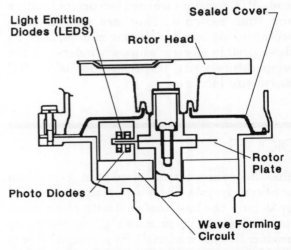

Figure 7-7 Photo diodes fitted as part of a distributor pulse generator.

through the diode's window, the diode becomes conductive in the reverse direction. When the light source is removed, the diode blocks the reverse current flow. A common automotive application for a photo diode, when used with a slotted rotating disc, is a pulse generator in distributors (see Figure 7-7) and speedometers.

LIGHT-EMITTING DIODE (LED)

A light-emitting diode acts in a similar manner to a 'normal' diode. This diode will block reverse current flow, but will allow current to flow in a forward direction (see Figure 7-8). When a forward current flow occurs, the diode emits light of a predetermined colour. When a reverse current is applied, the emitted light ceases. An

LED can be switched off and on at an extremely high rate (frequency) without being damaged.

Light-emitting diodes are suitable for 'flashing' system malfunction codes, so that the service technician can detect the fault.

TRANSISTOR

The NPN transistor is constructed with a 'P' material sandwiched between two pieces of 'N' material. Each piece of material is connected to a terminal. One of the end terminals is called the Collector (C) and the other is the Emitter (E). The centre terminal is the Base (B).

The circuit (see Figure 7-9) shows the positive terminal of a battery connected through a globe to C, and through a switch and resistor to B. The negative terminal of the battery is connected to E of the transistor. When the switch is open, the globe is 'off'. When the switch is closed, the globe will be illuminated. If the components in the circuit are selected correctly, the current flow through the switch and the resistor could be 100 times less than that through the globe. This implies that a sensitive switching device,

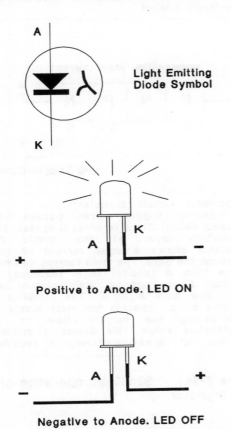

Positive to Anode. LED ON

Negative to Anode. LED OFF

Figure 7-8   Light emitting diode (LED) in forward and reverse bias.

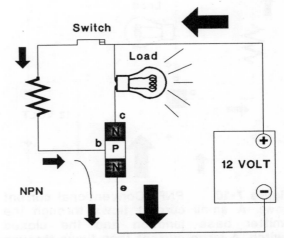

Figure 7-9   NPN (Conventional current flow). A small current flows, via the closed switch, through the base emitter junction. A large current then flows through the transistor from the collector to the emitter to operate the load.

operating at very low current flow, could be used to control a work unit requiring much higher current flow.

The PNP transistor is constructed with an 'N' material sandwiched between two pieces of 'P' material. Each piece of material is connected to a terminal. One of the end terminals is called the collector (C) and the other is the emitter (E). The centre terminal is the base (B).

The circuit (see Figure 7-10) shows the battery terminals connected in reverse to that of the NPN transistor. Except for the current flow going in the reverse direction through the transistor, the operation is the same as for the NPN transistor.

Both of these transistors can be used to amplify voltage. When a voltage of 0.1 V applied across B and E causes the transistor to conduct, the voltage across C and E could be as high as 10 V. This means the transistor has increased (amplified) the voltage by 100 times, provided the circuit is correctly designed.

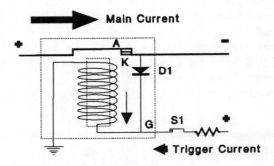

Theoretical 'Latching' relay:
Switch S1 is closed and current flows to terminal G. The solenoid coil is energised and the relay points A and K are closed, allowing the main current to flow. Opening S1 will not open points A and K because a small current is being supplied via D1 to keep the relay closed. Points A and K will not open until the main current flow is removed or decreased in value to an ineffective level.

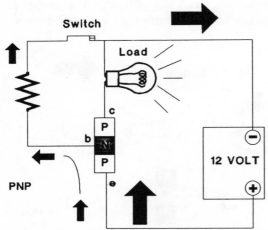

Figure 7-10 PNP (Conventional current flow). A small current flows through the emitter base junction and the closed switch. A large current then flows through the transistor from the emitter to the collector to operate the load.

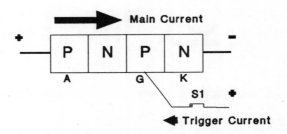

Electronic switch (thyristor):
A pulse or 'trigger' current passes from closed switch S1 to terminal G (gate). The depletion layer at the middle PN junction allows a small current to flow through the gate. The main current can now flow from A (anode) to K (cathode). If switch S1 is opened, main current will still flow from A to K. The thyristor is turned off by opening the main supply or decreasing the current flow to an ineffective value. This device is suitable where high speed switching is required.

Figure 7-11 Simplified operation of a P type thyristor.

THYRISTOR

An SCR, constructed with a PNPN layer configuration, has an anode (A) terminal attached to the outer P layer and a cathode (K) terminal attached to the outer N layer. A gate (G) terminal is attached to the P layer nearest the cathode. When a circuit that includes a voltage source is connected to the SCR so that the A is positive and K is negative, an extremely high resistance exists in the SCR and a very small current flow occurs in the circuit (see Figure 7-11). The SCR will not conduct until the gate is triggered by a small current flow. Once the gate has been triggered, the SCR conducts and the current flow in the circuit will be limited by the circuits resistance. When gate current is removed, the SCR will continue to conduct. The SCR can be turned off by opening the circuit with a switch or by placing a short circuit across the A and K terminals of the SCR. Current will not flow through the SCR until a voltage is again applied to the gate. SCR circuits can be constructed to operate with a DC or an AC supply.

EXHAUST GAS OXYGEN SENSOR

There are two common types of exhaust gas sensors, the zirconium dioxide type and the titanium type; both are designed to detect the concentration of oxygen in the exhaust gases.

ZIRCONIUM DIOXIDE TYPE

When the sensor reaches operating temperature, it generates a variable voltage signal depending on the amount of oxygen detected in the exhaust.

A zirconium dioxide 'thimble' coated with platinum inside and outside, is at the 'hot' end of the sensor. The outside surface of this 'thimble' is exposed to the exhaust gases and its inside surface is in direct contact with the 'outside' air.

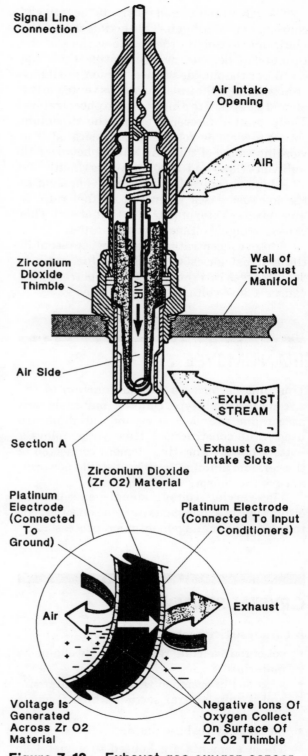

Figure 7-12   Exhaust gas oxygen sensor.

The zirconium dioxide attracts 'negatively' charged ions of oxygen. This action causes the platinum coating on the inside surface of the thimble to become more negative than the coating on the outside surface. The ion imbalance occurs because the inner surface is vented to the air and is exposed to a high oxygen concentration. The potential difference across the zirconium causes a variable voltage to be generated. The working range of this unit will be between 10 millivolts (low) and 1000 millivolts (high).

A low concentration of oxygen is present in the exhaust gases when the air/fuel ratio is 'rich' (excess ratio of petrol to that of air). This causes a 'high' voltage to be generated.

A high concentration of oxygen is present in the exhaust gases when the air/fuel ratio is 'lean' (excess ratio of air to that of petrol). This causes a 'low' voltage to be generated.

TITANIUM TYPE

This unit performs a similar function to the zirconium dioxide type. Different materials such as ceramic titania and ceramic alumina are used in its construction. This type employs a battery powered heating element connected to the ceramic alumina to ensure stable performance of the sensor.

The varying voltage signal generated by either type of exhaust gas oxygen sensor is sent to the engine management microprocessor.

CRYSTAL SENSORS

A piezoresistive sensor consists of a:
- set of four sensing resistors made of a piezo material;
- silicon chip designed with a thick outer rim and a thin central section which forms a diaphragm;
- metal or pyrex base plate.

These elements are housed in a metal or plastic container. The sensing resistors are connected into the circuit as shown in Figure 7-13.

A constant DC voltage is supplied to the input of the sensor. As the current flows through the sensing resistors, a voltage drop occurs such that the output voltage between A and B is zero. When a pressure is applied to the diaphragm section of the sensor, the sensing resistors change their value proportionally to the pressure and the output voltage rises. An amplifier is used to raise the output voltage to a usable level.

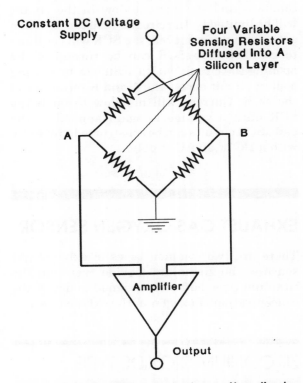

Piezo-resistive Sensor Circuit Connections (Diaphragm And Construction Not Shown)

Figure 7-13 Without strain on the diaphragm, the resistors are balanced in value and the voltage difference between A and B will be zero. Changes in manifold pressure cause the diaphragm to act on the resistors. Two resistors reduce in value and the remaining two increase in value. The variation in resistor value has the effect of changing the voltage difference between A and B.

By adding two capacitors to the circuit, the sensor becomes a frequency dependent device. A constant input frequency is changed to a variable output frequency, which is in proportion to the pressure applied to the diaphragm.

A piezo-electric sensor consists of a:

- piezo crystal sandwiched between two metal plates;
- steel block which contacts one of the plates;
- wire attached to each plate.

The piezo material may be made from quartz,

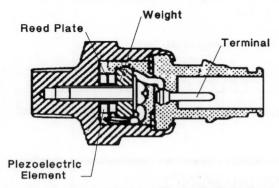

Figure 7-14 Detonation sensor based on a piezoelectric element.

rochelle salt or barium titanate. The device is enclosed in a combination plastic and metal housing.

When the steel block is subjected to a shock, the block moves and squeezes the crystal between the two plates. The squeezing action on the crystal produces a small potential difference (voltage) across the two plates. An amplifier increases the small voltage to a usable level.

OPEN AND CLOSED LOOP CONTROL

A simple process consists of an input, a conversion or decision and an output, e.g. When opening a door, the input is turning the handle followed by the mechanism converting movement and the output is the unlatching of

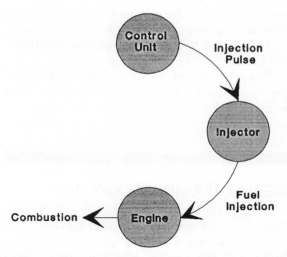

Open Loop: The output has no effect on, and does not modify the input.

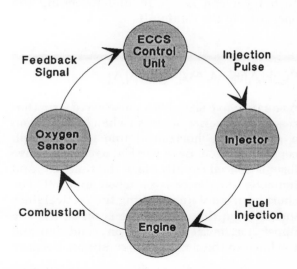

Closed Loop: The output has an effect on, and modifies the input.

Figure 7-15 Open and closed loop engine systems.

the door. This is called an **open loop control** process.

The output has no effect on and **does not modify the input.**

Considering the above example of the door, if the door did not open on the first try, the person would change the input by applying more force on the handle or turning the handle in the opposite direction. The action of modifying the input to achieve the correct output is called a **closed loop control process**.

The output has an effect on and **modifies the input**.

The principles of open and closed loop control are applied to the electronic processes of a motor vehicle.

SIGNAL TERMINOLOGY

To explain the operation of solid state devices requires the use of terms which may not be familiar to the reader. Minimal use of these terms has occurred but a basic knowledge of each term can assist in understanding the operation of the total unit.

SQUARE WAVE SIGNAL

A square wave signal is a pulse wave form that resembles a series of blocks when portrayed on a graph. The horizontal line of the graph represents time. Each vertical part of the wave form is spaced equally along the time axis and indicates the device is switched on or off. The short horizontal lines linking the vertical lines indicate the time the device is on or off. The upper horizontal lines are 'on' condition and the lines on the zero are 'off' condition. Because

the line lengths are equal the wave form is at constant frequency.

CLOCK SIGNAL

The 'clock' signal originates from an integrated circuit (chip) within the central processing unit (CPU). This signal is a continuous train of pulses at constant frequency, i.e. the pulses are equally spaced. Clock signals are used to control internal timing or synchronisation of the computer.

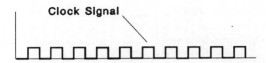

Figure 7-17   Constant pulse clock signal.

DIGITAL SIGNAL

Consider a graph representing a light that is turned 'on' then 'off' (see Figure 7-18). This 'on—off' action could be plotted along a horizontal line representing time. By considering the vertical line as voltage either 'present' or 'not present' the graph can be drawn in a 'square' wave as shown. However many digital applications may not occur at constant frequency because the device is switching on for varying time intervals. In these applications the on/off signal will appear as a block but is not a square wave. For digital control

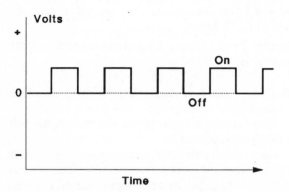

Figure 7-16   Square wave signal.

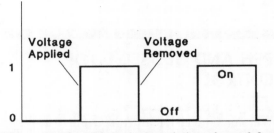

Figure 7-18   Digital signal developed from a square wave.

applications the voltage 'on' may appear as '1'; voltage 'off' will be '0'. This form of signal is useful in many switching applications.

DUTY CYCLE SIGNAL

A 'duty' cycle refers to a cyclic type operation (a process that is continually repeating itself).

Consider the illustration (Figure 7-19) as a fuel injector solenoid; when the injector valve is open, spraying fuel, the signal is 'on', when the injector valve is closed the signal is 'off'. One complete cycle consists of the time the injector is open plus the time the injector is closed. The duty cycle can then be expressed as a percentage of the complete cycle. A fuel injector which was open for exactly half of the complete cycle has a 50 per cent duty cycle.

Example

Set 50 per cent duty cycle as being the theoretical point at which correct air-fuel ratios are obtained from the injector system. To lean the air-fuel mixture would require a shorter duty cycle, i.e. the injector valve opens for less time. The shorter duty cycle would be expressed

as a percentage which would be less than 50 per cent.

The same signal terminology can be applied to any unit with a cyclic operation.

DIRECT CURRENT SIGNAL

Steady DC voltage is represented by a straight line on a graph. Battery voltage would appear as a horizontal line at a point (amplitude) approximately 12 volts positive above the zero line.

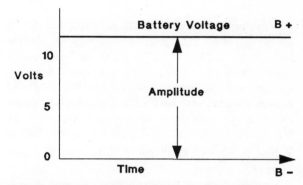

Figure 7-20   Steady DC voltage signal.

ANALOGUE SIGNAL

The most common form of analogue wave form is the alternating current sine wave. Figure 7-21 shows the vertical (amplitude) change in the wave form is not abrupt. In alternating current applications (as shown), the wave form passes through the zero voltage line and, depending on the point checked, the voltage may be positive or negative.

Analogue devices can also produce signals varying from zero to the system voltage. Consider a throttle position sensor which has a 'wiper' contact that can 'tap' along a resistor. The voltage available is a 'regulated voltage', the value of which is determined by the 'tap' point. The throttle body sensor is an analogue device sending a DC voltage signal to the ECU.

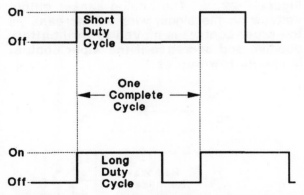

Figure 7-19   The ON time of the short duty cycle is less than 50 per cent of the complete cycle. The long duty cycle ON time is greater than 50 per cent of the complete cycle.

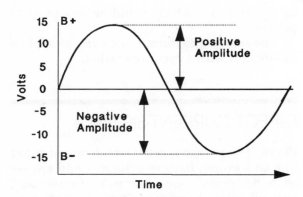

Figure 7-21   AC voltage signal.

Analogue signals are not suitable for direct switching of operating devices because the device would not be fully 'on' or fully 'off' for major parts of the wave form. For this reason their use in automotive applications is mainly confined to *input* signals to the ECU.. The ECU contains an A to D (analogue to digital) convertor to transform the incoming signal. The *output* signal from the ECU is in digital form.

RECTIFIED AC WAVE FORMS

The common AC wave form is the sine wave and, in some cases, it is necessary to change it to a DC wave. The best example of this procedure is the alternator's stator output which must be changed from AC to DC so that a battery can be recharged.

Changing an AC wave to DC is called 'rectification'. Rectifying an AC wave is achieved by preventing the wave from dropping below zero volts.

The simplest method is half wave rectification (see Figure 7-23) which results in the negative side of each wave being replaced with a straight line at zero volts. This effect is produced by placing a single diode into an AC circuit.

Full wave rectification (see Figure 7-24) results in the negative side of each wave being changed to positive. It has the effect of 'flipping'

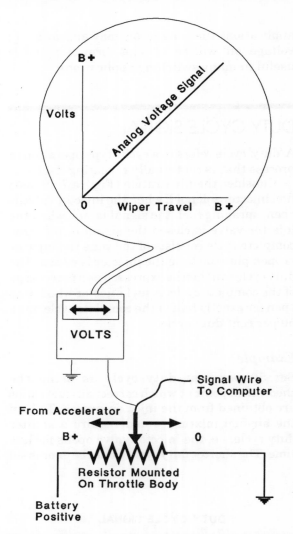

Figure 7-22   The analog sensor output voltage on the signal wire will increase as the wiper contact is moved toward battery positive and decrease if the wiper contact is moved towards earth.

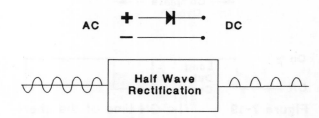

Figure 7-23 A single diode rectifier allows only half the wave cycle through with a resultant drop in output.

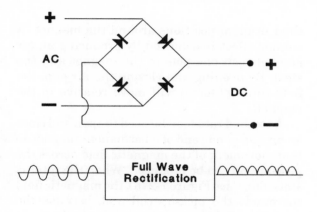

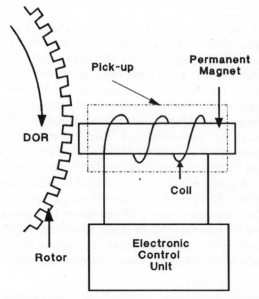

Figure 7-24 Four diodes form a rectifier bridge allowing the use of both half cycles of the AC wave form.

Figure 7-25 The basic construction of an inductive type pulse generator.

the negative section of the wave to the positive side of the zero voltage line. This action requires four diodes, connected in a special pattern, to be placed into an AC circuit.

Both methods of rectification produce pulsating DC waves which are undesirable as a supply for electronic devices. Full wave rectification is best suited to this application but it must first undergo a 'smoothing' process. A simple 'smoothing' method is obtained by placing a capacitor across the DC output which reduces the pulsations to very small ripples.

PULSE GENERATORS

INDUCTIVE MAGNETIC PULSE GENERATOR

CONSTRUCTION

The two parts of an inductive type magnetic pulse generator are the toothed rotor and the pick-up.

The rotor is a metal ring or disc that has tooth-like projections cut in or attached to its perimeter. This rotor is attached rigidly to a

suitable point on a rotating shaft. It is extremely important that the rotor runs concentric with the centre of the shaft to ensure perfect balance and to eliminate variations in the air gap between the tooth projections and the pick-up.

The pick-up consists of a coil wound onto a permanent magnet. The ends of the coil are connected by wires to the input stage of the electronic control unit. This pick-up is attached to a non-rotating part near the shaft so that the magnet is aligned and adjacent to the teeth on the rotor. The rotor tooth to magnet gap is very small (0.4–1.0 mm) and must be maintained to ensure efficient signal production.

OPERATION

As with all voltage generating mechanical devices, a magnet, a conductor and relative movement of the magnetic field across the conductor are essential to produce an AC current in a conductor.

A revision of the basic principles related to permanent magnet theory will assist the reader in understanding magnetic pulse generator operation. All permanent magnets have a magnetic field which consists of a huge number of lines-of-force (see Figure 7-26). These invisible lines-of-force emerge at the north pole and travel in an elliptical path to the south pole of the magnet. It is easier for them to travel through metal rather than through air. They do not cross but they can be bent, shaped and concentrated into an extremely tight pattern. It must be remembered that lines-of-forces completely surround the permanent magnet and that they cannot be 'switched off' or broken.

A piece of steel, shaped like a 'U', is placed across the ends of the magnet (see Figure 7-26). The magnetic lines-of-force move to a position in which almost all of them travel through the steel. When the steel piece is removed from the ends of the magnet, the lines-of-force return to their original positions around the magnet. A similar effect is achieved by producing an air gap between one pole of the magnet and the steel. By opening and closing the air gap, the field (lines-of-force) will move relative to the magnet.

A toothed rotor can be used to open and close an air gap at one end of a permanent magnet so that movement of the magnetic field across the pick-up coil can be produced. When the rotor is stationary (see Figure 7-27a), the magnetic field surrounds the pick-up coil and, because the magnetic field is stationary, no voltage is generated in the coil.

When the rotor turns, a tooth will align with the end of the magnet (see Figure 7-27b) and,

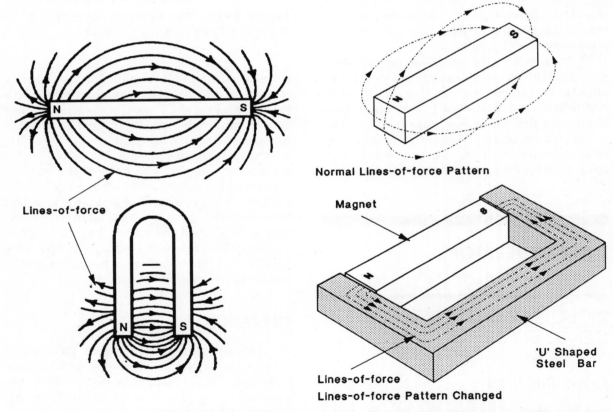

Normal Lines-of-force Pattern

Lines-of-force Pattern Changed

(a) A graphical representation of the invisible lines-of-force around permanent magnets.

(b) The lines-of-force around a magnet can be moved to a new position by placing a 'U' shaped steel bar across its ends.

Figure 7-26 The characteristics and behaviour of the lines-of-force around permanent magnets.

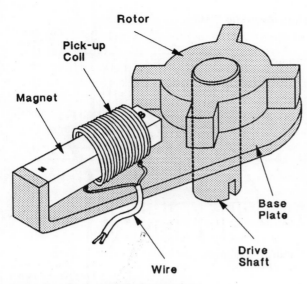

Magnet
Pick-up Coil
Rotor

Base Plate
Drive Shaft
Wire

Inductive Type Pulse Generator

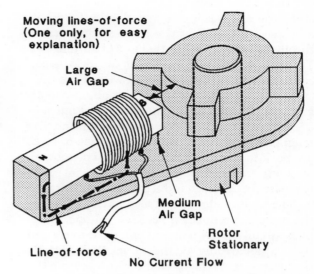

Moving lines-of-force (One only, for easy explanation)

Large Air Gap
Medium Air Gap
Rotor Stationary
Line-of-force
No Current Flow

(a) The Line-of-force enters the baseplate and jumps across the medium air gap to return to the magnet. The Line-of-force is not moving across the coil windings so no current is produced.

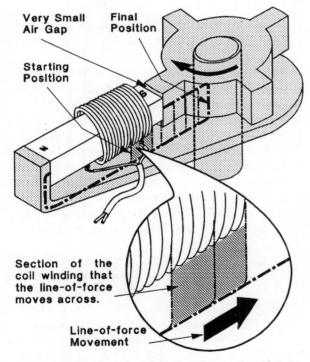

Very Small Air Gap
Final Position
Starting Position

Section of the coil winding that the line-of-force moves across.

Line-of-force Movement

(b) Voltage is induced into the coil winding, only while the line-of-force is moving across the coil.

REMEMBER *that there is vast number of lines-of-force acting on this coil, in the same manner as the one described in this figure.*

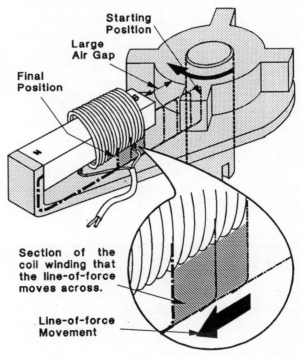

Starting Position
Large Air Gap
Final Position

Section of the coil winding that the line-of-force moves across.

Line-of-force Movement

(c) Voltage is induced into the coil winding in the opposite direction, only while the line-of-force is moving across the coil.

Figure 7-27 Generating pulsating voltage in a coil winding by opening and closing an air gap at the end of a magnet (Magnetic pulse generator).

since the air gap is very small, the field will move so that most of the lines-of-force travel through the rotor to the other end of the magnet. The moving field cutting across the pick-up coil causes a voltage to be induced within the coil.

Further movement of the rotor aligns a tooth gap with the end of the magnet (see Figure 7-27c). The air gap is now very large and the magnetic field moves so that the lines-of-force are back in their original position. The moving magnetic field cuts across the pick-up coil in the opposite direction to induce a reverse voltage.

The leading edge of the tooth, as it approaches the magnet produces a positive voltage in the pick-up coil. At the moment when the tooth and the magnet are aligned the induced voltage falls towards zero. As the tooth moves away from the magnet (into the gap between the teeth) a negative voltage is induced into the pick-up coil. The change between positive and negative voltages causes an AC signal to be delivered from the coil. This AC signal is not a pure sine wave. Size, shape and spacing of the teeth on the rotor determine the characteristics of the wave form. However, regardless of the wave form, the frequency is the most important aspect of this AC signal. The frequency of the AC signal will vary with the speed of the rotor. When the rotor speed increases, the AC frequency will increase. It is also important to note the AC voltage increases with increase in the rotor speed.

The AC signal is delivered to the input stage of the ECU where it is modified for use by the central processing unit (CPU).

HALL GENERATOR

Constant supply current 'Iv', passing through a semi-conductor 'H' (the Hall layer) is acted upon by a magnetic field 'B' positioned at a right angle to the current flow. A force, known as Lorentz force, is developed within the Hall layer and causes the moving electrons to drift towards the negative terminal plate 'A1'. A surplus of electrons occurs at 'A1', therefore 'A1' becomes negative. The terminal area at 'A2' now has a deficiency of electrons and

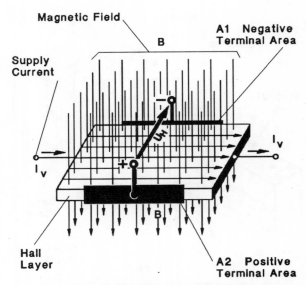

Figure 7-28 Hall layer exposed to a magnetic field.

becomes positive. The difference in potential between 'A1' and 'A2' is known as the Hall voltage 'Uh'.

Increasing the strength of the magnetic field will increase the value of the Hall voltage. Applying, then removing the magnetic field 'B' will cause voltage 'Uh' to rise to a maximum then drop to a minimum value. If the magnetic field apply and remove is timed by mechanical means then the voltage wave form will be in phase with the mechanical device. This voltage is only in the millivolt range and to be usable within an automotive ignition system the Hall voltage must be amplified and shaped into a digital signal.

In the Hall generator, the amplification and shaping operations are encapsulated into a Hall integrated circuit chip (see Figure 7-29). The IC is then fitted into an application package consisting of the magnet, chip mounting, wires and mechanical timer device. This type of assembly is used within distributors, engine speed sensors, vehicle speed sensors and crankshaft position sensors.

A widely used application of the Hall effect IC is within a distributor. In this application package a magnet and a Hall IC are separated by an air gap through which the vanes of a rotor move.

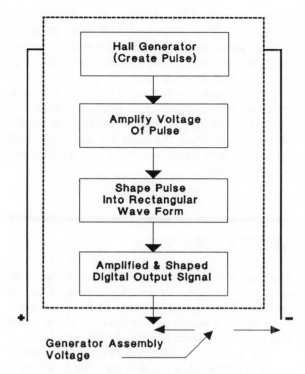

Figure 7-29 Simplified block diagram of a Hall IC chip.

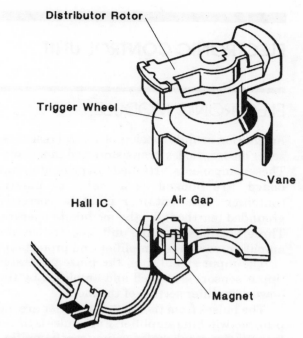

Figure 7-30 Hall generator assembly.

When the vane is not in the air gap, the Hall layer is exposed to the magnetic field. The Hall voltage will rise to its maximum and the Hall IC is switched 'on'.

With further shaft rotation the vane enters the air gap shielding the Hall layer from the magnetic field. The Hall voltage will fall to a minimum and the Hall IC is switched 'off' (see Figure 7-31).

A switching function in phase with the number of vanes and the r.p.m. of the shaft is available from the output stage of the Hall IC. The distributor was chosen for this example, however it is possible to have different application packages. A similar switching function can be achieved by mounting magnets on to a spinning disc and placing a Hall IC in close proximity to the disc edge. The movement of the magnet towards, then away from the Hall IC will cause the IC to switch 'on' and 'off'.

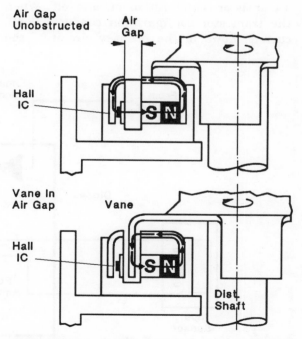

Figure 7-31 When the vane is in the air gap, most of the magnetic flux is intercepted by the vane metal.

ELECTRONIC CONTROL UNIT

ELECTRONIC MODULE

An ignition module consists of several resistors, zener diodes, diodes, transistors and capacitors. These components, attached to a printed circuit board, are housed in a metal or plastic container. The container must be correctly grounded (earthed) for the module to operate. Three sections in this unit are the power amplifier, switching amplifier and protection.

The input signal from the pulse generator (input sensor) is shaped and amplified by the power amplifier section of the module.

The pulses from the power amplifier are fed into the switching amplifier. Each pulse is timed to the engine crankshaft position (see the ignition section in Chapter 11 for a detailed description). After further amplification, each pulse switches a transistor configuration 'on' and 'off'. When the transistor configuration conducts ('on'), current flows in the primary circuit of the

ignition coil. The instant the transistor configuration is switched 'off', the primary current stops and ignition occurs in the selected cylinder (see the ignition section in Chapter 11 for a detailed description).

Zener diodes are connected between the ignition coil terminal and ground inside the electronic module. They protect the transistor configuration from the high voltage generated in the primary circuit, when the primary current is interrupted.

MICROCOMPUTER

The service technician collects information, analyses and processes the information by referring to diagnostic charts and service data. A decision is made on the correction required and action is taken to remedy the problem.

The microcomputer operates in a similar manner to the service technician. The computer collects information from specific inputs, performs calculations, compares the results to service data, determines and performs the

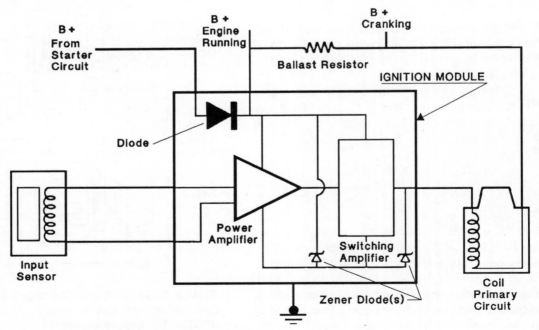

Figure 7-32 Simplified electronic ignition module.

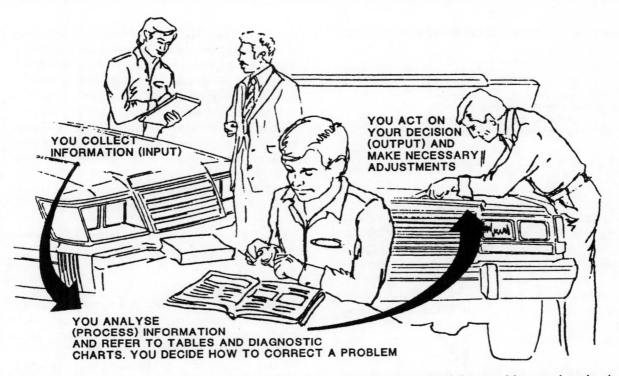

Figure 7-33 The stages of information input, processing, decision making and output action are a normal part of every service operation.

required action. These operations can be grouped into three sections, which are:

- Input;
- Processor and Storage;
- Output.

INPUT SECTION

The input section is housed in a metal container called the 'computer case' and formed on a printed circuit board (mother board). External connections are made via a multi-pin plug located on the outside of the computer case. The input section selects, conditions and amplifies the input signals.

CONSTRUCTION

The input section comprises an input conditioner and a multiplexer.

An input conditioner has two parts: the amplifier (AMP) and the analogue to digital (A/D) convertor. These parts are constructed on the very large scale integration (VLSI) design which means all the components are microminiature. Several thousand transistors, diodes, resistors and capacitors are formed in an area smaller than a pin head.

The input conditioner is connected to input wires from the various sensors and connected to the microprocessor through the printed circuit board.

OPERATION

The 'input' signals can be analogue or digital. An analogue signal is one which can vary in value. A temperature sensor is a device that produces an analogue signal, e.g. a variable voltage signal throughout its range of temperature changes. A digital signal is one which determines specific conditions—on or off, high or low, yes or no.

All analogue signals are sent to the analogue to digital convertor where they are changed to

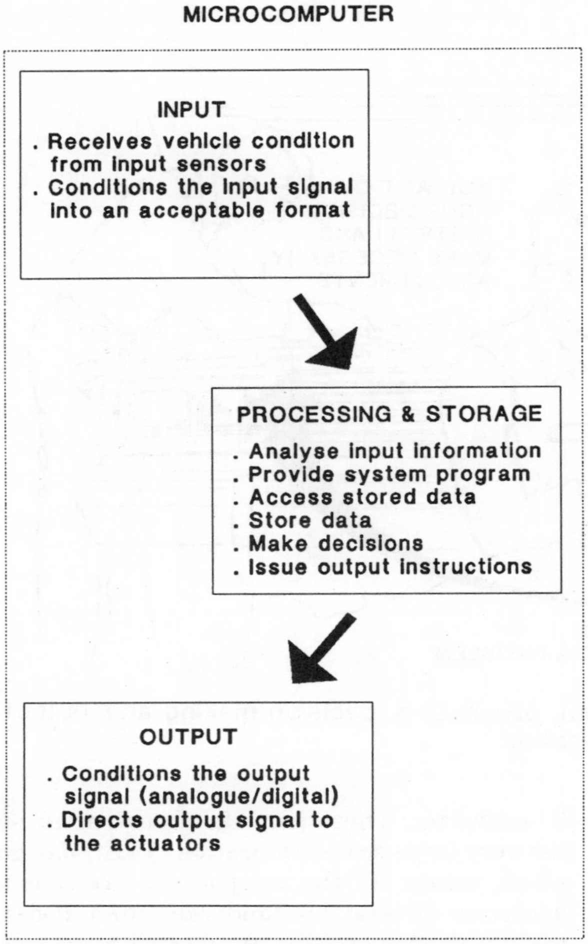

MICROCOMPUTER

INPUT

. Receives vehicle condition
  from input sensors
. Conditions the input signal
  into an acceptable format

PROCESSING & STORAGE

. Analyse input information
. Provide system program
. Access stored data
. Store data
. Make decisions
. Issue output instructions

OUTPUT

. Conditions the output
  signal (analogue/digital)
. Directs output signal to
  the actuators

Figure 7-34 The microcomputer, assisted by the sensors and actuators, operates in a similar manner to the service technician.

digital signals and increased (amplified) to a level that can be identified by the microprocessor. These converted signals enter the microprocessor.

Input signals, which are digital, bypass the A to D convertor and are directed to the amplifier section where they are increased to a level that the microprocessor can identify.

Input signals can be passed through a multiplexer. The multiplexer is a device which can be compared to a multi-input and single output mechanical switch. When the computer requires information from a system sensor, the multiplexer connects the required sensor to the microprocessor. Not all the input signals from the system sensors pass through the multiplexer. There are signals that a microprocessor needs to monitor in very short periods of time, such as a throttle position sensor. Those sensors that do not require continuous monitoring, can pass through the multiplexer and be selected when the microprocessor is not busy with the critical sensors. An example would be the fuel level indicator.

Multiplexers reduce the number of inputs that the microprocessor needs to service.

PROCESSOR AND STORAGE SECTION

In most cases, the processor is a microcomputer. A microcomputer consists of a microprocessor and a bank of memory (storage). It is constructed on the printed circuit board housed in the computer case. Connections are made through the printed circuit board.

CONSTRUCTION

The microprocessor or central processing unit (CPU) is an IC constructed on the very large scale integration (VLSI) design which means all the components are microminiature. A microprocessor consists of many sections such as a timer, decoder, program counter, an arithmetic-logic unit and registers. These sections allow the microprocessor to fetch, interpret and execute instructions.

There are three main types of computer memories, ROM, KAM and RAM. These are IC devices housed in a plastic block which has many prongs moulded into it to permit the IC to be connected to the other microcomputer components.

OPERATION

The operation of a microprocessor is a complex subject so the description will be extremely simple.

The microprocessor requires a set of logic

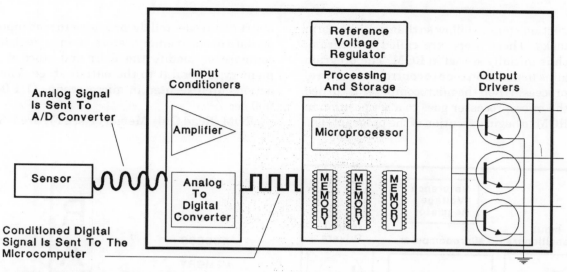

Figure 7-35 The illustration shows, in simplified form, the input, processing and storage and output sections of a computer.

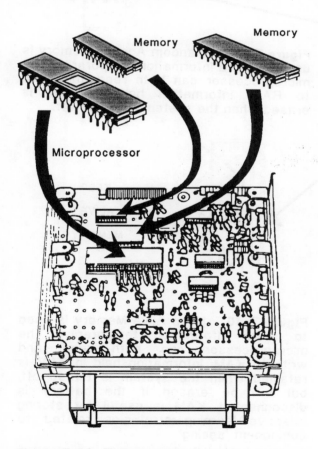

Figure 7-36 The processor and memory are contained in IC chips.

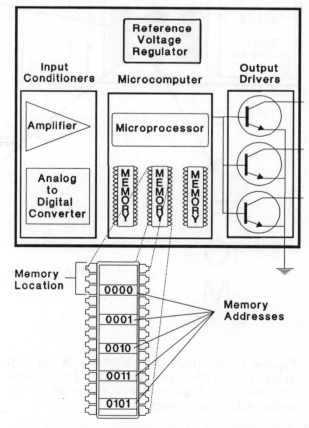

Figure 7-37 Each memory location, within the chip, is identified by a memory address.

instruction steps to follow so that it can perform its tasks. These steps are called a program, which is initially stored in ROM. The program contains instruction code to control the action of the process. When the microcomputer is turned on, the microprocessor goes to a set position in the ROM to begin execution of the program. The instruction codes tell the processor to read inputs or data from memory, store it in a register, compare or modify the data and store it in memory or send it to the output stage. These actions are executed in microseconds (1/1 000 000 sec.).

ROM (Read Only Memory) (see Figure 7-38)

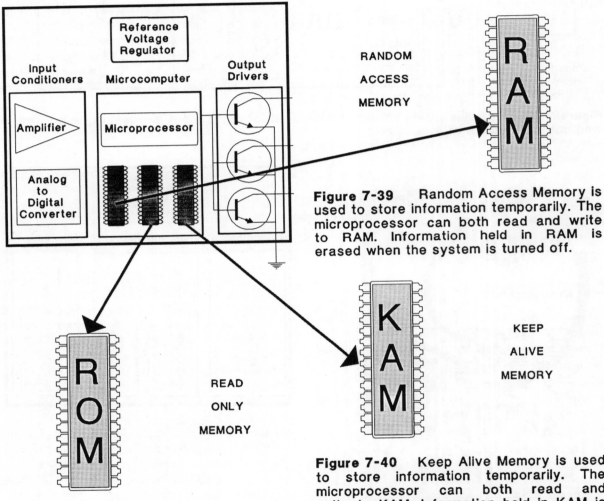

Figure 7-39 Random Access Memory is used to store information temporarily. The microprocessor can both read and write to RAM. Information held in RAM is erased when the system is turned off.

Figure 7-38 Read Only Memory is used to store information permanently. The stored information will be retained even when the battery is disconnected.

Figure 7-40 Keep Alive Memory is used to store information temporarily. The microprocessor can both read and write to KAM. Information held in KAM is retained when the system is turned off, but will be erased if the battery is disconnected. KAM is useful for storing adaptive learning strategies related to component 'ageing'.

is programmed information which is stored permanently and cannot be changed or erased by normal use of the computer. A program containing specific instructions, look up tables and vehicle component specifications is placed in ROM.

EPROM (Erasable Program Read Only Memory) is a special type of ROM, in which the program can be erased or changed. By using an EPROM the manufacturer can configure the computer with specifications to suit another engine.

RAM (Random Access Memory) (see Figure 7-39) is used as a non-permanent storage area (work area) when the ROM program is executed. The information is constantly changing and being updated, for example sensor inputs, which constantly change would be in RAM.

KAM (Keep Alive Memory) (see Figure 7-40) is similar in operation to RAM. The main difference is; the RAM loses its information when the vehicle's ignition is switched off. KAM retains its information when the vehicle's ignition is switched off but loses its information when the battery is disconnected.

OUTPUT SECTION

When the microprocessor decides to sends an output, the correct output has to be selected. Three common types of paths available to the microprocessor are:

- direct to an output driver;
- via a digital to analogue (D/A) convertor, to an output driver;
- via a demultiplexer, to an output driver.

These components are constructed on the printed circuit board housed in the computer case. Connections are made through the printed circuit board between the components. A multi-pin plug provides the external connection point for the output devices.

CONSTRUCTION

The output driver generally has two large transistors connected in series and a few resistors. This device is called a Darlington amplifier. The most important aspect of an output driver is that it must be mounted on a heat sink for protection.

The digital to analogue (D/A) convertor and the demultiplexer are integrated circuits (IC) designed on the very large scale integration (VLSI) method. These circuits contain a vast array of transistors, diodes, resistor and capacitors. Each device has a set of small prongs to provide a method of connecting them to the printed circuit board. The prongs are inserted into a base that has been previously mounted on the printed circuit board or the IC may be surface mounted by soldering its prongs to the board.

OPERATION

The digital to analogue (D/A) convertor changes a digital signal into a stepped output signal (analogue). This means that a digital signal can be used to increase or decrease the voltage or current flow applied to a work unit. An output driver provides the link with the work unit.

A digital signal can be connected directly to an output driver. Assuming the output driver is a Darlington amplifier, when the digital signal is 'High' at the base of the power transistor it conducts and voltage is applied to the base switching transistor. At the instant current flows in the base circuit of the switching transistor, it conducts between the collector and emitter to 'switch on' the work unit (see Figure 7-43).

When the digital signal is 'Low' at the base of the power transistor, there is no current flow in its base circuit so the transistor goes to or remains in the 'switched off' state. Current will not flow between the collector and the emitter. This action prevents the switching transistor from becoming or remaining conductive so the work unit is 'switched off' (see Figure 7-44).

The demultiplexer is similar to a rotary switch that has one input and several outputs. The number of outputs will depend on the range of work units that has to be controlled by the microprocessor. Each output is connected to the

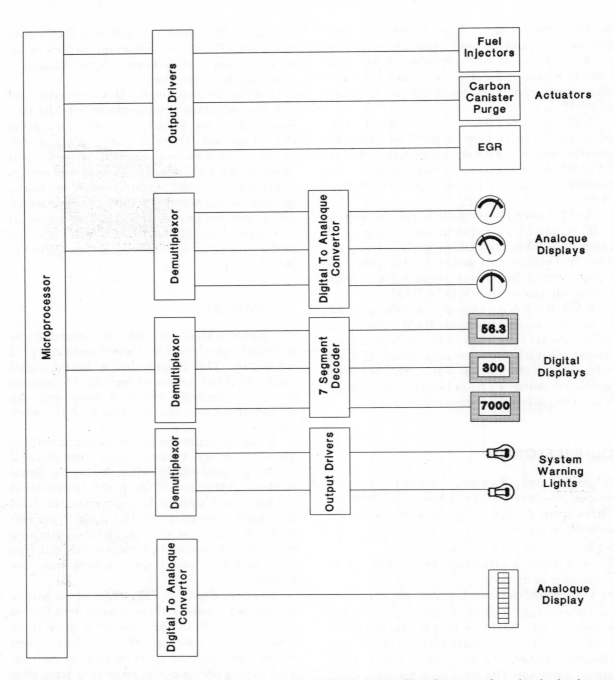

Figure 7-41   The illustration shows, in simplified form, the types of output devices which can be controlled by the microprocessor. The microprocessor does not supply the current to the output actuators. These devices are connected to the battery and switched to earth through the output drivers as required by the microprocessor.

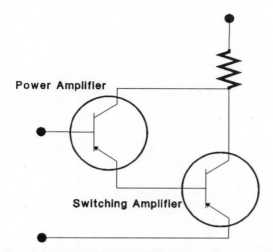

Figure 7-42 Darlington amplifier based on two PNP transistors. The amplifier can be separate transistors or two transistors contained in a single case. PNP or NPN types may be specified.

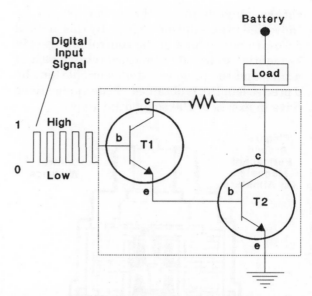

Figure 7-44 When a LOW (off or 0) input signal is present at the base of power transistor T1, the transistor turns OFF. Current cannot flow through T1 to reach the base of T2, therefore T2 is OFF. Load current is not able to flow through T2 and out to earth as a path from C to E of T2 is not available.

input by the microprocessor in any order. These outputs can be switched so fast that the signal appears to be continuous. Each output needs to be connected to an output driver.

STEPPING MOTOR

A stepping motor (see Figure 7-45) has an armature containing permanent magnets with a series of spaced 'fingers' around the circumference. Four control windings encircle the armature. The control windings are energised by signals from the ECU. In response to the magnetic field, formed when a control winding is energised, the armature can be made to rotate slightly. By energising alternate control windings the armature is rotated forward or backward. Linking the armature to a suitably threaded valve plunger allows the

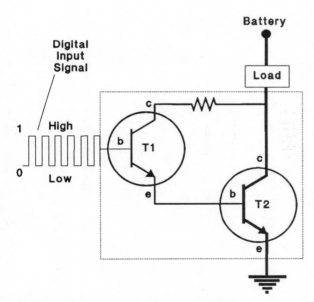

Figure 7-43 When a HIGH (on or 1) input signal is applied to the base of power transistor T1, the transistor turns ON. A current flows through T1 from C to E. This current then flows to the base of T2, turning T2 ON. The load current (which could be the ignition coil) flows through T2 from C to E and out to earth.

rotary movement of the armature to be converted to an axial movement. By this method accurate positioning of the control plunger, in sequential steps (the armature rotates part of a turn and the plunger extends or shortens by a precise length), is obtained. Some applications have in excess of 200 sequential steps.

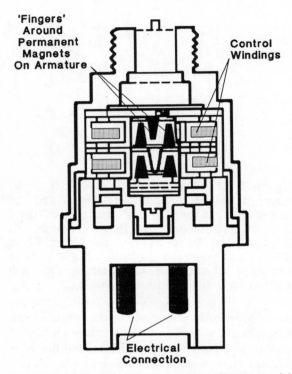

Figure 7-45 Four control windings enable the stepper motor to be rotated in both directions.

8

BRAKE SYSTEMS

INTRODUCTION

Regardless of the special features included in most vehicle brake systems, such as: dual hydraulic circuits, four wheel disc brake units and proportioning valves, they share one basic fault: that of skidding or wheel lock-up. The tendency of a vehicle to skid when it is braking is increased by the braking force, load and road to tyre conditions.

Wheel lock-up (skidding) is very dangerous, particularly when it occurs on the front wheels of a vehicle. The driver experiences severe lack of control of the vehicle's steering.

The pioneers of motor car racing overcame this problem by pumping the brake pedal with a high pulsating force. This action reduced the tendency for the wheels to lock-up, and it also increased the braking efficiency. In effect, the instant before the wheels locked, the brakes released and the wheels returned to the vehicle's speed. This method of braking is still practised by competitive motor sports drivers who have developed a keen awareness of their vehicle's braking and handling characteristics. Most motorists do not have these skills, and must rely on the vehicle brake systems being adequate to cope with normal braking conditions.

Anti-lock brake systems have been fitted to prestige motor vehicles for several years. These systems prevent wheel lock-up under severe braking conditions.

The names that have been used for anti-lock brake systems are:

- anti-lock brake system—ABS;
- anti-lock brake—ALB;
- anti-skid braking system—ABS;
- automatic wheel-slip brake control system—AWB.

The main advantages of an anti-lock brake system are that:

- the driver can steer the vehicle while braking severely;
- stability can be maintained;
- the braking distance is reduced;
- the ability to avoid an accident is increased.

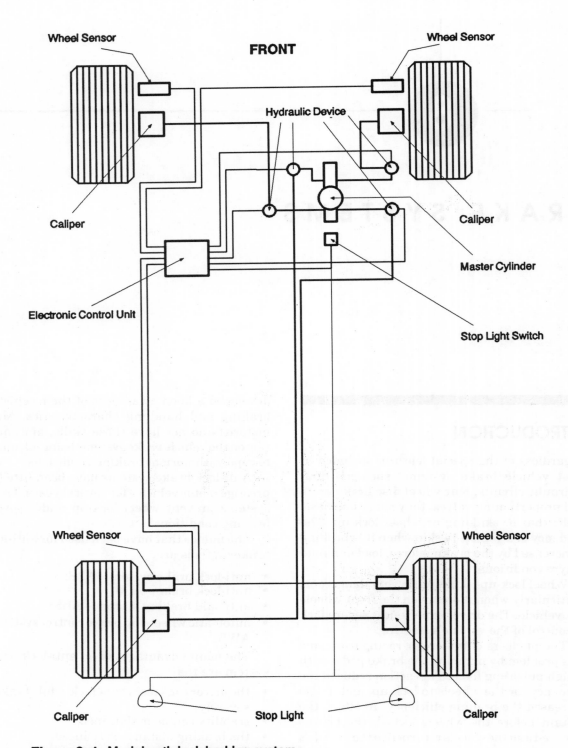

Figure 8-1 Model anti-lock braking system.

MODEL ANTI-LOCK BRAKE SYSTEM

This model is not an actual ABS, but it is presented to allow the reader to gain an understanding of the principles associated with the design, construction and operation of an actual system.

The model ABS has been designed to:

- apply the brakes until a wheel (or wheels) begins to lock and then release the brakes so that the wheel (or wheels) regains speed;
- repeat the above action very quickly until the vehicle has stopped or the brake pedal has been released.

Components chosen for the model (see Figure 8-1) are:

- a diagonally split hydraulic brake system comprising a dual master cylinder, a caliper and disc pads at each wheel, hydraulic pipes and hoses and a stop light switch;
- four hydraulic devices—one located in each brake caliper line;
- four electronic wheel speed sensors connected to the control unit;
- an electronic control unit connected to the wheel sensors, the stop-light switch, the battery and the hydraulic devices.

These components are models only, which means that they may not be found in an actual ABS.

COMPONENT CONSTRUCTION— MODEL

HYDRAULIC DEVICES

Each hydraulic device (see Figure 8-2) consists of:

- a cylinder that has a spring-loaded metal diaphragm attached to the plunger (iron core) of a solenoid and a spring-loaded valve stem;

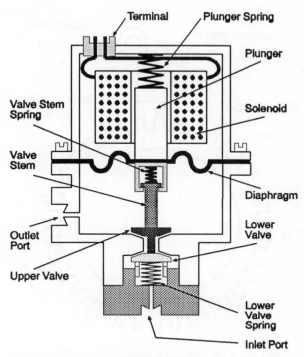

Figure 8-2   Hydraulic device.

- an upper valve which is spring-loaded through the valve stem;
- a spring-loaded lower valve located in an inlet port union;
- an inlet port which is connected via a line to the master cylinder;
- an outlet port which is connected via a line to the caliper pipe;
- an insulated electrical terminal which has its pins connected to the solenoid.

The solenoid of each hydraulic device is connected by wires to the electronic control unit (ECU).

WHEEL SPEED SENSORS

The wheel speed sensors (see Figure 8-3) are induction type pulse generators. Each pulse generator consists of a toothed rotor which is attached to the wheel hub and of an induction pick-up attached to the axle support. The pick-up coil of each pulse generator is connected by wires to the electronic control unit. The signals

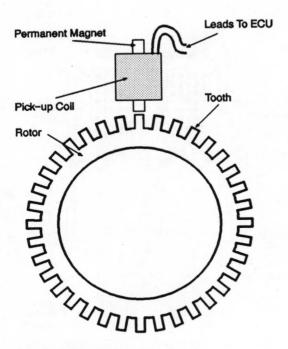

Figure 8-3 Wheel sensor.

produced by the wheel speed sensors are AC waves which change in frequency with variations in the wheel speeds.

ELECTRONIC CONTROL UNIT

The electronic control unit (ECU) consists of two input stages, a central processing unit (CPU), a memory and an output stage. One input stage shapes, divides and converts the wheel sensor pulse generator signals into digital signals that can be stored in memory and processed by the CPU. The other input stage changes the signal from the stop-light switch to a digital signal and sends it to the CPU where it is processed. The CPU processes the digital input signals, decides whether the wheels are changing speed too rapidly, calculates the number of hydraulic fluid pressure pulses that are needed at each caliper and sends the information to the output stage. The output stage selects and switches the solenoids, in the hydraulic devices, on and off at the correct time

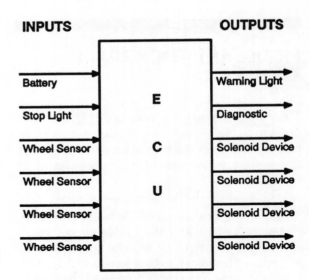

Figure 8-4 Electronic control unit.

and rate. Output signals are also supplied to a warning light and a diagnostic link. The electronic control unit is connected to the vehicle's battery (see Figure 8-4).

OPERATION—MODEL

MEDIUM BRAKING— NO LOCK-UP

The brake fluid pressure from the master cylinder is transferred to the wheel calipers through the open valves and the chamber of each hydraulic device (see Figure 8-5). The electronic control unit receives a signal from the stop-light switch that the brakes have been applied and starts to monitor the speed of each wheel. Each speed input value is stored in memory (RAM) and is constantly replaced with new values, after the old and new values have been compared. The CPU decides that the wheel speeds are not changing too rapidly and does not activate the hydraulic devices. The upper and lower valves in the hydraulic devices are held open by the valve stems because the diaphragms have not moved against spring pressure.

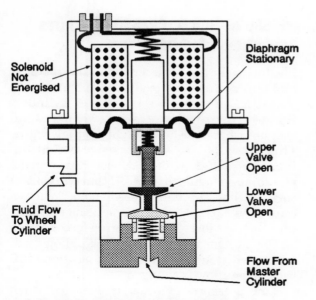

Figure 8-5 Normal operating position.

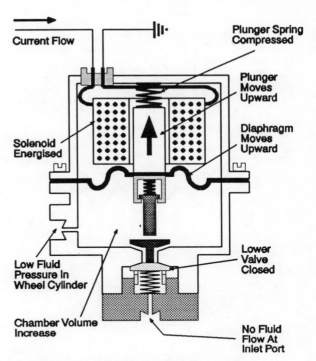

Figure 8-6 Pressure reduction mode.

MEDIUM BRAKING—ONE WHEEL LOCK-UP

Let us assume, under medium braking, one wheel contacts a patch on the road that causes a sudden reduction in the tyre to road friction, i.e. the wheel starts to lock-up. This action produces a rapid reduction in the pulse (frequency) output from the wheel sensor. The input stage of the ECU receives the signal and converts it for use by the CPU. When the CPU compares the new and old speed input signals, it decides that there is a large difference in the two values and sends a signal to the output stage.

The output stage connects the solenoid of the selected hydraulic device to the battery (see Figure 8-6). Current flow through the solenoid produces a strong magnet. The magnetic attraction causes the plunger to move against the spring pressure. As the plunger moves inside the solenoid, the diaphragm and the valve stem move away from the inlet port. The slightest upward movement of the valve stem allows the spring to close the lower valve. The instant the valve seats, the chamber is isolated from the master cylinder. Further upward movement of the diaphragm increases the volume of the

chamber and causes a sudden pressure drop in the wheel cylinder. The brake releases and the wheel regains speed.

The sudden rise in the wheel speed causes an increase in the pulse (frequency) output from the wheel sensor. The input stage of the ECU receives the signal and converts it for use by the CPU. When the CPU compares the new and old speed input signals, it decides that there is a large difference in the two values and sends a signal to the output stage.

The output stage connects the solenoid of the selected hydraulic device to the battery, in the reverse direction (see Figure 8-7). Current flow through the solenoid produces a strong magnetic force in a downward direction. This causes the plunger to move with spring pressure. As the plunger moves inside the solenoid, the diaphragm is forced towards the inlet port to decrease the volume of the chamber. The slightest downward movement of the stem pushes the upper and lower valves against the lower spring pressure to seat the upper valve. The chamber is isolated from the master cylinder

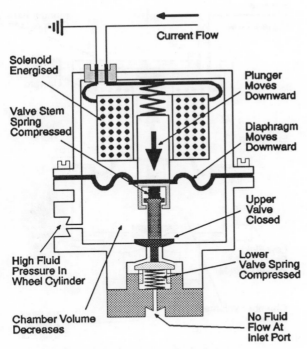

Figure 8-7 Pressure increase mode.

and the seated upper valve prevents the fluid from returning to the master cylinder. The valve stem spring starts to compress to provide an extra force on the upper valve. Further downward movement of the diaphragm decreases the volume of the chamber and causes a sudden rise in pressure in the wheel cylinder. The brake is applied and the wheel speed is reduced.

These two modes continue until the driver releases the brake pedal—the input signal from the stop-light is switched off—or the vehicle stops.

SEVERE BRAKING—ALL WHEELS LOCK-UP

When this situation occurs, the electronic control unit monitors the four wheel speed sensors and operates the four hydraulic devices in a sequence that will ensure maximum braking effect. The brake on each wheel will be operated independently of the other wheels but the overall braking action will be smooth and precise.

PRESSURE-HOLD (PRESSURE-RETENTION) MODE

The hydraulic device described in the 'medium braking' section will continue to produce a pulsating pressure in the wheel cylinder. However, there will be occasions when the fluid pressure should be held at a high or low value. This is called brake pressure-retention. Pressure-retention can be achieved in the model by adding a module to the ECU that will control current flow through the solenoid. When a controlled current flows through the solenoid, the magnetic force and the spring forces acting on the plunger will equalise. The plunger will stop moving and be held in that position until the current flow is changed.

Low pressure-retention is achieved when the lower valve is closed and the diaphragm is held up against the plunger spring pressure (see Figure 8-8).

High pressure-retention is achieved when the upper valve is closed and the diaphragm is held down against the valve stem spring and

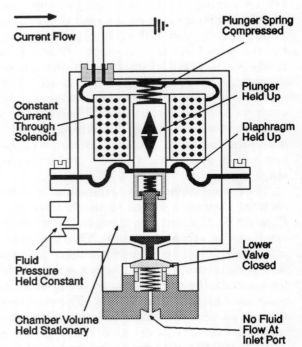

Figure 8-8 Pressure retention mode (low pressure).

lower valve spring pressures (see Figure 8-9).

When the current flow ceases, the diaphragm and the valves return to their initial position.

Four main principles of ABS that the model has demonstrated are:

- normal braking;
- pressure-increase;
- pressure-reduction;
- pressure-retention.

Normal braking occurs when the wheels reduce their speed at an acceptable rate.

Pressure-increase is achieved by reducing the chamber volume, inside the hydraulic device, while the fluid return valve is closed.

Pressure-reduction results from the chamber volume of the hydraulic device being increased while the fluid inlet valve is seated.

High or low pressure-retention is obtained in the hydraulic device when the chamber volume is held at a given value.

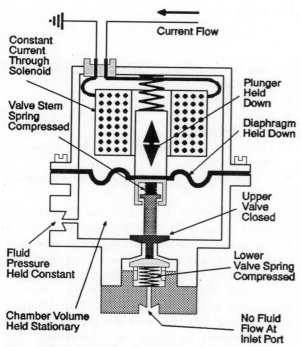

Figure 8–9 Pressure retention mode (high pressure).

ANTI-LOCK BRAKE SYSTEMS

The major problem with the theoretical model is related to the use of a solenoid to apply force on the diaphragm. The solenoid cannot generate enough force on the area of the diaphragm to increase the fluid pressure to a level suitable for operating a braking system. To obtain a very high pressure, the diaphragm would be so small that the distance it would have to travel would be excessive. Alternatively, a very large solenoid would have to be used to obtain the required pressure.

Manufacturers overcome the problem with hydraulically actuated pistons and solenoid actuated valves. The number of components in an ABS must be increased to support an additional, high pressure, hydraulic system.

An anti-lock brake system (see Figure 8-10) requires:

- a hydraulic brake system comprising a dual master cylinder, a caliper and disc pads at each wheel, hydraulic pipes and hoses and a stop-light switch;
- a high pressure hydraulic system consisting of several hydraulic devices, a pump, pressure switches, an accumulator and several solenoid control valves;
- four inductive wheel speed sensors connected to the control unit.;
- an electronic control unit connected to: the wheel sensors, the stop-light and/or park brake switch, the battery and the hydraulic devices;
- electrical components including wires, switches, relays and a battery.

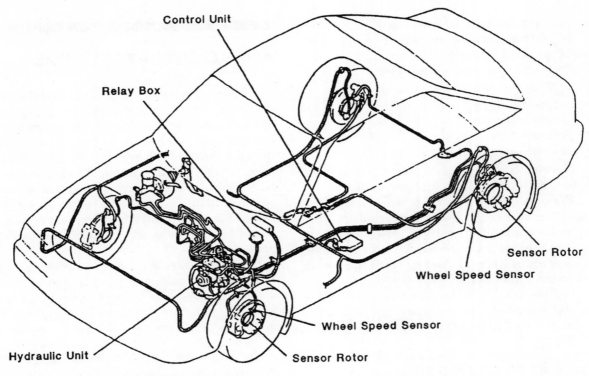

Figure 8-10 Anti-lock brake system.

VEHICLE BRAKE SYSTEM

Either a diagonal or a front and rear split configuration is used as the standard brake system (see Figure 8-11). The diagonal split brake system is common on front wheel drive vehicles. Conventional rear axle drive vehicles are generally fitted with a front and rear wheel split brake system.

Normally, the proportioning valves are an integral part of the master cylinder in a standard system, but they are now located in the hydraulic device of the ABS.

HIGH PRESSURE HYDRAULIC SYSTEM

The high pressure hydraulic system comprises:

- four control pistons and cut-off valve assemblies or stroke switches;

- two, three or four pairs of solenoid valve assemblies;
- a high pressure pump;
- an accumulator;
- a fluid reservoir;
- two proportioning valves or two pressure control valves;
- a high and low pressure switch or a pressure switch.

Two different high pressure hydraulic systems are in wide use. Both systems have a primary objective of pressure decrease, retention or increase to the wheel cylinders, as described in the system model. One manufacturer has an integrated hydraulic unit, whereas another integrates four control pistons, two solenoid valve assemblies, pressure control valves and a reservoir into a modulator unit. The integrated hydraulic unit will be described first, followed by the modulator unit.

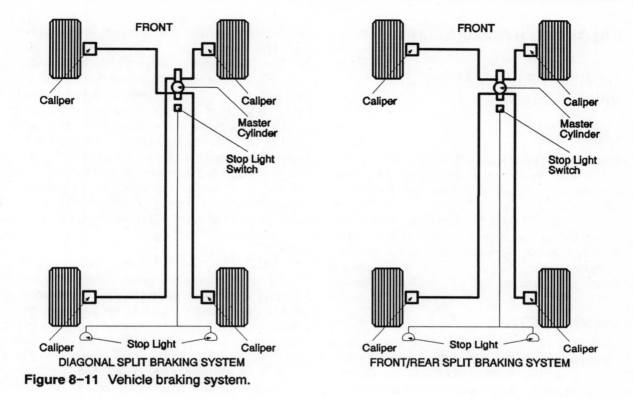

Figure 8–11 Vehicle braking system.

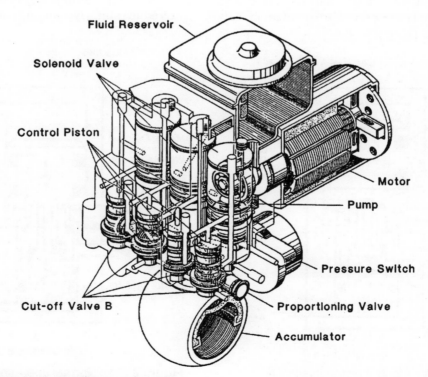

Figure 8-12 Integrated hydraulic unit.

INTEGRATED HYDRAULIC UNIT

CONTROL PISTON AND CUT-OFF VALVE ASSEMBLY

LOCATION

Four control pistons and cut-off valve assemblies are housed in the hydraulic unit. All these assemblies are identical to the following description.

Both ends of the bore of each assembly are connected by passages to a solenoid valve unit. Another two passages leave the bore and end at external unions, to provide connection points for the wheel and master cylinders respectively.

CONSTRUCTION

Each control piston is fitted with two sealing rings and is positioned into a bore (see Figure 8-13). The section in the bore above the control piston is known as chamber A and the section below the control piston is known as chamber B. Chamber A is connected to a solenoid valve

assembly. Chamber B is connected to a wheel cylinder.

The lower section houses the cut-off valve assembly. This assembly comprises a cut-off valve A, a cut-off valve B and a spring. Cut-off valve B is fitted with a sealing ring at one end and a circular seal at its spigot end. The spigot end of the valve protrudes into chamber B to allow the circular seal to seat. Cut-off valve B is large enough to house cut-off valve A and the spring. The stem of cut-off valve A protrudes through the spigot end of cut-off valve B to allow the spring to seat the valve head on a seal.

OPERATION

In the normal operating position, high pressure fluid is supplied to chamber A. This pressure forces the control piston against the stem and the spigot of the cut-off valves (see Figure 8-14). The control piston holds both cut-off valves off their seats. This action connects the master cylinder directly to a wheel cylinder.

Just before the wheel locks, the ECU and the solenoid valve allows the pressure in chamber A to reduce slightly, causing the control piston to

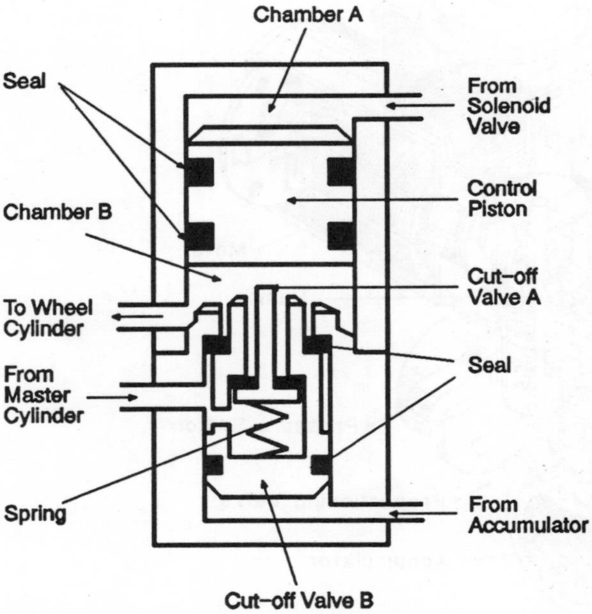

Figure 8-13 Control piston assembly.

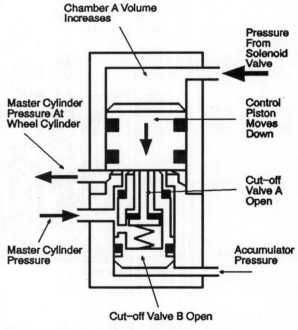

Figure 8-14 Normal operating position.

move to a position that will allow the cut-off valves to seat (see Figure 8-15). The master cylinder is isolated from the wheel cylinder. Further movement of the control piston away from the cut-off valves increases the volume in chamber B and reduces the pressure in the wheel cylinder. The brakes are released slightly.

When the control piston has moved to a point where the pressures in the chambers A and B are equal (see Figure 8-16), the braking effort will remain constant.

If the ECU senses that the wheel is not stopping quickly enough, the ECU and the solenoid valve allow more pressure to be applied to chamber A which forces the control piston towards the cut-off valves (see Figure 8-17). The slight movement of the control piston reduces the volume in chamber B and increases the fluid pressure in the wheel cylinder. The increased braking effort slows the vehicle more rapidly.

This cycle repeats itself until the vehicle stops moving or the brake pedal is released.

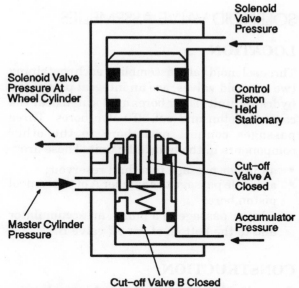

Figure 8-16   Pressure retention mode.

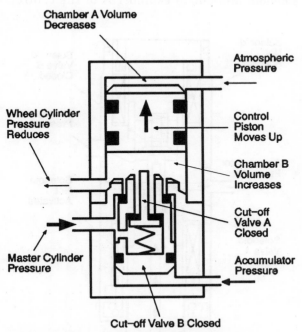

Figure 8-15   Pressure reduction mode.

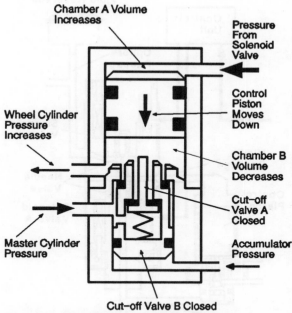

Figure 8-17   Pressure increase mode.

SOLENOID VALVE ASSEMBLIES

LOCATION

Three solenoid valve assemblies, each containing two solenoid valves, are an integral part of the hydraulic unit. Their bores are located near the control piston and cut-off valve bores. Three passages connect each bore to the other components in the hydraulic unit. These are:

- an upper passage to a fluid reservoir;
- a middle passage to chamber A of the control piston bore;
- a lower passage to a pump, an accumulator and to the bottom of cut off valve B.

CONSTRUCTION

Two solenoid valves are located in each bore; one at each end (see Figure 8-18). They are labelled solenoid A (lower unit) and solenoid B (upper unit). Each solenoid valve comprises:

- a solenoid;
- a needle valve and seat;
- a spring.

One end of each solenoid coil is electrically connected to a common external terminal. The other end of each solenoid coil is electrically connected to separate external terminals.

A spring, needle valve and seat is placed inside each solenoid so that the seat forms an opening to a lower passage. The spring holds the needle valve on its seat, when a low accumulator pressure exists in the passage.

OPERATION

In the normal braking mode, the ECU has switched off the voltage to both solenoids, and the high accumulator pressure holds solenoid needle valve A off its seat (see Figure 8-19). Fluid flows in the open valve, through the solenoid and out to chamber A of the control

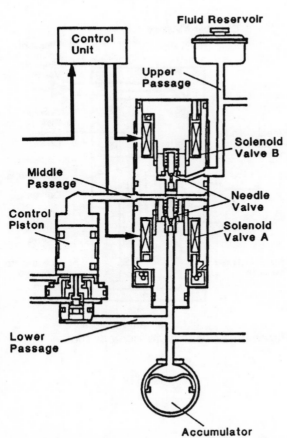

Figure 8-18 Solenoid valve assembly.

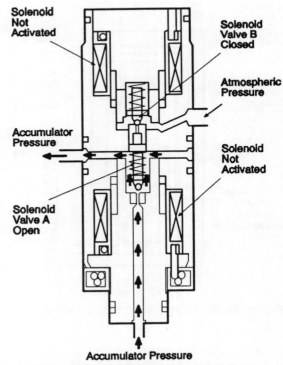

Figure 8-19 Normal operating mode.

piston. Chamber A now receives hydraulic pressure from the accumulator. The resulting control piston movement pushes the cut-off valve off its seat and the master cylinder is connected to the wheel cylinder.

When the ECU applies voltage to solenoid A, the current through the coil creates a strong magnet which forces the needle onto its seat (see Figure 8-20). This action closes the passage from the accumulator and the control piston is held stationary. Pressure-retention is achieved.

Pressure-reduction is obtained when the ECU applies voltage to solenoid B while solenoid A is energised. Solenoid needle valve B is pulled away from its seat by the strong magnet created as current flows through the coil. This action connects chamber A of the control piston to the reservoir. The pressure in chamber A is reduced to atmospheric pressure and the wheel cylinder pressure in chamber B forces the control piston upwards. As the control piston moves away from the cut-off valve it increases the volume in chamber B; the pressure in the wheel cylinder reduces and the brake is released.

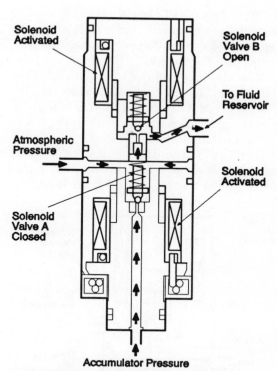

Figure 8-21 Pressure reduction mode.

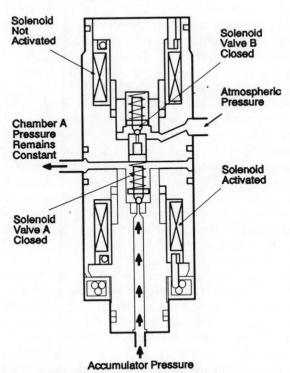

Figure 8-20 Pressure retention mode.

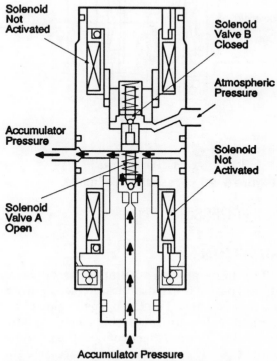

Figure 8-22 Pressure increase mode.

Immediately the ECU switches off the applied voltage to both solenoids, valve B is seated by the spring and valve A is opened by accumulator pressure (see Figure 8-22). High pressure from the accumulator is supplied to chamber A of the control piston. The piston is forced towards the cut-off valve reducing the volume of chamber B. This action increases the pressure in the wheel cylinder and the brake is applied.

The valve at solenoid B also acts as a pressure relief valve. It dumps excessively high pump pressure to the reservoir (see Figure 8-23).

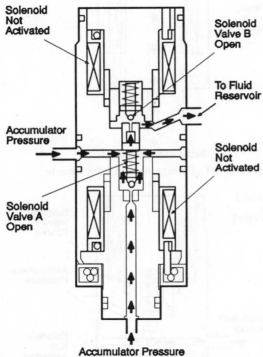

Figure 8-23   Pressure relief mode.

HIGH PRESSURE PUMP

LOCATION

The high pressure pump housing is an integral part of the hydraulic unit (see Figure 8-24). It is positioned on the opposite side of the solenoid valve units to the control piston assemblies. Two passages connect the pumping element to the reservoir and the accumulator respectively.

CONSTRUCTION

The high pressure pump comprises:
- an electric motor;
- a cam-eccentric bearing;
- a spring-loaded plunger;
- a spring-loaded inlet valve;
- a spring-loaded outlet valve.

A DC electric motor is connected to the battery via a relay. The motor's shaft is fitted with a cam constructed from a small crank and a roller bearing. In some cases, a reduction gear set is installed between the motor shaft and the small crank. The cam bears on the end of the spring-loaded plunger so that the plunger will move back and forth in its bore during each motor revolution. The inlet and outlet valve may be either of a ball or of a plunger type.

OPERATION

When the motor is connected to the battery by the relay, the armature spins the eccentric bearing to produce a reciprocating motion at the end of the plunger. With each upward movement (see Figure 8-25) of the plunger the pressure in the pump chamber drops to below atmospheric pressure. The spring force and accumulator pressure seats the outlet valve. Atmospheric pressure forces brake fluid through the inlet valve to fill the pump chamber. With each downward movement (see Figure 8-25), the plunger pressurises the brake fluid in the pump chamber. The spring force and fluid pressure seats the inlet valve. The brake fluid is forced out through the outlet valve into the accumulator. The resulting pumping action is repeated for every revolution of the eccentric bearing. System pressures of more than 21 000 kPa can be achieved by the use this type of plunger pump.

Pressure relief is provided by the needle valves and seats in each solenoid valve assembly. Normally, needle valve A, in each solenoid assembly, is held off its seat by the accumulator pressure so pressure relief is controlled by needle valve B (see Figure 8-26). Extremely high accumulator pressure forces needle valve B off its seat and excess brake fluid is dumped into

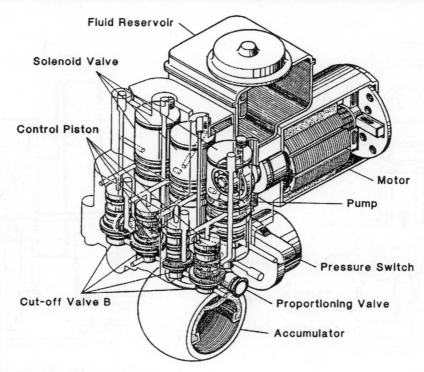

Figure 8-24 High pressure pump.

the reservoir to protect the control valves and other components in the system.

Typical pump working pressures are in the range of 19 620 to 21 582 kPa (2844 to 3128 psi).

ACCUMULATOR

LOCATION

The accumulator housing is attached to the base of the high pressure pump housing. It is connected by a common passage to:

- the outlet port of the pump;
- needle valve A in each solenoid valve assembly;
- the lower end of cut-off valve B in each control piston assembly.

CONSTRUCTION

A thick walled metal sphere is fitted with a union to provide a method of attaching it to the pump housing. The interior of the sphere is divided into two chambers by a thick rubberised diaphragm. The lower chamber is initially pressurised with nitrogen to the value of 10 104 kPa (see Figure 8-27) . The upper chamber, connected to the pump, is designed to operate with brake fluid pressures of more than 21 000 kPa (3000 psi).

OPERATION

As the brake fluid pressure in the upper chamber increases above 10 104 kPa, the diaphragm is forced into the lower chamber, compressing the nitrogen. Fluid pressure will continue to force the diaphragm into the lower chamber until the two pressures (fluid and nitrogen) balance. At this point, the upper chamber becomes a storage area for the brake fluid. When the system's fluid demand becomes greater than the pump's capacity, the compressed nitrogen and the diaphragm will force the stored brake fluid into the system to maintain the pressure.

Typical working pressures of the accumulator are in the range of 19 620 to 21 582 kPa (2844 to 3128 psi).

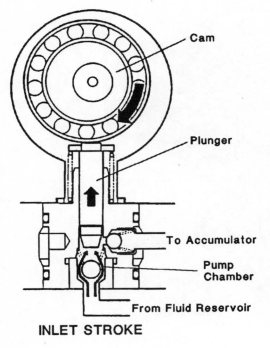

Cam

Plunger

To Accumulator

Pump Chamber

From Fluid Reservoir

INLET STROKE

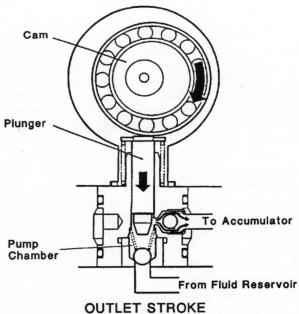

Cam

Plunger

To Accumulator

Pump Chamber

From Fluid Reservoir

OUTLET STROKE

Figure 8-25   Pump operation.

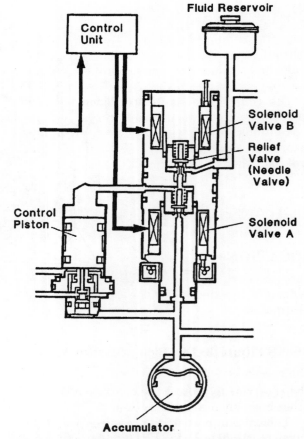

Control Unit

Fluid Reservoir

Solenoid Valve B

Relief Valve (Needle Valve)

Control Piston

Solenoid Valve A

Accumulator

Figure 8-26   Pressure relief valve.

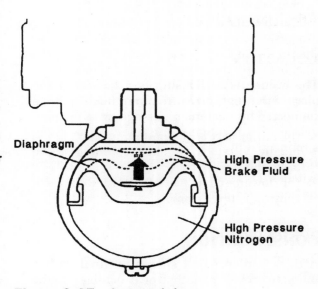

Diaphragm

High Pressure Brake Fluid

High Pressure Nitrogen

Figure 8-27   Accumulator.

PROPORTIONING VALVE

Two proportioning valves are located inside the hydraulic unit. Each valve is connected into the brake line between the master cylinder and the control piston assembly of each rear brake.

The construction and operation of these valves are similar to those of valves fitted to conventional brake systems. Their function is to prevent rear wheel lock-up under normal braking, when the ABS has failed.

HIGH AND LOW PRESSURE SWITCH

LOCATION

The pressure switch is mounted onto the hydraulic housing near the pump and the accumulator.

CONSTRUCTION

The pressure switch comprises a:

- heavy bourden tube;
- spring;
- contact switch;
- microswitch;
- lever;
- heavy metal housing.

The open end of the bourden tube is connected by a passage to the pump and the accumulator. The tip, fitted to the closed end of the bourden tube, bears on the end of a spring-loaded lever (see Figure 8-28). Positioned between the microswitch and the arm of the contact switch, the other end of the lever is designed to open and close both switches. One pole of each switch is connected to the ECU by a separate wire. A third wire grounds the common arm of both switches. These components are arranged inside the heavy metal housing.

OPERATION

Three operating positions are possible with this switch. These positions are:

- low pressure warning;

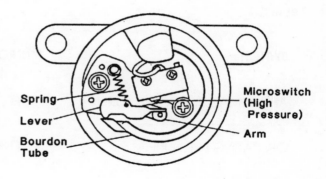

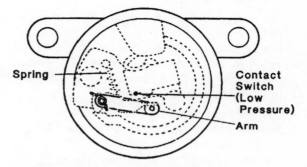

Figure 8-28 Pressure switch.

- normal operating pressure;
- cut-off (high) pressure.

When an abnormally low pressure exists in the hydraulic unit, the bourden tube bends, causing its tip to move inward. Inward movement of the tip is transferred to the lever which opens both switches. Immediately the switches open, current flow to the pump is interrupted by the relay and the warning light is illuminated. This action prevents the pump from running continuously when a failure has occurred.

At normal operating pressure, the bourden tube straightens slightly, and the spring moves the lever to a position that closes the contact switch and allows the microswitch to remain open. Pump operation resumes and the warning light is cancelled.

When cut-off pressure is reached, the bourden tube has straightened to a point where the spring and lever has closed the microswitch. The contact switch remains closed. Immediately the microswitch closes, the current flow to the

Table 8.1

SWITCH	MICRO	CONTACT	PUMP
ABNORMALLY LOW PRESSURE	OFF	OFF	OFF
OPERATING PRESSURE	OFF	ON	ON
CUT-OFF PRESSURE	ON	ON	OFF

pump is interrupted by the relay. Table 8.1 shows the relationship between the switch positions and the fluid pressure in the ABS. Typical pump pressures are:

- low pressure less than 12 263 kPa;
- normal operating pressure between 12 263 and 21 582 kPa;
- cut-off pressure between 19 620 and 21 582 kPa.

As indicated throughout this section, the operation of the integrated hydraulic unit relies on electronic componentry and circuits. These components are covered in the section following the description of the modulator hydraulic unit.

MODULATOR HYDRAULIC UNIT

Two control piston assemblies, two solenoid valve assemblies, two pressure control valves (proportioning valves) and a reservoir are combined into a modulator unit (see Figure 8-29). The high pressure pump is a separate unit and the pressure switch has been integrated with the accumulator to form another unit.

CONTROL PISTON ASSEMBLIES

LOCATION

The four control pistons are housed in two bores formed in the body of the modulator unit. Nine pipes and one hose connect the modulator unit to the other brake components. Four wires connect the stroke switches to the electronic control unit and another wire is connected to the fail-safe relay (see Figure 8-30).

CONSTRUCTION

Two bores are drilled through the modulator body. The open ends of each bore are stepped and threaded to take the stroke switches. Internal passages are used to connect the two bores at specific points and to connect them to the solenoid valve assemblies. Two control pistons, separated by a compression spring, are housed in each bore, which is sealed at its ends with a stroke switch. The control pistons in one bore relate to the front wheels of the vehicle and the other bore contains the rear wheel control pistons. To achieve a cross-braking configuration, the right side front control piston in one bore is positioned next to the left side rear control piston in the other bore. These piston assemblies are interconnected through a pressure control valve (not shown in illustration) which is housed in a small bore formed between the two larger bores. The pressure control valve has similar construction and operating features to a proportioning valve located in a standard braking system.

Each control piston (see Figure 8-31) comprises:

- a piston with a head at one end of a stem and a thread at the other end. The stem has three different diameters. They reduce in

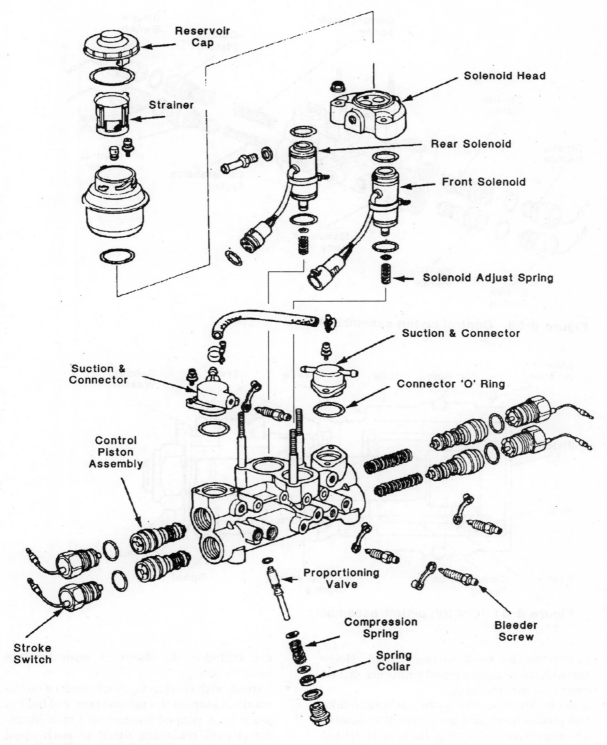

Figure 8-29   Modulator hydraulic unit.

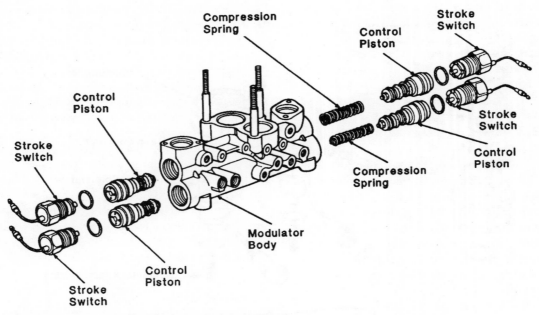

Figure 8-30 Control piston assemblies.

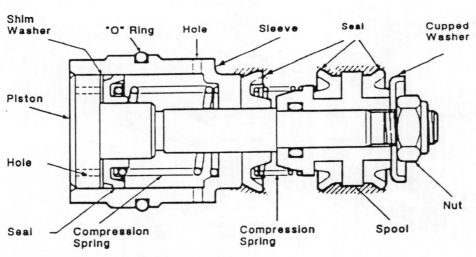

Figure 8-31 Control piston assembly.

size from the head to the thread. Holes through the head at several points are drilled near its circumference;

• a sleeve finished with a smooth bore to take the piston head and an external groove to accommodate an 'O' ring. Its base is drilled to allow the piston stem to pass through the sleeve and move freely back and forth. Holes are drilled in the sleeve at several points near its base;

• a spool, with two lands, that is located on the smallest step on the piston stem and held in place by a cupped washer and nut. Mono-directional seals are fitted to each spool groove;

• four mono-directional seals. One against the

piston head. One on the piston stem against the sleeve base. Two on the spool but facing in opposite directions;

- two compression springs. One between the sleeve base and the piston head seal. The other spring is located between the inner end of the spool and the seal which is pressed against the sleeve base;

- a shim steel washer, positioned between the piston head and the seal.

When two piston assemblies, a compression spring and two stroke switches are installed into a bore, seven chambers are formed. A common chamber is formed in the compression spring area of the bore between the ends of the two pistons. Three more chambers are formed by one control piston assembly and the another three chambers are associated with the second control piston.

When the modulator is completely assembled, the four control pistons in their respective bores form fourteen chambers. These chambers are connected to one another and to the system components by a network of passages and brake pipes.

Since the construction and operation of all four control pistons are similar, the construction and operation of only one control piston will be described in the following sections.

The common chamber (chamber 1) is connected to the inlet valve section of a solenoid valve assembly (see Figure 8-32).

Chamber 2 is formed in the area between the inner spool seal and the sleeve base seal. A pipe connects this chamber to a chamber in the master cylinder.

A separate brake line connects chamber 3 to its respective wheel cylinder (calipers). This chamber is formed inside the bore of the sleeve by the control piston seal.

Chamber 4 is an atmospheric pressure chamber formed between the end of the piston head and the stroke switch. It is connected by a brake line to the master cylinder reservoir.

OPERATION

It must be stressed that, in this unit, the master cylinder chambers are not, and cannot be, directly connected to the wheel cylinders (calipers).

Under light and medium braking conditions, chamber 1 is connected through the solenoid valve assembly to the reservoir of the modulator

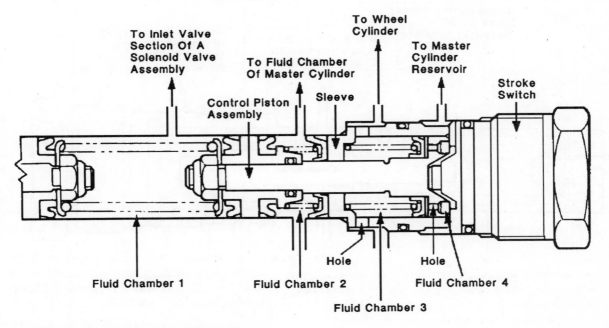

Figure 8-32 Piston and modulator unit.

unit. The fluid in chamber 1 is subjected to atmospheric pressure throughout this braking sequence.

When the brake pedal is depressed, master cylinder pressure is developed in chamber 2 (see Figure 8-33). This high pressure forces the control piston away from the stroke switch and displaces fluid from chamber 1 into the modulator reservoir. The piston head movement in the sleeve is such that the volume in chamber 3 is reduced and the resulting pressure increase is transferred to the wheel cylinder. The brake is applied. The piston head movement also causes an increase in the volume of chamber 4 which is kept at atmospheric pressure by fluid flowing into the chamber from the master cylinder reservoir. The normal braking mode is achieved.

When the brake pedal is released, the pressure in chamber 2 drops to atmospheric pressure and the compression springs return the control piston to its initial position. As the

piston returns rapidly, the piston head seal collapses to allow fluid from chamber 4 to flow through the piston head holes into chamber 3 to balance the pressure. This action ensures air is not drawn into the wheel cylinder circuit and also compensates for any slight fluid loss from the wheel cylinder.

Under medium to heavy braking conditions that will produce wheel lock-up (skidding), the ECU through the solenoid valve assembly determines the braking action that will minimise brake lock-up.

When the brake pedal is depressed, the transfer of pressure to the wheel cylinder is the same as the method previously described. Immediately the ECU senses that the wheel is slowing too rapidly and likely to skid, it isolates chamber 1 from the modulator reservoir and connects chamber 1 to the accumulator (see Figure 8-34). High pressure (accumulator pressure) is applied to the spool end seal which forces the control piston towards the stroke

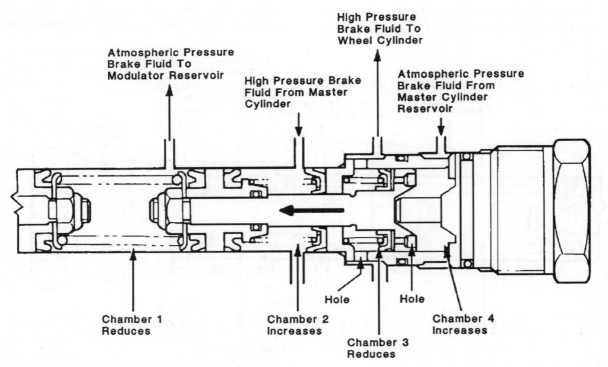

Figure 8-33 Normal operating mode.

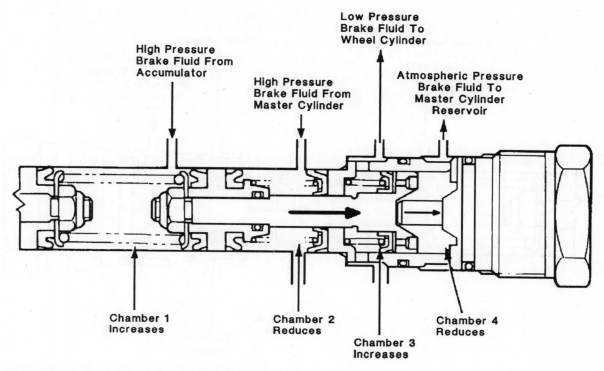

Figure 8-34 Pressure-reduction mode.

switch. The movement of the piston inside the sleeve increases the volume in chamber 3 and reduces the pressure at the wheel cylinder (caliper) to release the brake. The control piston movement also reduces the volume in chamber 2, which increases the master cylinder pressure. The pressure-reduction mode results from the above events.

By monitoring the wheel speed, the ECU detects when the wheel speed has been sufficiently restored. Immediately the ECU isolates the accumulator and opens chamber 1 to the modulator reservoir (see Figure 8-35). Under the influence of the high master cylinder pressure in chamber 2, the control piston is forced away from the stroke switch to reduce the volume in chamber 3. The volume reduction in chamber 3, increases the pressure applied to the wheel cylinder (caliper) and the brake is reapplied. The pressure-increase mode is achieved by these actions.

The complete cycle is repeated until the brake pedal is released or the vehicle stops moving.

Circumstances will arise during the braking cycle that will require the brake pressure to be held constant. The pressure-retention mode is achieved when signals from the ECU completely seal chamber 1 (see Figure 8-36). With chamber 1 sealed, the piston will be held stationary by the balanced forces provided the brake pedal pressure remains constant. The pressure applied to the wheel cylinder from chamber 3 will be maintained and the resulting braking effect will be relative to this pressure. Provided the brake pedal pressure remains constant, the pressure-retention mode will be sustained.

SOLENOID VALVE ASSEMBLIES

LOCATION

The two solenoid valve assemblies are clamped into recesses formed in the top of the modulator body (see Figure 8-37). They are positioned between the relative bores of the control piston

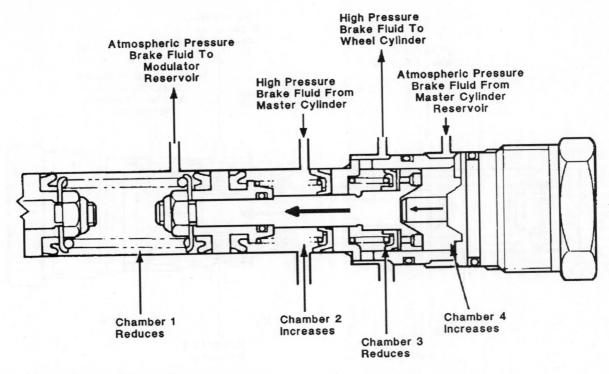

Figure 8-35 Pressure-increase mode.

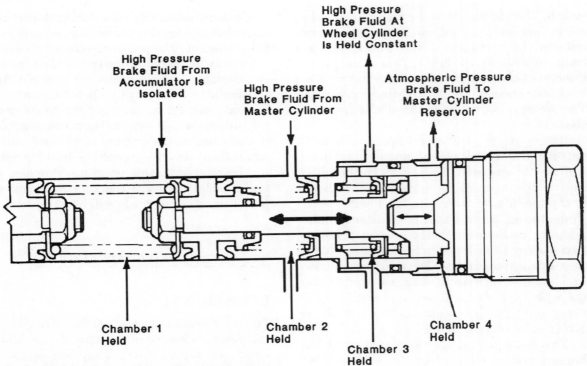

Figure 8-36 Pressure-retention mode.

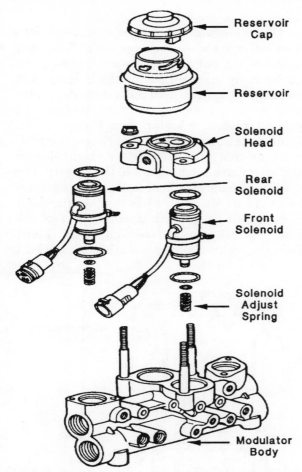

Figure 8-37 Solenoid valve assemblies.

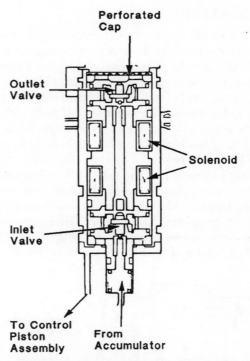

Figure 8-38 Solenoid valve assembly.

assemblies and the modulator reservoir. Each solenoid valve assembly is connected into the electrical circuit by three short wires which terminate in a plug.

CONSTRUCTION

Although one solenoid valve assembly operates the front brakes and the other one operates the rear brakes, they are identical in construction.

A solenoid valve assembly (see Figure 8-38) consists of:

- a cylindrical housing to which all the other components are fitted;
- two coils, one at each end of a centrally drilled core piece which is located at the midpoint of the housing;
- an outlet valve that is positioned at one end

(top) of the core piece which has the valve seat inserted into its drilling;
- an inlet valve that is located at the other end of the core piece;
- an inlet valve seat formed in a small stem piece that is held rigidly in the housing. The flange section of the stem piece is perforated to provide a fluid passage and its hollow spigot is grooved to take an 'O' ring;
- a perforated cap that is held rigidly in the outlet valve end of the cylinder.

The recesses in the modulator body are stepped to accept the spigot end of the solenoid valves (see Figure 8-38). A passage from the bottom of the smallest recess bores (spigot bores) terminates at the accumulator union. Each of the largest recess bores are connected by a passage to the respective control piston bores.

A solenoid head is used to clamp the two solenoid valves firmly into the modulator body. A moulded plastic fluid reservoir attaches to the upper side of the solenoid head and provides fluid to the modulator through the solenoid valves.

OPERATION

Under light and medium braking conditions, the outlet valve is held 'open' by a spring and the inlet valve is held 'closed' by a spring (see Figure 8-39). In this position, the accumulator is isolated from the control piston by the closed inlet valve, and the fluid reservoir is connected to chamber 1 of the control piston cylinder (see Figure 8-33). Fluid, displaced from chamber 1 when the brake pedal in depressed, passes through the perforated flange and the inlet valve mounting. It travels up the centre of the solenoid core, passes through the open outlet valve and the perforated cap into the reservoir. Since solenoids have not been energised by the output stage of the ECU, normal braking mode is achieved.

Under medium to heavy braking conditions that will cause wheel lock-up, the ECU sends a signal to the output stage which connects the outlet valve solenoid coil to the battery. The solenoid's magnetic field pulls the outlet valve onto its seat, isolating the reservoir from the control piston cylinder (see Figure 8-40). Immediately, the outlet valve closes, the ECU sends a signal to the output stage which connects the inlet valve solenoid coil to the battery. The solenoid's magnetic field pulls the inlet valve away from its seat allowing accumulator pressure to be applied to chamber 1 of the control piston cylinder (see Figure 8-34). Fluid, from the accumulator, passes through the inlet

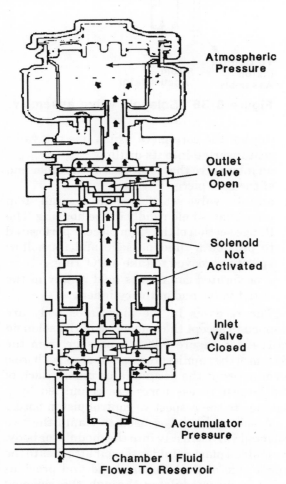

Figure 8-39 Normal braking.

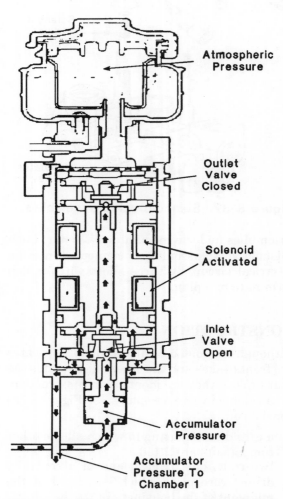

Figure 8-40 Pressure-reduction mode.

valve and the perforated flange into chamber 1 of the control piston assembly. The resulting pressure rise forces the piston in a direction that will cause a reduction in the fluid pressure applied to the wheel cylinder. This action results in the pressure-reduction mode.

Sensing that the wheel speed is increasing, the ECU switches off the current flow to the inlet and outlet valve solenoids. The springs cause the outlet valve to open and the inlet valve to close (see Figure 8-41). With both solenoids deactivated, the pressure in chamber 1 drops to atmospheric pressure and master cylinder pressure in chamber 2 forces the control piston away from the stroke switch (see Figure

8-35). Fluid, displaced from chamber 1 by the control piston movement, passes through the perforated flange and the inlet valve mounting. It travels up the centre of the solenoid core, passes through the open outlet valve and the perforated cap into the reservoir. The decrease in volume of chamber 3 increases the wheel cylinder pressure to apply the brake—provided the brake pedal remains stationary. The pressure-increase mode is achieved.

The pressure-retention mode is achieved when the ECU switches off the current flow to the inlet valve solenoid and applies current to the outlet valve solenoid (see Figure 8-42). The magnetic force closes the outlet valve and the

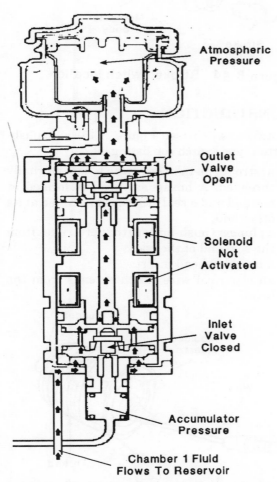

Figure 8-41 Pressure-increase mode.

Atmospheric Pressure

Outlet Valve Open

Solenoid Not Activated

Inlet Valve Closed

Accumulator Pressure

Chamber 1 Fluid Flows To Reservoir

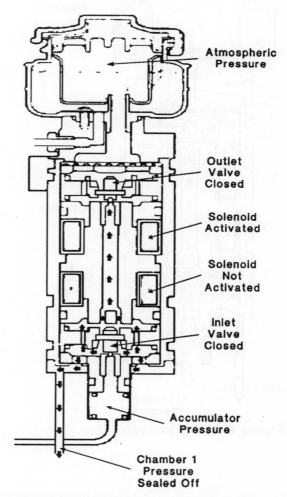

Figure 8-42 Pressure-retention mode.

Atmospheric Pressure

Outlet Valve Closed

Solenoid Activated

Solenoid Not Activated

Inlet Valve Closed

Accumulator Pressure

Chamber 1 Pressure Sealed Off

spring closes the inlet valve. With both valves closed, chamber 1 is completely sealed and the control piston will be held stationary by the balanced forces (see Figure 8-36).

When the brake pedal has been released, the ECU switches off the current flow to the outlet valve solenoid coil which allows the spring to open the outlet valve. With both solenoids deactivated, the pressures in chamber 1 and chamber 2 drop to atmospheric pressure and the springs return the control piston to its initial position (see Figure 8-43).

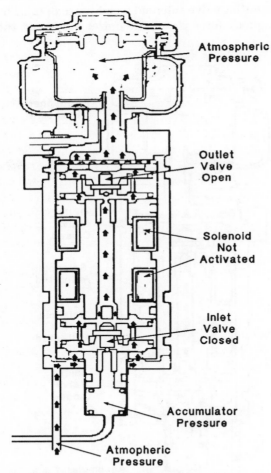

Figure 8-43　Brake released.

STROKE SWITCHES

LOCATION

Each of the four stroke switches is screwed into an end of a control piston bore. The four switches are connected by wires to the electronic control unit (see Figure 8-44).

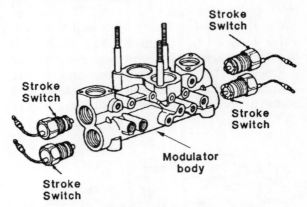

Figure 8-44　Stroke switch location.

CONSTRUCTION

A stroke switch (see Figure 8-45) is a push-button type switch consisting of:

- a strong metal body which is externally threaded. A hexagonal head is formed at one end and a seal groove is machined at its other end;
- a plunger (push-button) that protrudes from the seal end of the body;
- a moulded head insert;
- an electrical wire which extends from the head.

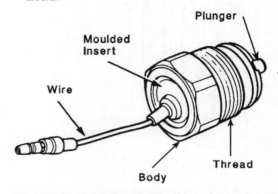

Figure 8-45　Stroke switch.

OPERATION

When the control piston is in its initial position in its bore, the switch plunger is fully depressed by the head of the control piston. The switch is 'on' when its plunger is depressed. If a condition exists where the piston travels more than 13.2 mm away from the switch (see Figure 8-46), the plunger will be extended to a point that will turn 'off' the switch. The electronic control unit, sensing that the switch is 'off', illuminates the brake warning light.

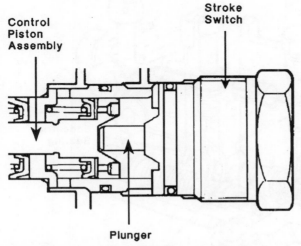

Figure 8-46 Stroke switch plunger extended.

HIGH PRESSURE PUMP

The high pressure pump is identical in construction and operation to the one described in the integrated hydraulic unit section. Its maximum pressure may reach the value of 22 897 kPa (3271 psi).

ACCUMULATOR

The accumulator is identical in construction and operation to the unit described in the integrated hydraulic unit section (see Figure 8-25).

PRESSURE SWITCH

CONSTRUCTION

There is a subtle difference in the construction of the pressure switch compared to the one previously described (see Figure 8-28). It is not designed to monitor abnormally low pressures in the system. The contacts have been eliminated.

OPERATION

When a low pressure exists in the accumulator, the bourden tube bends causing its tip to move inward. Inward movement of the tip is transferred to the lever which opens the microswitch. Immediately the switch opens, the control unit closes the relay which connects the pump to the battery. Pressure then starts to increase in the system. When a high pressure exists in the accumulator, the bourden tube straightens and the spring moves the lever to a position that closes the microswitch. Immediately the switch closes, the control unit opens the relay which disconnects the pump from the battery. Pressure then start to drop with each brake application. Table 8.2 shows the relationship between the microswitch positions and the fluid pressure in the ABS.

When a failure occurs and high pressure cannot be achieved, the computer will interrupt the pump operation after the low pressure warning light has been 'on' for more than 120 seconds. The typical operating pressures are:

- low pressure within the range of 19 800 to 20 500 kPa;
- high pressure within the range of 21700 to 22000 kPa.

Table 8.2

PRESSURE	SWITCH POSITION	PUMP
LOW	OFF	ON
HIGH	ON	OFF

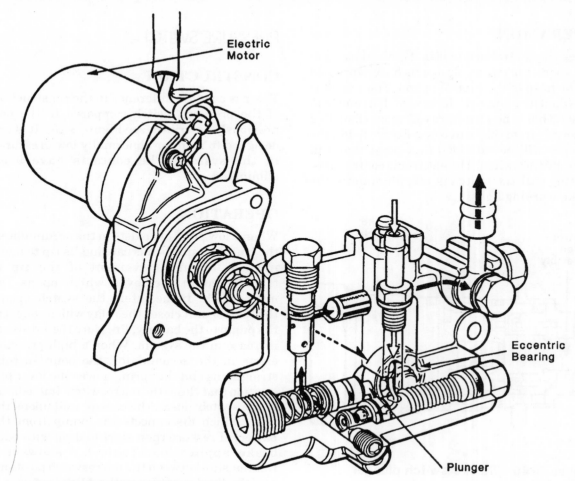

Figure 8-47 High pressure pump.

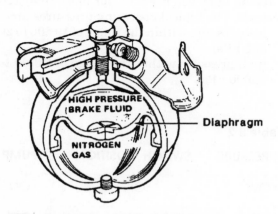

Figure 8-48 Accumulator.

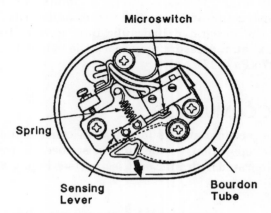

Figure 8-49 Pressure switch.

INDUCTIVE WHEEL SENSORS

LOCATION

As previously stated, one sensor is located at each wheel to detect wheel speed.

Some manufacturers use three sensors on their vehicles—one at each front wheel and one located on the final drive pinion shaft of the rear axle assembly. These vehicles are fitted with the front and rear split brake system and do not require a sensor on each rear wheel.

CONSTRUCTION

The two parts of a wheel sensor are the toothed rotor (pulser) and the pick-up (see Figure 8-50).

The rotor is a metal ring or disc that has tooth-like projections cut in its perimeter. The tooth width is the same dimension as the tooth space. This rotor is attached rigidly to a suitable point on the hub of the wheel. It is extremely important that the rotor runs concentric with the wheel hub centre to ensure perfect balance and to eliminate variations in the air gap between the tooth projections and the pick-up.

The pick-up consists of a coil wound onto a permanent magnet. The ends of the coil are connected by wires to the input stage of the electronic control unit. This pick-up is attached to a non-rotating part near the wheel hub so that its magnet is just above and in alignment with the teeth on the rotor. The rotor tooth to

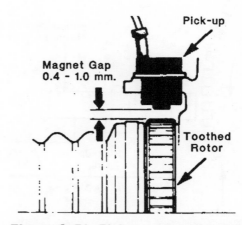

Figure 8-51 Pick-up magnet gap.

magnet gap is very small (0.4–1.0 mm) and must be maintained to ensure efficient signal production (see Figure 8-51).

OPERATION

As the leading edge of the tooth aligns with the pick-up, a positive voltage is induced into the pick-up coil. When the trailing edge of the tooth aligns with the pick-up, a negative voltage is induced into the pick-up coil. The change between positive and negative voltages causes an AC signal to be delivered from the coil. This AC signal will vary with the wheel speed. When the wheel speed increases the AC frequency will increase (see Figure 8-52). It is also important to note the AC voltage increases with increase in the rotor speed.

The AC signal is delivered to the input stage of the ECU where it is modified for use by the CPU.

ELECTRONIC CONTROL UNIT (ECU)

The ECU is a microprocessor that has been developed for use in an ABS. This unit consists of a main section which controls the operations of the ABS and a subsection which controls the pump motor and fail-safe functions.

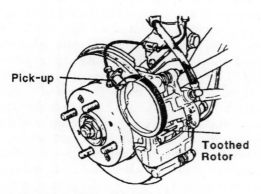

Figure 8-50 Wheel sensor.

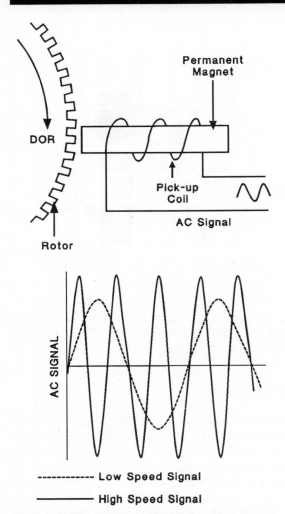

Figure 8-52 Wheel sensor operation.

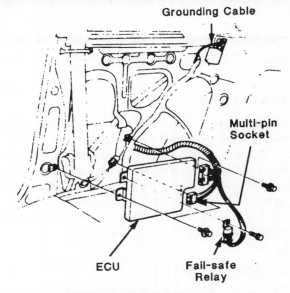

Figure 8-53   ECU location.

LOCATION

The ECU is attached rigidly to a metal panel inside the passenger compartment of the vehicle.

The grounding of the ECU is critical. This is achieved by returning the current flow through the metal case and mounting brackets, and also through the wiring harness. A large multi-pin socket provides the connection points for the inputs and outputs to the vehicle wiring harnesses.

CONSTRUCTION

The ECU consists of:

* a central processing unit (CPU);
* read only memory (ROM);
* random access memory (RAM);
* an input stage that converts signals into digital signals;
* an output stage that converts digital signals into pulses.

Figure 8-54 shows the inputs and outputs of the specific ABS electronic control unit.

This description is brief, however, the construction of a microprocessor has been covered in earlier chapters of this book.

OPERATION

The operation of an ABS is based on a slippage ratio determined from the coefficient of friction (tyre to road), the wheel speed and the vehicle speed. Research has shown that the most effective braking can be achieved in the range of an 8 to 30 per cent slippage ratio (see Figure 8-55). A rolling wheel has a zero slippage ratio, whereas a locked wheel has a 100 per cent slippage ratio. The ECU can calculate the slippage ratio from the signals received from the wheel speed sensors, and then look up a table to decide the action that is needed to reduce the chance of the wheel locking up. Assuming that each wheel is travelling at 100 km/h (vehicle speed) before braking and then one wheel suddenly reduces to 80 km/h as the brake is applied, the slippage ratio would be 20 per cent. After the computer has calculated this value, a look-up table would

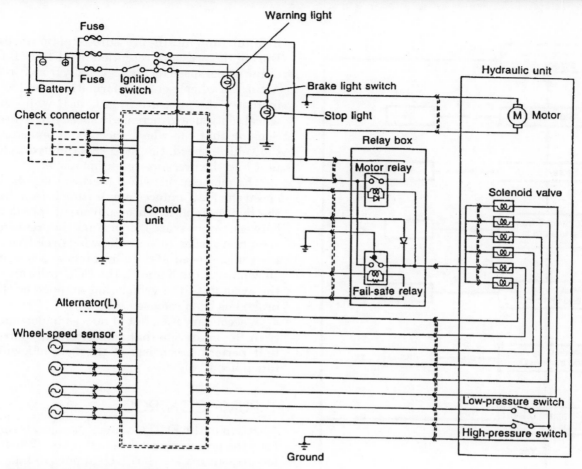

Figure 8-54 ABS Electronic layout.

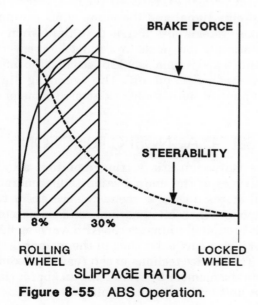

Figure 8-55 ABS Operation.

be used to determine the action to be taken, which, in this case, would be zero—the slippage ratio being less than 30 per cent. Another example would be when the vehicle speed is 90 km/h and a wheel speed is suddenly reduced to 15 km/h which results in a 83 per cent slippage ratio. Immediately, the computer looks up this value in the table, its resulting action would be to release the brake on this wheel. The very basic steps the computer would take for this case are as shown in Figure 8-56.

Remember that these steps would be completed by the computer in few microseconds (one millionth of a second).

Some ECUs are capable of using the wheels' acceleration or deceleration to calculate a projected vehicle speed that would be expected after a given time. At that time, if the project value is not obtained, the computer would decide

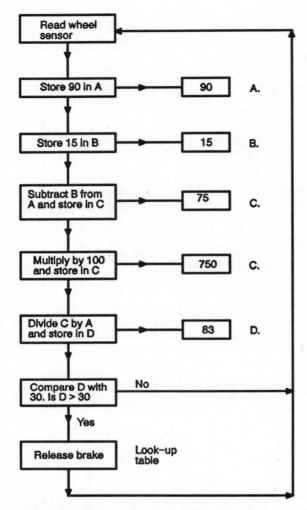

Read wheel sensor		
Store 90 in A	90	A.
Store 15 in B	15	B.
Subtract B from A and store in C	75	C.
Multiply by 100 and store in C	750	C.
Divide C by A and store in D	83	D.
Compare D with 30. Is D > 30	No	
Yes		
Release brake	Look–up table	

Figure 8–56   Basic computer steps.

the action to be taken to correct the situation. Figure 8-57 illustrates a typical example of the combination of the three critical conditions that are influenced by the ECU signal for one wheel.

At point O, the brakes are applied and the hydraulic pressure builds up rapidly to A which reduces the wheel speed to a value (1) below the vehicle speed. Sensing the wheel speed situation, the ECU calculates a high slippage ratio and signals the hydraulic unit to release the pressure. The pressure drops from A to B while the wheel continues to lose speed from 1 to 2.

At point 2, the ECU, sensing that the wheel speed is not reducing as rapidly, sets the

hydraulic unit into the pressure-retention mode. During the pressure-retention mode from B to C, the wheel speed increases rapidly from 2 to 3. Acting on wheel sensor information, the ECU increases the pressure from C to D while the wheel continues to increase in speed from 3 to 4. Sensing that the wheel speed is approaching the vehicle speed, the ECU sets the hydraulic unit into the pressure-retention mode.

The moment the wheel reaches the vehicle speed the ECU induces a pulsating action into the hydraulic unit. These pulsations oscillate between the pressure-increase and the pressure-retention modes to slow the wheel relative to the vehicle speed. If the wheel tends to lock, as shown at points 5 and E, the ECU will repeat the cycle until the vehicle has stopped or the brake has been released.

A secondary role, but of no less importance than the main role that the electronic control unit performs, consists of the four following functions:

1 PUMP CONTROL

A section in the ECU monitors the signals sent from the pressure switches in the ABS. When a low pressure exists, the ECU supplies voltage to the pump relay which activates the pump. When a given pressure is reached, the ECU interrupts the relay signal and the pump stops running. Some systems are designed so that when an abnormally low pressure cannot be recovered within a given time (say 120 seconds) the pump will be switched off, the warning light illuminated and a trouble code stored in memory.

2 SELF-DIAGNOSTIC FUNCTION

The self-diagnostic section monitors all the activities of the main section then generates and stores a code in memory once every two seconds. When an abnormal condition occurs, this section illuminates the ABS warning light and logs a particular code in the memory.

The service technician can read these codes from the memory by connecting an appropriate test unit to the diagnostic plug.

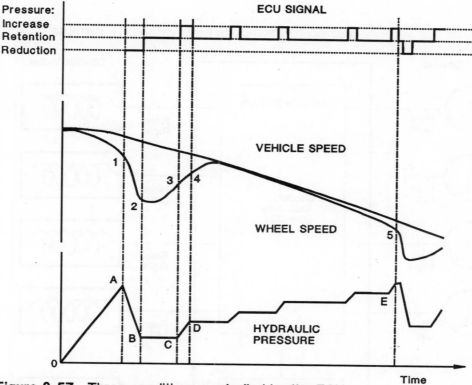

Figure 8-57 Three conditions controlled by the ECU.

Since the codes used by manufacturers vary considerably, it is advisable to obtain the relevant workshop manual to help assist in decoding the problems.

3 FAIL-SAFE FUNCTION

When a defect occurs in the ABS, the fail-safe relay is instructed to cut the power supply to the ECU. Immediately, the braking system reverts to the normal operation mode.

The conditions for which fail-safe occur will vary between vehicles. Some manufacturers set a fail-safe condition related to the tyre diameter. A small tyre will produce a high wheel speed and when one wheel speed is 12 per cent greater than the other wheels the fail-safe condition will be activated.

4 BACK-UP FUNCTION

Some ABS have a back-up function that pulses the rear brakes when a defect occurs in the main section of the ECU. The reason for pulsating the rear brakes is that they lock-up more easily than the front brakes.

INPUTS AND OUTPUTS

Figure 8-58 shows the inputs and outputs that the electronic control unit is designed to service. An ECU may have different inputs and outputs to the ones shown in the diagram. The diagram is only a representation of what could be expected when diagnosing a system.

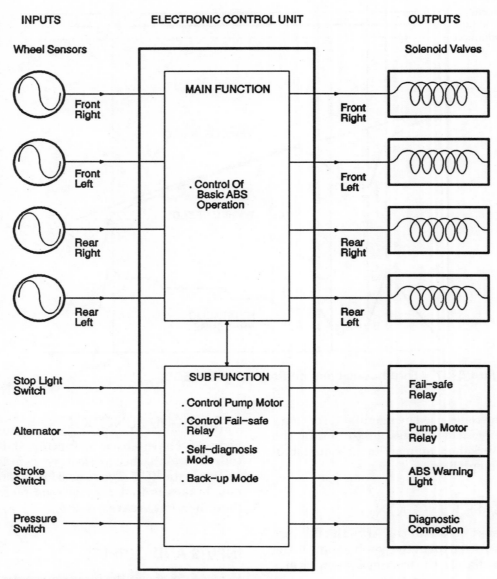

Figure 8-58 ECU Inputs and outputs.

9

ELECTRONICALLY CONTROLLED VEHICLE STEERING SYSTEMS

Power assisted steering is not a recent development and has been fitted to vehicles for many years. This chapter is not intended to explain the many power steering systems which have been developed but will deal with two recent developments in vehicle steering systems. These are:

- Vehicle speed sensing power assisted steering;
- Electronically controlled four wheel steering systems.

The systems chosen for explanation are not the only types available in mass produced vehicles. However they are excellent examples of the manner in which electronics can be applied to vehicle systems.

VEHICLE SPEED SENSING POWER ASSISTED STEERING

Many changes have occurred, as designers have tried to introduce 'road feel' into the power assist system. The steering effort required at parking speeds is markedly different to the effort required at highway speed. The 'perfect' system is one which requires minimum steering effort when parking, yet maintains 'road feel' at highway speeds. It is unlikely this 'perfect road feel' steering will ever be developed because drivers have individual preferences related to steering effort input. Prior to electronics the major design effort has concentrated on:

- hydraulic characteristics of the system, particularly the flow control valve;
- mechanical changes to the worm shaft (rack teeth profile in some systems) to introduce a variable steering ratio as the steering is moved towards either steering lock.

A number of excellent systems have resulted from these design changes but most were still a compromise because the major system inputs are obtained from steering effort and engine (pump) speed.

It is assumed the reader is familiar with the concepts of power-assisted steering and steering box design. This explanation will concentrate on the changes introduced to provide vehicle speed sensing as an input to power assistance control.

HYDRAULIC CONTROL

Figure 9-1 depicts a system very similar in appearance to other rack and pinion type power-assisted steering systems. The changes to the steering box are in the reaction valve area, otherwise the box construction is almost unchanged. A control valve is placed between the oil pump and the steering box. The oil pump has three pipes (two on conventional systems) (see Figure 9-2).

OIL FLOW AND PIPE APPLICATION

Oil from the pump enters the control valve at the upper right inlet and leaves at the lower right outlet. This oil is sent to the steering box valve where it is used for power assistance in the normal manner. Steering box return oil is routed back to the pump reservoir.

The pipe from the upper left of the control valve is used to supply or exhaust oil from a 'reaction force chamber' within the steering box. The pipe located at the upper centre of the control valve allows oil, when exhausted from the reaction force chamber, to be returned to the pump.

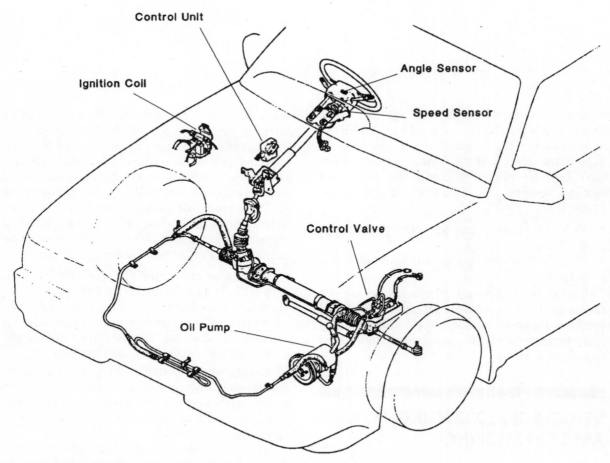

Figure 9-1 Vehicle speed sensing power steering.

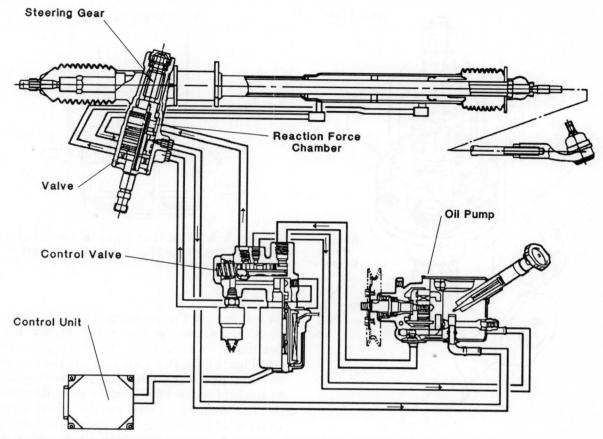

Figure 9-2 The computer (control unit), control valve and the reaction force chamber are the components which combine to produce the change in steering effort.

The function of the control valve is to control the pressure of the oil sent to the steering box reaction force chamber.

REACTION FORCE CHAMBER

Conventional torsion bar actuated valves rely solely on the torsional characteristics of the bar to determine how far the valve ports will be opened. High steering wheel input effort causes the ports to fully open, low steering wheel effort will only cause minimal port opening. This electronically controlled system contains refinements of the basic concept. The relative movement between the control valve sleeve (outer) and input shaft valve sections (which is the movement area when the torsion bar twists) is controlled by the hydraulic force on four small pistons (see Figure 9-3). The pistons are housed in a reaction force chamber formed on the lower part of the control valve sleeve. The input shaft section of the valve has two small projections, against which the pistons can press.

At low vehicle speeds, pressure from the externally mounted control valve is not sent to the steering box reaction force chamber. The pistons are not forced against the projections and the input shaft valve section can move against torsion bar resistance with minimum input effort (see Figure 9-4). This action will open the steering valve ports, allowing high oil flow to the steering rack. The effort required at the steering wheel is at a minimum.

At high vehicle speeds, pressure from the externally mounted control valve is sent to the steering box reaction chamber. The pistons are

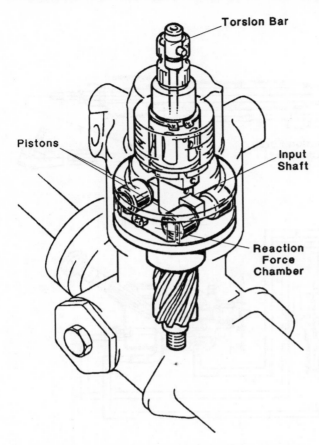

Torsion Bar

Pistons

Input Shaft

Reaction Force Chamber

Figure 9-3 The lower part of the control valve contains a reaction force chamber and four pistons.

Low Speed Driving

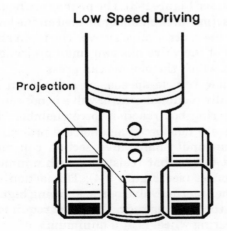

Projection

Figure 9-4 The pistons are not forced against the projections and steering effort is at a minimum.

High Speed Driving

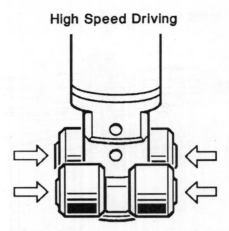

Figure 9-5 The four pistons are forced against the projections and the steering effort is increased.

forced against the projections and the input shaft valve section requires a much higher input twisting force to open the steering valve ports (see Figure 9-5). The basic action of the torsion bar to resist the twisting force is supplemented by the load applied from the pistons within the reaction chamber. A high steering wheel input effort would be required to open the steering valve ports, therefore under normal conditions oil flow to the steering rack would be low. The hydraulic force exerted by the pistons is not on/off but is progressively controlled by the externally mounted control valve. The effort required at the steering wheel is maintained at a value matched to the vehicle road speed (see Figure 9-6).

CONTROL VALVE

The control valve contains an internal valve, solenoid and a pressure switch. The rod will be moved in or out of the orifice by the solenoid. The distance the rod moves will depend on the current value supplied to the solenoid from the electronic control unit. The rod has a vast number of operating positions; however, the description will concentrate on the valve action with the rod fully out (vehicle stationary) or fully in the orifice (high speed).

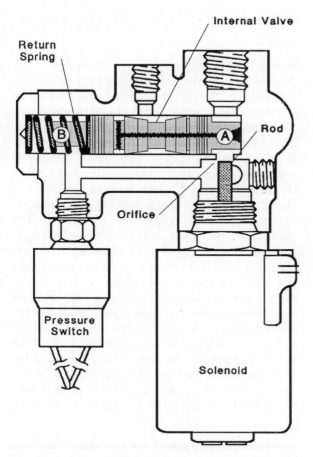

Figure 9-6 Major parts of the control valve. The rod, moved by the solenoid, controls oil flow through the orifice. The change in oil flow causes a change in the pressure differential between chambers A and B.

STEERING ACTION—VEHICLE STATIONARY

If the steering wheel is turned with the vehicle stationary, the ECU will supply maximum current to the control valve solenoid. The solenoid rod is withdrawn from the orifice causing the opening to become large. Oil flows through an internal passage to chamber B at the spring end of the internal valve. The difference in pressure between chambers A and B is minimal and the return spring positions the internal valve to the right (see Figure 9-7). The valve position prevents oil being supplied

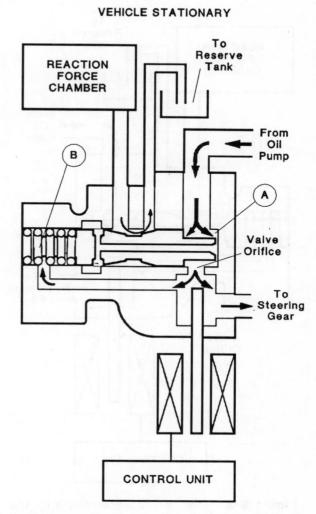

Figure 9-7 With the solenoid rod withdrawn from the orifice the pressure differential between chambers A and B is small. The spring positions the valve, opening a passage to exhaust the reaction force chamber oil to the reserve tank.

to the reaction force chamber in the steering box. At the same time the valve position allows any pressure in the reaction force chamber to exhaust past the valve back to the reserve tank. The reaction force chamber pressure will be at zero and steering box valve opening will be controlled solely by the torsion bar. The steering will 'feel' very light.

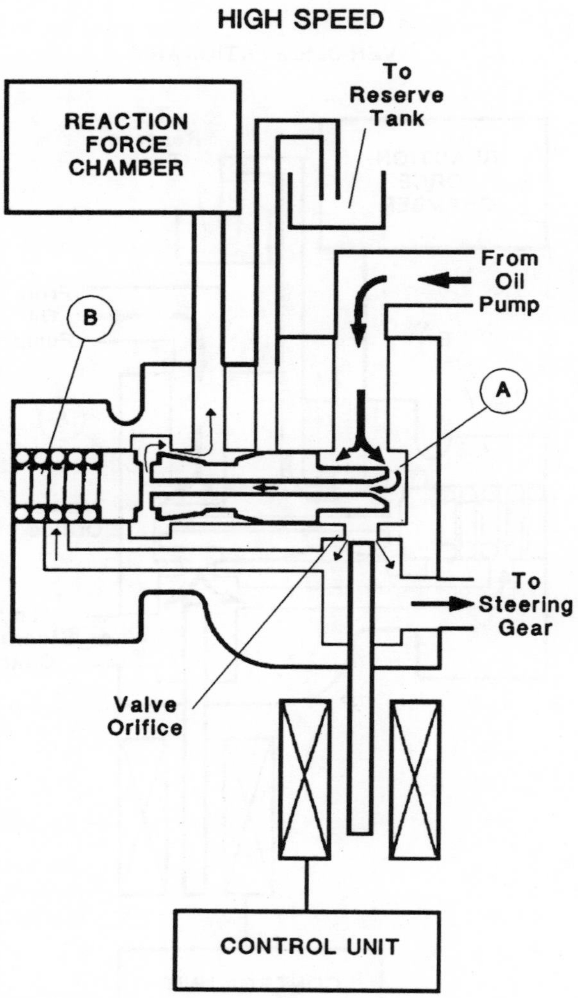

HIGH SPEED

REACTION
FORCE
CHAMBER

To
Reserve
Tank

B

From
Oil
Pump

A

To
Steering
Gear

Valve
Orifice

CONTROL UNIT

Figure 9-8 The rod is extended into the orifice, restricting the flow and causing a large difference in pressure between chambers A and B. The valve is positioned opening an oil supply passage to the re-action force chamber. The valve position also closes the reaction chamber exhaust to the reserve tank.

STEERING ACTION—VEHICLE AT HIGH SPEED

When the steering wheel is turned at high vehicle speeds, the ECU will supply minimum current to the control valve solenoid. The solenoid rod is moved into the orifice causing the opening to become small. The oil quantity flowing through the internal passage to chamber B at the spring end of the internal valve is restricted. A pressure difference exists between chambers A and B. The greater pressure in chamber A causes the internal valve to move against return spring force (see Figure 9-8). The valve position allows oil to be supplied through the centre of the valve to the reaction force chamber in the steering box. At the same time the valve position closes the line to the reserve tank preventing reaction force chamber pressure from exhausting past the valve back to the reserve tank. The reaction force chamber pressure will be at maximum and steering box valve opening will be controlled by the torsion bar supplemented by the chamber pistons. Steering wheel operation will be heavier in both the straight ahead position and when turning.

ELECTRONIC CONTROL

An electronic control unit (see Figure 9-9) receives inputs from the ignition coil, steering angle sensor and the vehicle speed sensor.

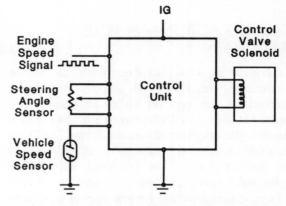

Figure 9-9 Control unit input and output signals.

The steering angle sensor (see Figure 9-10) is a potentiometer located on the steering shaft under the steering wheel. Care should be taken when installing this unit that alignment marks are correctly positioned.

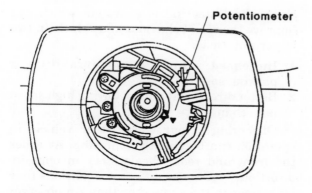

Figure 9-10 Steering angle sensor.

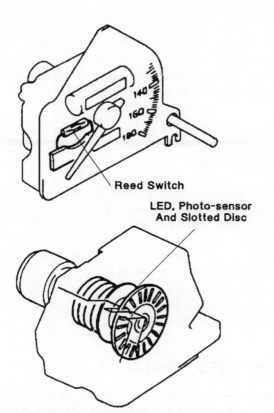

Figure 9-11 Two different types of speed sensors are used, depending on the type of speedometer fitted. (analogue or digital)

The vehicle speed sensor (see Figure 9-11) is located within the speedometer cluster and sends a pulse signal to the control unit.

Based on these input sensors, an output current value is established which is then supplied from the control unit to the control valve solenoid (Figure 9-9).

To enable the steering effort to be progressive and suited to all vehicle operations it is necessary to vary the position of the solenoid rod within the control valve. Figure 9-12 illustrates how the control valve current is determined. Input values from the vehicle speed sensor and the steering angle sensor are used by the ECU to determine the solenoid valve current. The solenoid valve can therefore produce a reaction force chamber pressure which results in power-assisted steering that is progressive in application and retains excellent steering 'feel'.

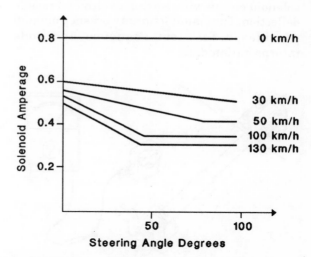

Figure 9-12 A combination of steering angle and vehicle speed are used by the control unit to supply a suitable current value to the control valve solenoid.

FAIL SAFE

The system contains a number of fail-safe options, e.g. a failure of the vehicle speed sensor causes the ECU to operate on data from the engine speed sensor only. Although the steering effort may increase under a system failure condition there is still manual steering capability. In common with other types of torsion bar steering valves, a mechanical link is always present between the input shaft and the steering pinion.

SELF-DIAGNOSIS FUNCTION

A voltmeter can be connected to the power steering test connector (see Figure 9-13). The vehicle should be driven at a speed above 10 km/h with the steering wheel in the straight ahead position. Any faults present in the system will cause the voltmeter needle to deflect in a coded sequence. Open circuits in the vehicle speed sensor, engine speed, angle sensor or control valve and their associated wiring will cause voltmeter needle deflection. An out-of-position angle sensor or a short in the control valve solenoid/circuit will also cause voltmeter needle deflection. The manufacturers workshop manual provides a full test sequence and code interpretation data.

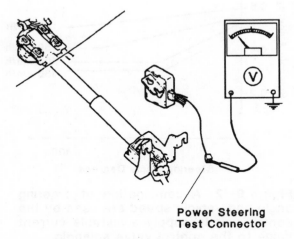

Power Steering Test Connector

Figure 9-13 Diagnostic codes can be retrieved from the control unit with an analogue voltmeter.

ELECTRONICALLY CONTROLLED FOUR WHEEL STEERING SYSTEMS

Four wheel steering is a system which can improve the road behaviour of a vehicle in two important areas:

1 Increased vehicle manoeuvrability at parking speeds.
2 Better directional control during high speed lane changes.

Observing a four wheel steer vehicle in operation can be slightly confusing. At times the front and rear wheels steer in opposite directions, yet under different conditions they both steer in the same direction. An observer may note that the front and rear wheel steering action is in opposite directions, when the vehicle is manoeuvring at low speeds. At highway speeds the steering action of both front and rear wheels is in the same direction. The following text will explain why both types of steering action are necessary and how electronics are being used to make the system vehicle speed related.

OVERVIEW

To understand the benefits of four wheel steering it is necessary to examine what happens when a vehicle is turning at both low and high speeds. Because this concept may be new to many readers, an explanation of road wheel and steering action is provided for:

- low speed turning;
- high speed turning;
- front/rear steering phase.

LOW SPEED TURNING

Figure 9-14 shows a two wheel steer vehicle moving through a sharp right-hand turn. A line extended through the centre of the rear axle is intersected by lines drawn at right angles to the front wheel turning angle. The intersection of

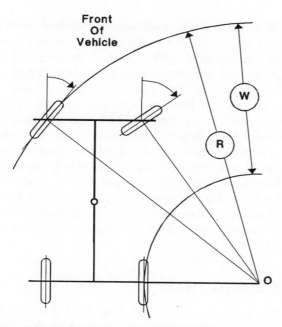

Figure 9-14 Two wheel steer vehicle turning to the right at low speed.

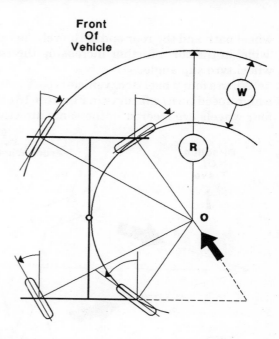

Figure 9-15 Four wheel steer vehicle turning to the right at low speed.

these lines is point O: the centre of the turn. R is the minimum turning radius and W is the difference between the inner and outer wheel paths.

Figure 9-15 depicts a four wheel steer vehicle negotiating a turn. The right angle lines drawn from each wheel turning angle intersect at the turn centre O. The front and rear wheels are turned towards opposite 'locks' placing the turn centre much closer to the vehicle. By comparing the size of R and W with the two wheel steer vehicle (Figure 9-14) it is possible to see the superior turning capability of four wheel steering. This feature is particularly useful when the vehicle is engaged in small radius turns, e.g. parking.

HIGH SPEED CORNERING AND TURNING

When a wheel assembly is viewed from above during a high speed turn the forces, shown in Figure 9-16, act upon the wheel. As vehicle speed increases, a greater cornering force is necessary and the tyre side-slip angle increases.

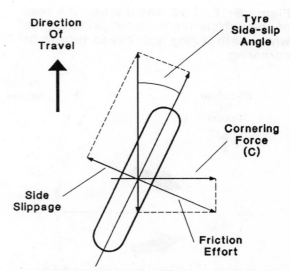

Figure 9-16 Forces acting on a wheel during high speed turning or cornering.

Figure 9-17 shows the action of one front and one rear wheel of a two wheel steer vehicle during left-hand cornering. The tyre side slip angle increases as the steering is turned into the curve. The vehicle tends to yaw about the wheel path and the rear end of the vehicle runs 'wide' in the turn, further increasing the rear wheel tyre slip angle.

When a four wheel steer vehicle is performing a high speed turn (left curve in Figure 9-18), the four wheels are steered in the same direction.

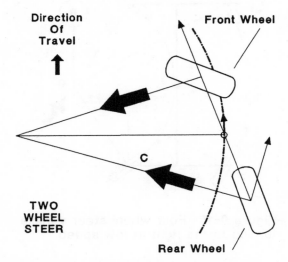

Figure 9-17 Two wheel steer: the rear of the vehicle tends to run 'wide' of the wheel path during high speed turning or cornering.

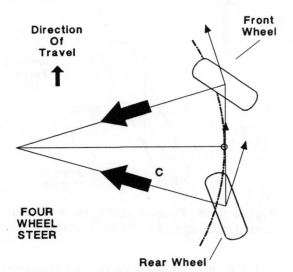

Figure 9-18 Four wheel steer: the rear of the vehicle runs closer to the wheel path during high speed turning or cornering.

This steering action reduces the tendency of the rear wheels to yaw and the vehicle runs closer to its intended wheel path. This feature is particularly useful during a lane change manoeuvre. A two wheel steer vehicle tends to cross the lane markers, run slightly wide at the rear end and requires a slight steering correction (towards the opposite lock then back to straight ahead) to align the vehicle into the direction of travel. A four wheel steer vehicle does not require this correction to align the vehicle. The reduced length of time of rear end yawing and attitude change provides the driver with greater steering precision, steering wheel response and overall cornering balance.

STEERING PHASE

When the front and rear wheels of the vehicle angle towards the same steering lock they are described as being in the same 'phase'. If the front and rear wheels move towards opposite locks, they are described as being in the opposite phase. The front and rear wheels are moving in the opposite phase at slow speeds but moving in the same phase at high speeds. Obviously there must be a transition point where the steering action moves from opposite phase to same phase. There is at least one mass production vehicle which achieves the transition in steering movement by mechanical means and is not vehicle speed related. The system described in this text is a combination of electronic, mechanical and hydraulic systems. This system is based on vehicle speed related principles for all steering actions of four wheel steering.

The maximum steering angle available at the rear wheels is approximately 5 degrees, therefore any reference to both wheels steering in the same phase (or opposite phase) does not mean both front and rear wheels are pivoting by the same number of degrees.

A full lock movement at the front wheel can only produce a rear wheel steering movement of 5 degrees even though the vehicle may be stationary (see Figure 9-20). As the vehicle speed increases from rest, the front to back wheel steering ratio is reduced so less rear wheel steering movement occurs on turns.

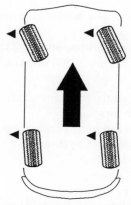

IN PHASE - All wheels steer in same direction.

OPPOSITE PHASE - Front and rear wheels steer in opposite directions.

Figure 9-19 In-phase and opposite-phase steering actions.

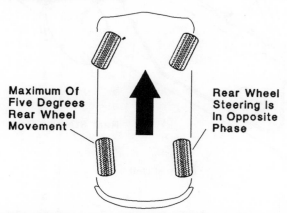

Maximum Of Five Degrees Rear Wheel Movement

Rear Wheel Steering Is In Opposite Phase

Figure 9-20 The opposite phase, rear wheel steer action is at a maximum of five degrees when the vehicle is stationary. The rear steer action reduces as the vehicle approaches 35 km/h.

At approximately 35 km/h, the opposite phase movement has been cancelled out (see Figure 9-21).

With further speed increase the same direction phase is in operation and the front to rear steering ratio starts to increase (see Figure 9-22). At traffic speeds, the maximum rear steering movement is again available with the front and rear steering in the same phase.

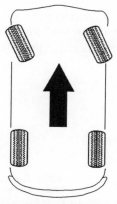

Figure 9-21 At the phase transition point (approx. 35 km/h), the rear wheels remain in the straight ahead position during both left and right turns.

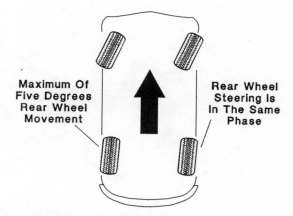

Maximum Of Five Degrees Rear Wheel Movement

Rear Wheel Steering Is In The Same Phase

Figure 9-22 The same phase, rear wheel steer action is at a maximum of five degrees when the vehicle is at traffic speeds. The rear steer action started to increase from zero after the vehicle speed passed 35 km/h.

FOUR WHEEL STEERING SYSTEM

A typical four wheel steering system consists of a relatively conventional power-assisted rack and pinion front steering system which performs the main steering function. The rear system is an electronically controlled hydraulic unit. The rear steering wheels are steered in response to steering wheel angle and vehicle speed. This system is equipped with 'fail-safe' operation. If an electrical or hydraulic malfunction prevents rear steer operation the rear wheels are moved to the straight ahead position by a self-centring spring. During fail-safe conditions the vehicle will operate in the same manner as any two-wheel steer vehicle.

The major components of the rear steer system (see Figure 9-23) are the:

- oil pump;
- steering angle transfer shaft;
- speed sensors;
- rear steering gear containing the:
  —phase control system;
  —stepper motor;
  —control valve;
  —power cylinder;
- control unit;
- fail safe solenoids.

4 WHEEL STEERING (4WS) SYSTEM

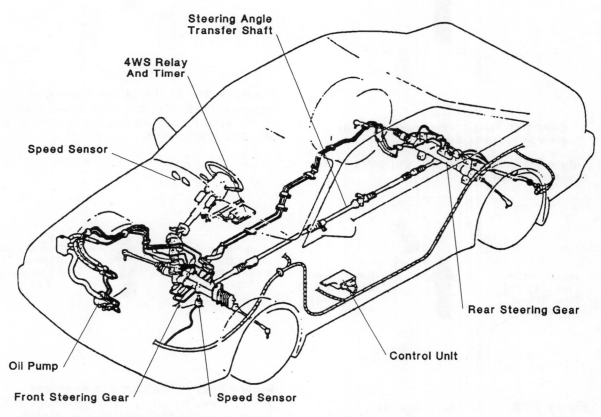

Figure 9-23 Electronically controlled four wheel steer system.

OIL PUMP

The oil pump is a belt driven unit supplied with fluid from a single remote-mounted reservoir. The pump contains two vane type pumping chambers. One chamber supplies the front system and the other chamber is for the rear system. The pump also contains two flow control valves, one for each system (see Figure 9-24).

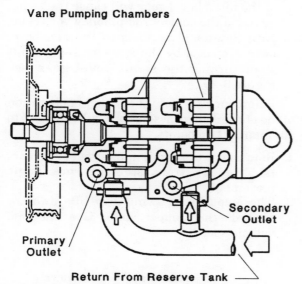

Figure 9-24 One, dual chamber, pump supplies oil to both front and rear steering systems.

STEERING ANGLE TRANSFER SHAFT

The steering angle transfer shaft (see Figure 9-25) is geared to the front rack and connects to the phase control system within the rear steering unit. The phase control unit is aware of how many degrees the front wheels have moved by the rotation of the transfer shaft.

SPEED SENSORS

Two speed sensors are fitted. The main unit is inside the instrument cluster (see Figure 9-26) and a back-up unit is installed at the transaxle. The speedometer cable rotation is detected by the sensors and vehicle speed is sent as an electrical signal to the control unit.

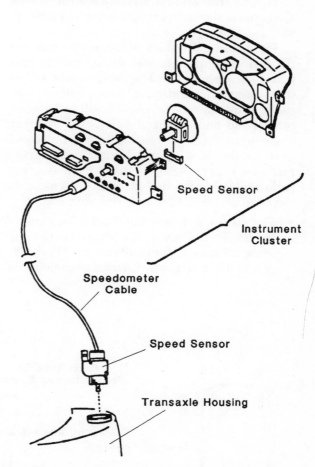

Figure 9-26 Location of speed sensors.

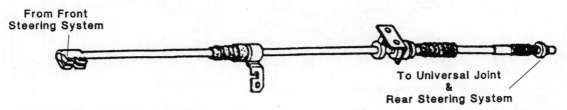

Figure 9-25 Steering angle transfer shaft.

REAR STEERING SYSTEM

CONSTRUCTION

The rear steering unit contains several sub-assemblies, each of which will be described. The rear section of the unit with the stepper motor and the control valve is the phase control system. This area is shown in greater detail in Figure 9-28.

Phase control system—this section controls the direction, and number of degrees, of rear wheel steering action. The components within this system are the stepper motor, control yoke, swing arm, main bevel gear, small bevel gear, control rod and a control valve. The steering angle input shaft is connected to the small bevel gear and provides input from the front steering rack (see Figure 9-28). A rear to front steering ratio sensor is attached through the casing to the rotation shaft of the control yoke.

- The **stepper motor** is bolted to one end of the housing. A bevel gear is fitted to the end of the stepper motor shaft (rotor). This bevel gear is in contact with a second gear mounted

on a worm shaft. Rotation of the worm shaft causes the **control yoke** to pivot and the **swing arm** will be pulled towards or pushed away from the stepper motor.

- The **control rod** forms a pivoting link between the swing arm and the **spool valve** located within the control valve assembly. The control rod passes through a hole in the **main bevel gear**. Any rotational movement of the main bevel gear will cause the swing arm and the control rod to follow a circular path. Rotational movement of the main bevel gear is caused by the steering angle input shaft turning the **small bevel gear** (see Figure 9-29).

Before moving on to the theory of operation of the phase control assembly it will be useful to summarise two points.

1. Rotation of the small and large bevel gears is based on front steering rack angle movement.
2. The amount the swing arm tilts away or towards the stepper motor is a result of vehicle speed.

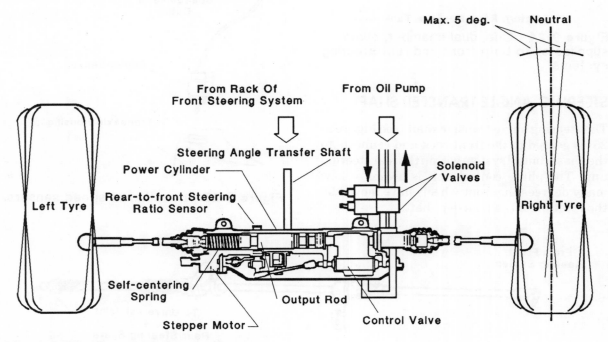

Figure 9-27 Rear steering system.

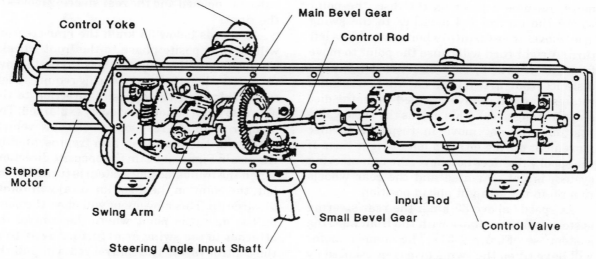

Figure 9-28 The phase control system consists of the parts shown in the illustration. The housing forms part of the rear steering rack assembly.

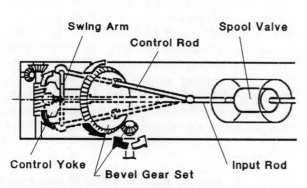

Figure 9-29 The steering angle input shaft is connected to the small bevel gear. Movement of the front steering system will cause the bevel gear set to rotate.

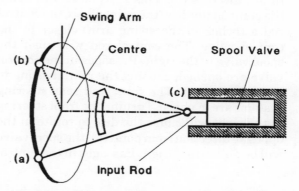

Figure 9-30 At 35 km/h the swing arm is at right angles to its axis. In this position, movement of the swing arm from (a) to (b) [or (b) to (a)] will not move the spool valve. Rear steering action will not be available.

OPERATION

During the description of the phase control system the reader must remember that movements caused by the stepper motor are directed to a spool valve and not a steering rack. The unit is hydraulic and left-hand movements may cause the rear steering to move to the right.

At 35 km/h the rear steering system is at the transition point from opposite phase to same phase, therefore rear steering is not available. Figure 9-30 represents the swing arm as a disc. The stepper motor, in response to vehicle speed, has positioned the swing arm at right angles to its axis. The main bevel gear is rotated by the front steering rack input. The control rod, passing through the bevel gear will be carried around its centre causing the swing arm to move. Assume a point on the disc (the point where the control rod attaches to the swing arm) moves from (a) to (b) when rotated for a left turn. A right turn will cause the point to move from (a) in the opposite direction to (b). The distance from (a) to (c) will not alter as the disc rotates, nor will it have changed when the disc stops moving at (b). This means the spool valve position will not move in response to disc movement during either a left or right turn. If the spool valve does not move, the rear steering system is not activated and the rear wheels remain in the straight ahead position.

At speeds above 35 km/h the rear steering system will be in phase with the front steering system (see Figure 9-31). The stepper motor will have tilted the swing arm (represented by the disc rim) to an angle determined by vehicle speed. When the front steering system is turned to the left, the point on the disc rim at (a) will rotate towards (b). This movement will bring the point on the disc (the point where the control rod attaches to the swing arm) closer to the control valve. The control rod will push the spool valve to the right. Passages in the control valve are opened, causing the rear steering to move to the left. The front and rear steering systems are in phase. Turning the front steering to the right will cause the disc to rotate in the opposite direction from point (a), the spool valve will be pulled to the left. Passages in the control

valve are opened and the rear steering moves to the right.

At speeds below 35 km/h the rear steering will be in opposite phase to the front wheels. This change will require the spool valve to move to the left when the front steering requires a left turn. This is in the opposite direction to the above 35 km/h action previously explained. The stepper motor in response to the low vehicle speed will tilt the swing arm (represented by the disc in Figure 9-32) in the opposite direction. When the front steering system is turned to the left, the point on the disc rim at (a) will rotate towards (b). This movement will move the point on the disc (the point where the control rod attaches to the swing arm) further away from the control valve. The control rod will pull the spool valve to the left. Passages in the control valve are opened, causing the rear steering to move to the right. The front and rear steering systems are in opposite phase. Turning the front steering to the right will cause the disc to rotate in the opposite direction from point (a), the spool valve will be pushed to the right. Passages in the control valve are opened and the rear steering moves to the left.

The **steering angle input shaft** is a mechanical drive from the front steering system. As previously explained, the control yoke, stepper motor and swing arm combine with the

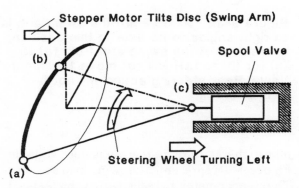

Figure 9-31 At speeds above 35 km/h the swing arm is tilted by the stepper motor. In this position, movement of the swing arm from (a) to (b) [or (b) to (a)] will move the spool valve. Rear steering action in the same phase is available.

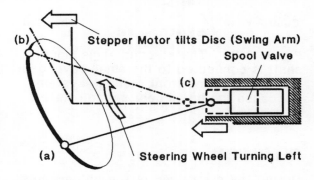

Figure 9-32 At speeds below 35 km/h the swing arm is tilted, by the stepper motor, in the opposite direction (refer Figure 9-31). In this position, movement of the swing arm from (a) to (b) [or (b) to (a)] will move the spool valve. Rear steering action in the opposite phase is available.

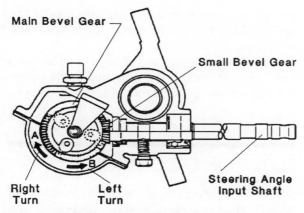

Figure 9-33 The main bevel gear moves toward A when the front steering system is turned to the right. The reverse action occurs when the vehicle is turning left.

control rod and spool valve to move the rear steering system. Several references were made to rotation of the theoretical disc formed by the connection point of the swing arm and the control rod. Figure 9-33 illustrates how rotation of this disc occurs. The input shaft turns the small bevel gear which acts on the main bevel gear. The control rod (see Figure 9-29), which is linked to the spool valve, passes through the large bevel gear. Main bevel gear movement carries the control rod around with it. Because the control rod and swing arm are joined, their connection point will follow the path of a disc. This arrangement only advises that the front system has turned left or right. The control yoke angle determines the direction and amount of rear steering movement.

SUMMARY OF THE PHASE CONTROL SYSTEM

- The spool valve moves to the right to cause the rear steering to turn to the left.
- The spool valve moves to the left to cause the rear steering to turn to the right.
- The stepper motor, in response to a speed related control signal, will move, positioning the control yoke and swing arm. This position will determine left, right or no rear steering movement
- Information indicating whether the front system is turning right or left is provided by the steering angle transfer shaft.

POWER CYLINDER

The spool valve is enclosed by a valve sleeve which contains the oil passages to link the phase control system to the power cylinder (see Figure 9-34). An output rod, fitted with a tie rod at each end, passes through the centre of the power cylinder. Two hydraulic chambers are formed at a central point and by introducing high pressure oil to either the left or right chamber, the output rod can be pushed in the required direction. It is the movement of the output rod that causes the rear wheels to steer. The self-centring spring and oil pressure in both chambers maintain the rear steering in the straight ahead position when steering action is not required. The spring will also return the rear wheels to the straight ahead position if an electrical or hydraulic failure occurs.

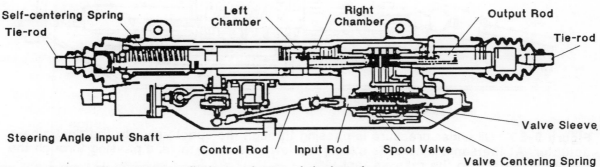

Figure 9-34 The power cylinder and associated parts.

CONTROL VALVE

The spool valve is positioned as required by the vehicle speed and the amount of front steering system movement. Figure 9-35 shows the spool valve moved to the left. In this position oil from the pump is directed via the sleeve passages, then through the ports opened by the spool valve. The oil is redirected back through the sleeve to the right side actuator chamber. At the same time oil from the left chamber can return to the reserve tank via the sleeve and spool valve. The unbalanced pressures between the left and right chambers will cause the output rod to move to the left, turning the rear wheels to the right. The valve sleeve is fitted to, and forms part of the moving output rod. When the rod moves to the left the sleeve will be carried with it, moving over the spool valve and changing the amount the ports are open. The output rod (and the valve sleeve) will stop moving when

the oil pressure and the combined force of the self-centring spring and cornering forces reach a balance.

The reverse action will occur when the spool valve is positioned to the right end of the sleeve for a left turning movement of the rear wheels.

ELECTRONIC CONTROL

System control is provided by a CPU, as depicted within the electrical system diagram. The diagram can be simplified very quickly when it is considered that the primary purpose of the CPU is to accept a vehicle speed sensor input and send a step signal to the stepper motor. The remaining inputs and outputs are to check, monitor and, if necessary, 'fail-safe' the rear steering system.

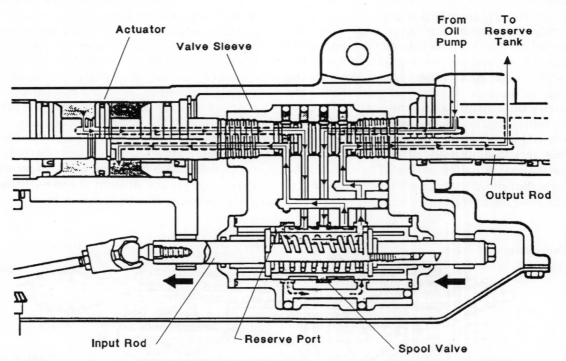

Figure 9-35 Oil under pressure is passing through the spool valve to the right side actuator. At the same time oil is exiting the left side actuator and passing through the spool valve back to the reserve tank.

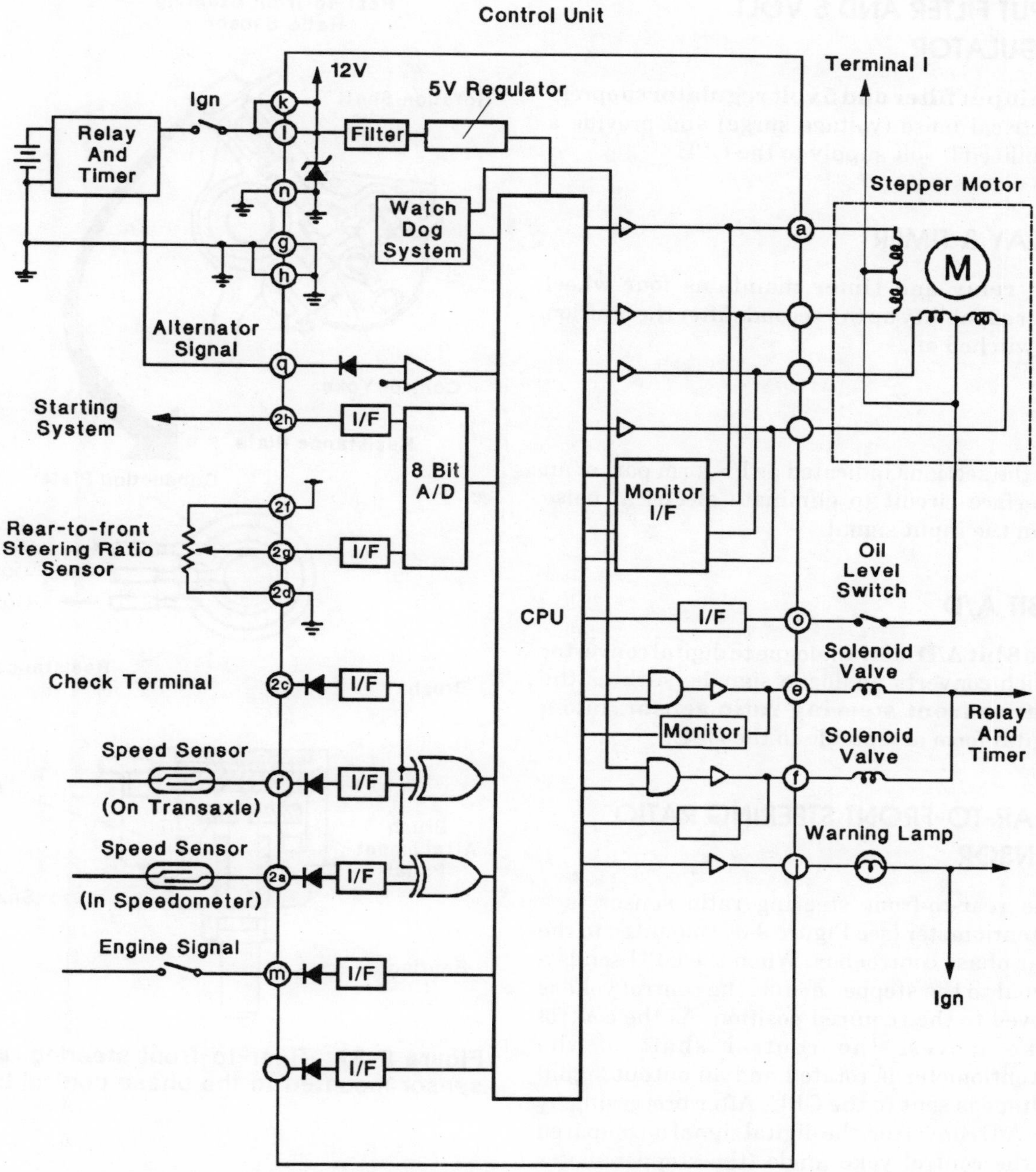

Figure 9-36 Electrical system diagram. An explanation for each of the major inputs and outputs is provided within the text.

INPUT FILTER AND 5 VOLT REGULATOR

The **input filter and 5 volt regulator** suppress electrical noise (voltage surge) and provide a stabilised 5 volt supply to the CPU.

RELAY & TIMER

The **relay** and **timer** maintains four wheel steer capability up to 7 seconds after the ignition is switched off.

I/F

All the sections indicated as I/F form part of an interface circuit to eliminate electrical noise from the input signal.

8 BIT A/D

The **8 bit A/D** is an analogue to digital convertor which converts analogue signals, such as the **rear-to-front steering ratio sensor**, into a digital form acceptable to the CPU.

REAR-TO-FRONT STEERING RATIO SENSOR

The rear-to-front steering ratio sensor is a potentiometer (see Figure 9-37) mounted to the rear phase control box. When the CPU sends a signal to the stepper motor, the control yoke is moved to the required position. As the control yoke moves, the central shaft of the potentiometer is rotated and an output signal voltage is sent to the CPU. After processing by the A/D convertor, the digital signal is compared to the control yoke angle (the stepper motor position signal) regulated from the CPU. If a variation exists, the CPU sends a signal to the stepper motor to compensate for the difference.

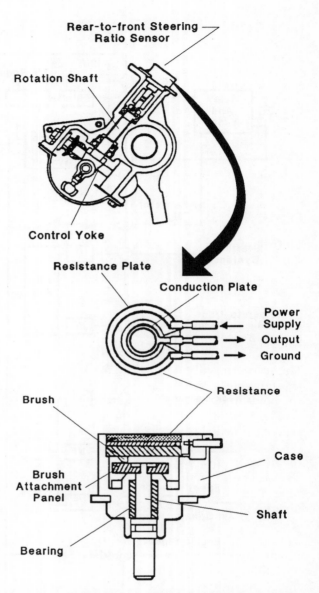

Figure 9-37 Rear-to-front steering ratio sensor mounted to the phase control box.

SPEED SENSORS

Vehicle **speed sensors** are located at the transaxle end and the instrument cluster end of the speedometer cable. Refer to Figure 9-26 for an illustration of the speed sensors.

MONITOR I/F

The **monitor I/F** checks the stepper motor for open circuits.

STEPPER MOTOR

The **stepper motor** operation and interconnection to the spool valve is described in the phase control operation section.

OIL LEVEL SWITCH

The **oil level switch** checks the power steering fluid in the reserve tank is maintained at a suitable level.

SOLENOID VALVES AND MONITORS

The **solenoid valves** and **monitors** are part of the fail-safe system. The monitors check the solenoids and wiring for open circuits.

The two solenoid valves are mounted at the rear of the vehicle near the rear steering system (see Figure 9-38). One solenoid valve is connected to the rear steering oil supply pressure pipe. The second solenoid is connected to the return pipe. The function of the solenoid valves is to divert the oil from the supply port of the rear steering system in the event of an electrical system failure. As can be seen from the illustration, only one valve would be required to 'fail safe' the system. The second valve is a back-up for the first valve. During normal operation, current is supplied to the solenoids from the CPU, the valve spools move against spring force and oil cannot flow through the valves. The CPU, upon detection of an electrical failure, cuts current to the solenoids. The springs push the valve spools back and supply pressure oil is diverted through the solenoid valves. The removal of oil under pressure from the rear steering supply port allows the self-centring spring to bring the rear steering system to the straight ahead position.

NORMAL

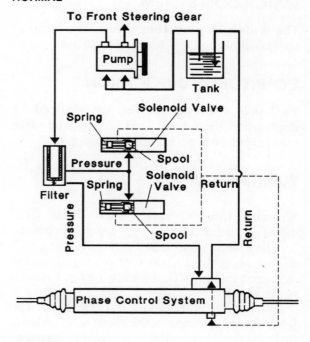

FAILURE

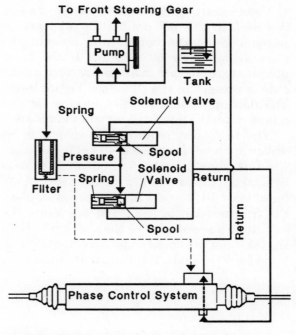

Figure 9-38 If a failure is detected, pressure oil is diverted through the solenoids and returned to the reserve tank.

WATCH DOG SYSTEM

The **watch dog system** checks the internal operation of the CPU for malfunctions.

CONNECTION DUPLICATION

Fail-safe operation is further assured by duplication of essential power supply and earth connections: e.g. terminals i, k and g, h.

WARNING LAMP

Warning lamp operation occurs if the CPU detects a malfunction in any of the check circuits.

The system is equipped with automatic self-diagnosis capability. If an electrical malfunction is 'logged' by the CPU, the warning light system is activated.

When the ignition key is turned 'on', the lamp should illuminate; the absence of the lamp illumination is an indication of globe, wiring or CPU malfunction.

Before starting the engine, ensure the key has been in the 'off' position for at least 10 seconds. Start the engine and observe the warning lamp for 60 seconds. If the lamp illuminates or flashes in a coded pattern a fault code is present in the CPU (see Figure 9-39). The flashing condition will continue for one minute then the lamp will remain illuminated.

The third test requires the vehicle to be driven at speeds in excess of 40 km/h; therefore the test should be performed on a chassis dynometer. Ensure the key has been turned 'off' for at least 10 seconds. Start the engine, place the transmission in gear, and accelerate the vehicle to a speed greater than 40 km/h. Hold the vehicle at this speed and observe the warning lamp for 60 seconds. If the lamp illuminates or

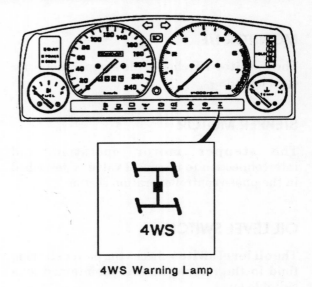

4WS Warning Lamp

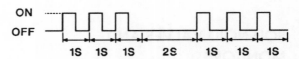

Figure 9-39 A dash mounted warning lamp advises of a failure by flashing in a coded sequence. The example shows a three-flash code repeated at two-second intervals.

flashes in a coded pattern, a fault code is present in the CPU. The flashing condition will continue for one minute then the lamp will remain illuminated.

The flashing code diagram (see Figure 9-40) is included to illustrate the range of faults which can be detected by the diagnostic features of the system. The manufacturer has designed this system with 'fail-safe' capability as a prominent feature. Repairs to the system should only be performed in accordance with the manufacturer's service manual and overhaul instructions.

Item		Check timing		Warning lamp	Reaction to failure
		Ign ON	Driving		
Speed sensor	Speed sensor in speedometer	-	*	Flashes 1 time [2 sec. period between cycles]	2WS
	Speed sensor on transaxle	-	*	Flashes 1 time [2 sec. period between cycles]	2WS
	Difference between above sensors	-	*	Flashes 1 time [2 sec. period between cycles]	2WS
Phase control system	Mis-stepping (Out of phase)	-	*	Flashes 2 times	2WS
Solenoid valve	Wiring circuit	*	*	Flashes 3 times	2WS
Stepper motor	Wiring circuit	*	*	Flashes 4 times	2WS
	Operation	-	*	Flashes 4 times	2WS
Rear-to-front steering ratio sensor	Output	*	*	Flashes 5 times	2WS
	Standard position	*	*	Flashes 6 times	2WS
Power steering fluid	Level	*	*	Flashes 7 times	2WS
Control unit	Program	*	*	Flashes 8 times	2WS
	Memory	*	*	Flashes 8 times	2WS
	Conversion from analogue to digital values	*	-	Flashes 8 times	2WS
	Computer error	*	*	OFF	4WS after computer reset
Power supply	Battery voltage	-	*	Stays ON	2WS

Note: After repairing a failure, turn off the ignition switch to cancel the warning lamp operation.

Figure 9-40 Warning lamp diagnostic chart.

10

ELECTRONICALLY CONTROLLED AUTOMATIC TRANSMISSIONS

INTRODUCTION

Earlier designs of automatic transmissions have been solely hydraulic-mechanical devices. The hydraulic section is used to select and shift the gears through the various speed ranges. The mechanical design of these transmissions provided either two or three gear ratios. In the late 1980s the need to improve the overall performance of vehicles has led to refinements in the automatic transmission. Significant improvements were gained by the introduction of electronic control and four speed ratios.

Initially, steps were taken to control the slippage that occurred in the torque convertor. Lock-up clutches which provide a precise one-to-one ratio input to the transmission have been incorporated into torque convertors. The lock-up action is controlled by an electronic system. Further refinements include a fourth gear, electronic control of the shift patterns and a driver selectable operating mode.

The following features are included in the modern transmission:

- four speed ratios;
- overdrive capability;
- lock-up convertor;
- power and economy modes;
- interface to the engine management system;
- microprocessor control for smooth gear change and engagement;
- self-diagnosis of electronic malfunctions;
- oil temperature monitoring system;
- interface with cruise control.

The introduction of these transmissions has allowed manufacturers to produce a vehicle with improved performance, greater fuel economy, and smooth gear changes, and has enabled the driver to select a transmission shift pattern suited to individual driving style.

Before the introduction of electronic control systems, automatic transmissions relied on the magnitude of hydraulic pressures for shift control.

A conventional automatic transmission is controlled by two inputs. The first input is driver controlled from the accelerator and the second input is developed from output shaft speed, therefore it is road speed related.

The accelerator pedal is mechanically linked to the butterfly valve of the carburettor or fuel injection throttle body. Information related to throttle position is sent to the transmission via a mechanical link, usually in the form of a cable (see Figure 10-1). An alternative method is to utilise a vacuum signal from the inlet manifold to take the place of the mechanical linkage. Either method is designed to produce a force on a throttle valve within the transmission valve body. Transmission line pressure passing through this valve is converted to throttle pressure. The throttle pressure has several functions, one of which is to modify the line pressure passing to the on/off shuttle valves (shift valves). The shift valve moves, in response to a shift signal, and the line pressure flows to

apply the friction elements (bands or clutches). The shift signal which moves the shift valve is a result of two other hydraulic pressures not shown in the illustration. Governor pressure, developed from tailshaft/road speed, overcomes throttle pressure developed from the amount of throttle opening and the resulting imbalance of forces causes the shift valve to move. In this way the transmission will 'shift gears' in response to engine load-throttle position or vacuum signal, balanced against road speed.

The introduction of electronic controlled automatic transmissions (ECAT) allowed a more precise control of the shift pattern. The engine speed and load inputs are supplied from electronic sensors (see Figure 10-1). A pulse generator located within the transmission senses vehicle speed and engine load information is supplied by a throttle sensor located on the throttle body. This information is modified within an electronic control unit (ECU) by signals received from other sensor inputs. The

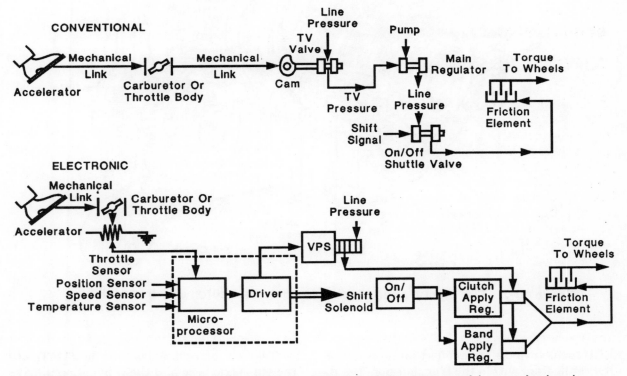

Figure 10-1 A conventional transmission control system compared to an electronic transmission control system.

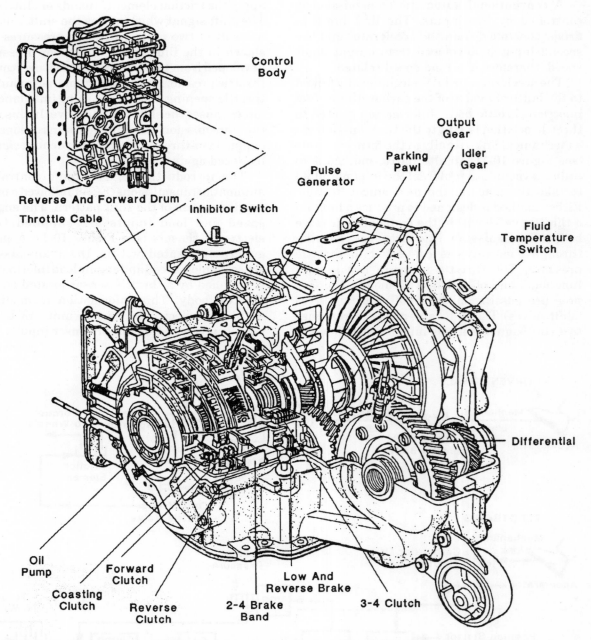

Figure 10-2 The layout of an electronic controlled automatic transaxle.

ECU turns on and off solenoids that redirect the hydraulic fluid within the transmission. Band, servo and clutch application then occurs, as directed by the solenoid valves.

Two common configurations for ECATs are the transaxle type and the conventional drive type. The abbreviation 'ECAT' has been widely used for both types of transmissions.

ELECTRONICALLY CONTROLLED AUTOMATIC TRANSAXLE

The electronically controlled automatic transaxle (ECAT) contains many features common to the conventional electronically controlled automatic transmissions.

The ECAT, being described, is representative of the available range that has four forward speeds and a reverse. The fourth speed is overdrive (OD). The torque convertor lock-up function can be applied in D range third, and overdrive.

Two shift programs, available from the ECU, offer the driver a choice of economy or power. The shift pattern in use is selected by the driver via a mode switch.

The common shift selection pattern of P, R, N, D, S and L, offered by these transaxles, is accessible to the driver. Selection of the D range shift pattern enables an automatic shift up or down between 1,2,3 (lock-up) and OD (lock-up).

When S range is selected, the transaxle will automatically shift up and down between 1,2, and 3 but it will not allow the torque convertor lock-up to be engaged. L range will only allow automatic shifts from 1 to 2 or 2 to 1.

Another driver selectable feature of the ECAT is the provision of a Hold mode which can be activated from a button on the shift selector lever. Hold mode can be accessed in D, S or L ranges but the actual selection of the gear is controlled so that engine overspeed can not occur. The Hold mode assumes priority over both the economy and the power modes.

COMPONENTS

An ECAT is a mechanical device which is operated by hydraulics that are controlled by electronics. The three main sections of an ECAT contain the:

- mechanical components;
- hydraulic components;
- electronic components.

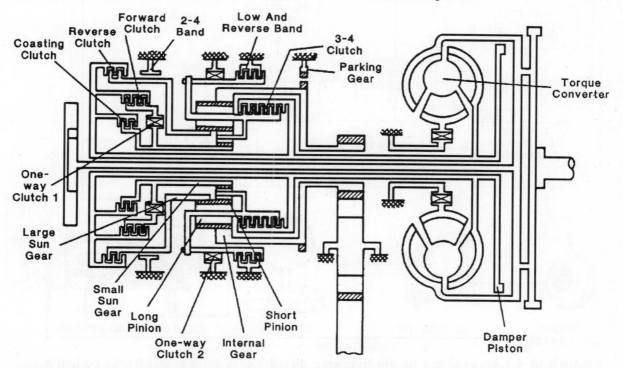

Figure 10-3 An example of the mechanical layout of a typical electronic controlled automatic transaxle (ECAT).

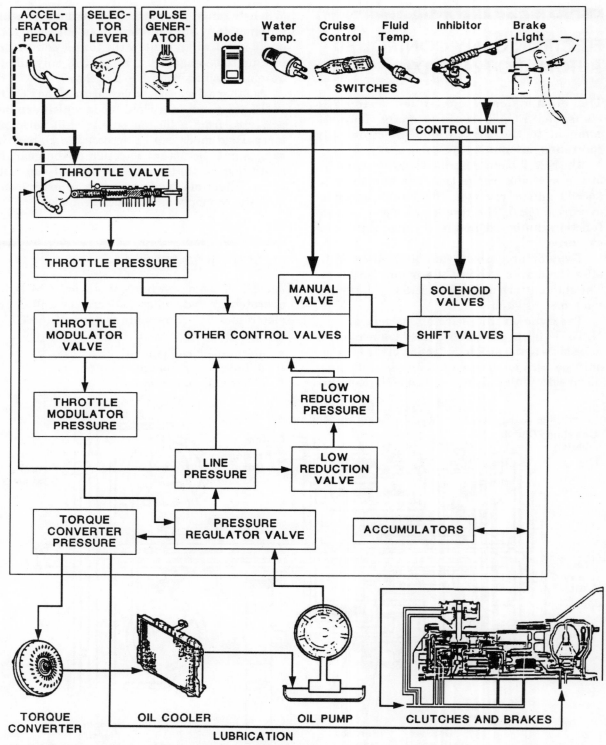

FIGURE 10-4 An example of the hydraulic system of a typical electronic controlled automatic transaxle (ECAT).

MECHANICAL AND HYDRAULIC COMPONENTS

The type and layout of the mechanical and hydraulic components is similar to the standard automatic transaxle.

The major components of the mechanical layout are:

- lock-up torque convertor;
- planetary gear sets;
- clutches and bands;
- input and output shafts;
- selector mechanism;
- bearings and bushings.

The major components of the hydraulic system are:

- variable capacity, vane type oil pump;
- torque convertor;
- clutch and band servos;
- shift valves;
- manual valve;
- pressure regulator valve;
- throttle valve;
- solenoid valves;
- accumulators;
- oil cooler.

It is not in the scope of this book to describe the construction and operation of these

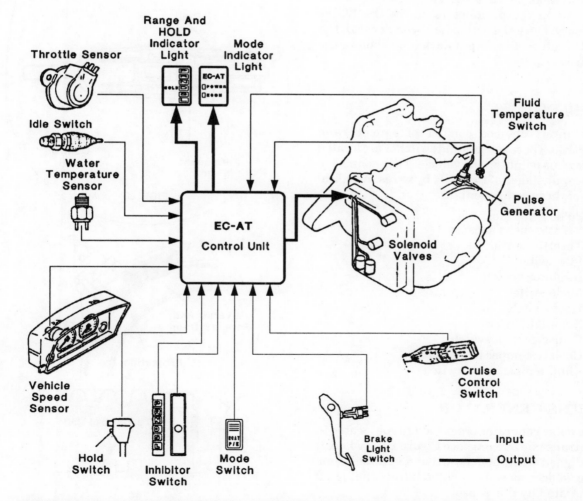

Figure 10-5 An example of the electronic layout of a typical electronic controlled automatic transaxle (ECAT).

components. Where the reader is not fully conversant with these components, it is recommended that the appropriate workshop manual be carefully studied.

ELECTRONIC COMPONENTS

The electronic components can be grouped under three subsections. These subsections are the:

- input devices;
- output devices;
- electronic control unit (ECU).

The input devices supply the ECU with information so that decisions can be made relative to the parameters set in the ECU's memory. Once the action has been decided, the ECU controls the output devices to achieve the desired results.

INPUT DEVICES

The input devices consist of sensors and switches. These sensors and switches are located at various points on the vehicle. They connect to the input stages of the ECU by wires. A typical list of these components is:

- Pulse generator;
- Vehicle speed sensor;
- Throttle sensor;
- Idle switch;
- Inhibitor switch;
- Mode switch;
- Hold switch;
- Stop-light switch;
- Cruise control switch;
- Coolant temperature switch;
- Fluid temperature switch.

PULSE GENERATOR

The pulse generator is a reluctor type located in the transaxle housing (see Figure 10-6) where it is aligned with the forward and reverse drum. The vehicle speed is calculated from the signal generated by this sensor.

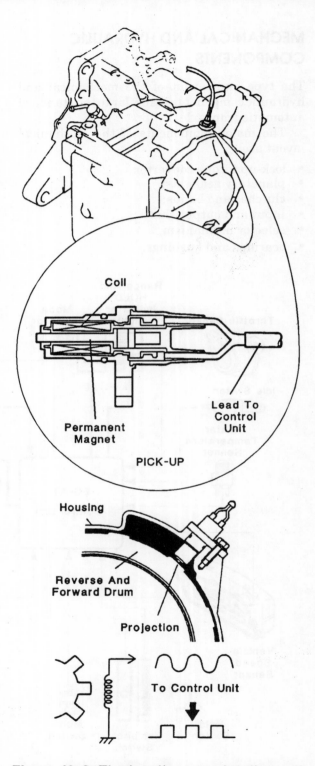

Figure 10-6 The location, construction and operation of a pulse generator.

Construction

The rotor is formed on the outer rim of the forward and reverse drum. It consists of twelve projections spaced at thirty degrees around the drum's perimeter. The rotor spins at the input speed of the transaxle.

The pick-up consists of a coil wound onto a permanent magnet. The ends of the coil are connected by two wires to the ECU.

Operation

When the rotor spins, the projections pass the end of the magnet which cause the magnetic field to cut back and forth across the pick-up coil. Moving the magnetic field across the coil induces a voltage into the coil windings. The voltage generated is an AC signal consisting of twelve pulses per revolution of the drum. This signal is sent to the ECU where it is converted to a square wave pulse and multiplied by a constant to obtain the vehicle speed. The constant is derived from the selected gear ratio, tyre diameter and the final drive ratio.

VEHICLE SPEED SENSOR

The vehicle speed sensor is a mechanically operated reed switch located at the base of the speedometer (see Figure 10-7). This sensor is a back-up unit should the pulse generator fail. Drive for the sensor is provided from the speedometer cable.

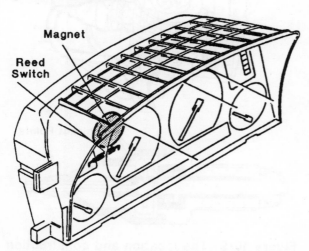

Figure 10-7 The location of the vehicle speed sensor.

Construction

The reed switch consists of a spring steel blade and two contacts. Each contact is connected to the ECU by a wire.

The operating mechanism is formed on a non-magnetic disc which is attached to the speedometer cable. This mechanism consists of four small magnets moulded at 90 degrees to one another on the rim of a non-magnetic disc.

Operation

When the disc is spun by the speedometer cable, each magnet aligns with the spring steel blade. The steel blade, attracted towards the magnet, moves in a direction that closes the contacts. For each disc revolution, the magnets attract the steel blade to open and close the contacts four times per revolution. This action produces a digital signal (square wave) at the input stage of the ECU. The ECU calculates the vehicle speed by multiplying the digital signal by a constant. This constant is derived from the speedometer gear ratio, tyre diameter and the final drive ratio.

THROTTLE SENSOR

The throttle sensor is a variable resistor which is attached to the throttle body and connected to the throttle shaft. Three wires connect the sensor to the ECU. The angle of throttle opening is determined by the ECU from the voltage signal that is received from the sensor.

Construction

A circular resistive element is positioned concentric to a wiper arm hub inside a small housing. The end of a wiper arm, attached to the hub, is positioned on the resistive element. Each end of the circular resistive element and the wiper arm are connected to terminals formed in a block on the housing.

Operation

A reference voltage is applied to one end of the resistive element and the other end is connected to ground. When the throttle is closed, the wiper arm is positioned near the grounded end of the

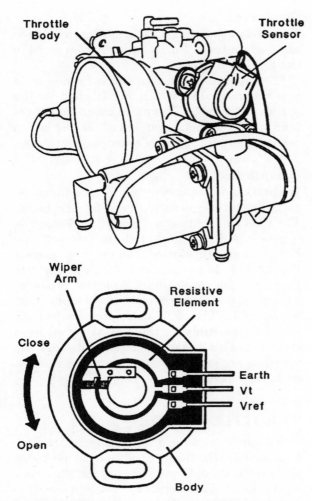

Figure 10-8 The location and construction of a throttle sensor.

resistive element. In this position the voltage supplied to the ECU through the wiper arm connection is very low. As the throttle starts to open, the wiper arm moves towards the reference voltage end of the resistive element and the voltage supplied to the ECU increases. At the full throttle position, the ECU receives a high voltage which is slightly less than the reference voltage.

An adjustment is provided on the sensor to allow the high voltage to be set to a value that the ECU will recognise as the full throttle position. The ECU can recognise the throttle plate opening angle from the sensor voltage.

IDLE SWITCH

The idle switch detects when the throttle butterfly is in its fully closed position. It is screwed into a bracket on the throttle body in a position that aligns it with the throttle cable lever (see Figure 10-9). A wire connects the switch to the ECU.

Construction

A spring-loaded plunger is housed in a hollow, threaded metal body. The plunger has a contact formed on its spring end and a dust seal groove at its other end. The body is closed with a metal cap that contains a fixed, insulated contact. The fixed contact is connected by a wire to the ECU. The plunger grounds its contact through the metal body.

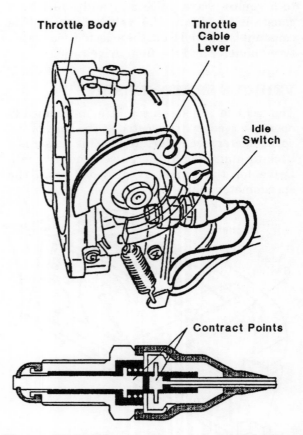

Figure 10-9 The location and construction of an idle switch.

Operation

When the throttle is fully closed (idle position), the throttle lever depresses the plunger against spring pressure to close the contacts. With the contacts closed, the circuit is completed through the plunger, switch body and the throttle body. The ECU senses the switch is closed by the voltage drop in the wire.

The instant the throttle cable moves the lever, the contacts are opened by the spring and the voltage is restored in the .wire. When the ECU senses a voltage in the wire, it monitors the throttle opening through the throttle sensor.

INHIBITOR SWITCH

An inhibitor switch is located on the transaxle case and positioned on the gear selection mechanism (see Figure 10-10). The switch is a multi-contact type and is connected by wires to the ECU.

Construction

A spring steel wiper arm is formed with a small hub at one end and a contact at its other end. The hub is attached to and moves with the gear selector shaft.

Five fixed contacts are positioned in a short arc around the centre of the hub. These contacts are insulated from one another and also the switch body. A wire is attached to each fixed contact.

Operation

When the driver moves the gear selector lever, the switch hub turns with the gear selector shaft. The wiper arm moves its contact to a position that will connect one of the fixed contacts to ground. The voltage drop in the wire indicates to the ECU the position of the manual valve as selected by the driver. When the gear selector is in Park or Neutral, the ECU will allow the engine to be started. In any other gear, the engine will not start. When the engine is running, the switch will indicate to the ECU the drive range selected.

MODE SWITCH

A push-button type switch is located on the dash panel near the driver (see Figure 10-11). The switch is connected by two wires to the ECU.

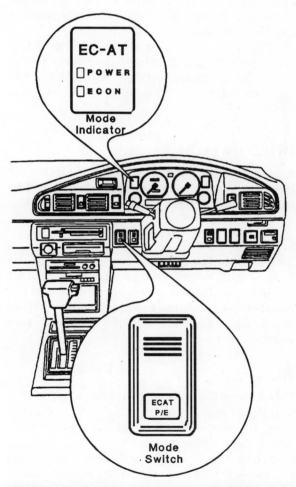

Figure 10-11 Typical locations for the mode switch and the mode indicator.

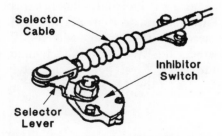

Figure 10-10 The gear selection mechanism and the inhibitor switch.

Construction

Two fixed contacts are located in the base of the switch where they are insulated from one another. A spring-loaded contact bridge is attached to an over-centre mechanism but insulated from the body. The over-centre mechanism is connected to a button.

Operation

When the driver pushes the button, the over-centre mechanism moves the contact bridge against spring tension to connect the two contacts together. The mechanism locks and holds the switch in the 'On' position. When the button is pushed, the mechanism is released and the contact bridge is forced away from the contacts by the spring. The switch is returned to the 'Off' position.

With the mode switch 'On', the ECU configures its operation to suit the power mode.

With the mode switch 'Off', the ECU configures its operation to suit the economy mode.

HOLD SWITCH

A push button switch is located in the handle of the gear selector lever (see Figure 10-12). The switch is connected by two wires to the ECU.

Construction

Two flexible fixed contacts are located in the base of the switch. A spring-loaded contact bridge, positioned between the two fixed contacts, is attached to a button. These components are housed in the moulded lever handle.

Operation

When the button is depressed, the contact bridge is moved towards the fixed contacts to connect them together. The switch is 'On' and the ECU activates the Hold mode.

When the button is released, the spring forces the bridge away from the contacts to open the switch. This action does not cancel the Hold mode.

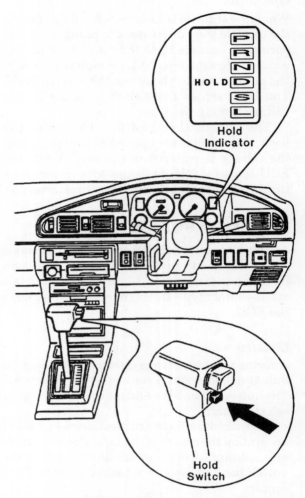

Figure 10-12 Typical locations for the hold switch and the hold indicator.

When the switch is depressed once the Hold mode is activated. When the switch is depressed a second time the Hold mode is cancelled.

The Hold mode restricts the shift to third and second with 'D' range selected. When 'S' is selected, the ECU locks the transmission in second gear. First gear is locked when the selector is in the 'L' range.

STOP-LIGHT SWITCH

The stop light switch is located on the upper section of the brake pedal (see Figure 10-13). It is a part of the brake system but it is also connected by a wire to the ECU of the ECAT.

Construction

A spring-loaded contact bridge is housed in the bore of the bakelite switch base. Two flexible fixed contacts are located at the open end of the bore. A plunger, housed in a metal body, joins with the contact bridge.

Operation

With the brake pedal in released position, the upper pedal depresses the switch plunger against spring force to hold the switch 'Off'.

When the brake pedal is depressed, the plunger and contact bridge is forced outward until the bridge joins the two contacts. The switch is 'On'. This action sends a signal to the ECU that the brake has been applied.

Torque convertor lock-up is inhibited when the switch is 'On'.

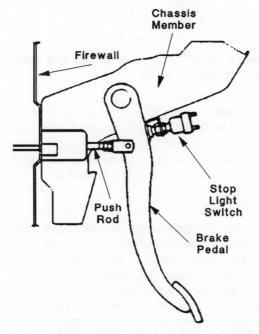

Figure 10-13 The location of the stop light switch.

CRUISE CONTROL SWITCH

This switch is a part of the cruise control system (see Figure 10-14). It is located in the cruise control unit and connected by a wire to the ECU.

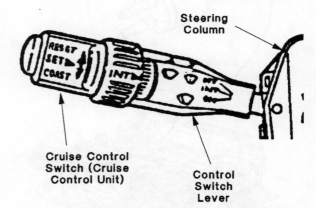

Figure 10-14 The location of the cruise control switch (cruise control unit).

Construction

The switch is one of the outputs from the cruise control unit.

Operation

When a signal is received from the Resume switch, Set switch or the vehicle speed drops 3 km/h below the set speed, the cruise control switch (within the cruise control unit) sends a signal to the ECU to inhibit overdrive.

COOLANT TEMPERATURE SWITCH

A bi-metallic type switch is screwed into the cylinder head (see Figure 10-15) of the engine. It is connected by two wires to the ECU.

Construction

One end of a bi-metallic strip is attached to the switch body and a terminal. A contact is formed on the other end of the strip. A fixed contact is aligned with the moving contact and connected to another terminal.

The switch body is threaded to allow it to be screwed into the cylinder head.

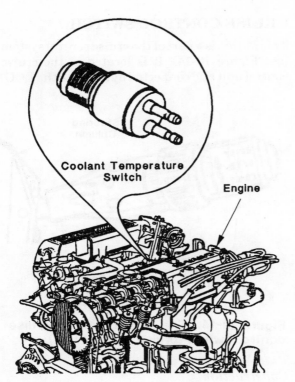

Figure 10-15 A typical location for a coolant temperature switch.

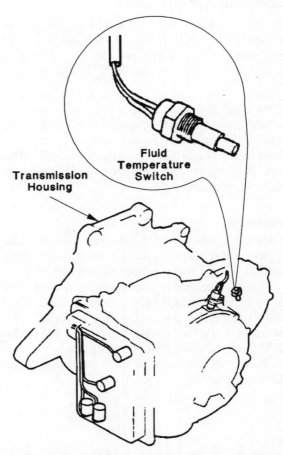

Figure 10-16 The location of the fluid temperature switch.

Operation

When the coolant temperature is below 65 degrees Celsius, the bi-metal strip is straight and the contacts are closed. A signal is sent to the ECU.

When the coolant temperature exceeds 65 degrees Celsius, the bi-metallic strip bends and the contacts open. There is no signal to the ECU when the contacts are open and overdrive operation becomes available.

FLUID TEMPERATURE SWITCH

A bi-metallic type switch is fitted to the cooler line from the transaxle. It is connected by two wires to the ECU (see Figure 10-16).

Construction

One end of a bi-metallic strip is attached to the switch body and a terminal. A contact is formed on the other end of the strip. A fixed contact is aligned with the moving contact and connected to another terminal.

The switch body is threaded to allow it to be screwed into the transaxle housing.

Operation

When the fluid temperature is below 150 degrees Celsius, the bi-metallic strip bends and the contacts open. There is no signal to the ECU when the contacts are open.

When the fluid temperature is above 150 degrees Celsius, the bi-metallic strip is straight and the contacts are closed. A signal is sent to the ECU, causing the torque convertor lock-up clutch to be engaged.

OUTPUT DEVICES

The output devices consist of four solenoids (see Figure 10-17). These solenoids operate a set of hydraulic valves which are timed by the ECU to open and close passages in the hydraulic system. Each solenoid is located on or near the valve body and is connected by wires to the ECU.

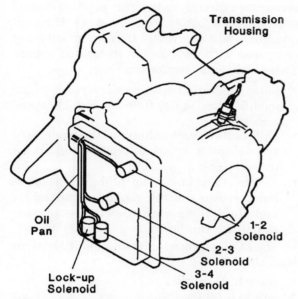

Figure 10-17 The location of the solenoids inside the transmission.

The following names are given to the solenoids:

- 1-2 solenoid;
- 2-3 solenoid;
- 3-4 solenoid;
- Lock-up solenoid.

Construction

Each solenoid valve is constructed in a similar manner (see Figure 10-18). A coil is wound around a cylindrical core which houses a spring and plunger. The plunger is formed with a rod tapered at its end. This tapered end is the spring-loaded section of a valve. The valve seat is formed in the lower section of the solenoid housing and provides access to a passage in the valve body near the shift valve. Provision is

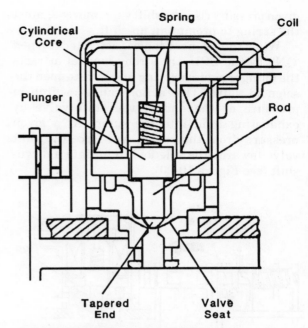

Figure 10-18 The construction of a solenoid valve assembly.

made on the solenoid body to clamp the solenoid housing to the hydraulic valve body.

Operation

When the solenoid valve is deactivated (turned 'OFF') by the ECU, the plunger spring forces the rod taper onto its seat. This action causes a rise in pressure at the end of the shift valve to oppose spring pressure (see Figure 10-19). At a

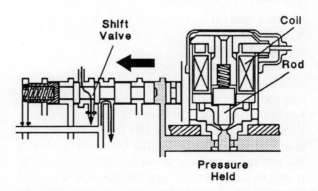

Figure 10-19 The solenoid valve in its OFF position.

given pressure rise, the shift valve moves against the spring to prompt an up-shift.

When the solenoid valve is activated (turned 'ON') by the ECU, the strong magnet attracts the plunger against spring pressure to open the solenoid valve. This action causes a drop in pressure at the end of the shift valve by exhausting fluid into the sump. At a given pressure drop, the spring will force the shift valve towards the solenoid to initiate a down-shift (see Figure 10-20).

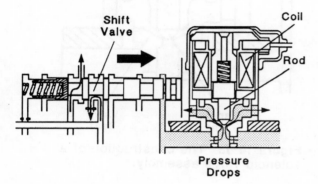

Figure 10-20 The solenoid valve in its ON position.

ELECTRONIC CONTROL UNIT

The electronic control unit (ECU) has all the features of the microprocessor described in Chapter 7. The ECU in the electronic controlled automatic transaxle has four functions. These functions are to provide:

- shift control;
- self-diagnosis;
- fail-safe mode;
- warning code display.

The ECU is located in the passenger compartment of the vehicle. Its precise location varies with the manufacturer. Common positions include, under the dash, behind the side kick panels or under the front seats. One of the critical aspects of mounting the ECU is that it must be correctly grounded to the metal structure of the vehicle.

SHIFT CONTROL FUNCTION

Four input stages provide information to the gear position control section in the ECU so that one of nine shift patterns can be selected. These input stages receive their information from the various sensors and switches associated with the automatic transaxle (see Figure 10-21). Signals from the pulse generator and the throttle sensor feed directly to the gear position control section. The other signals pass through additional sections where they are modified or compared with data stored in look-up tables. These additional sections are the:

- pulse generator back-up (which receives and modifies the signal from the vehicle speed sensor);
- shift pattern determination (the mode, hold and inhibitor switches provide signals for this section);
- overdrive and/or lock-up inhibitor (the brake light, cruise control, coolant temperature and fluid temperature switches provide signals for this section).

The ECU selects one of the nine shift patterns from the:

- three in D range;
- two in S range;
- two in L range;
- one from the P and N ranges;
- one in R range.

The pattern selected is in accordance with the status from the shift pattern determination section. Once the status has been determined, the ECU configures the output devices to provide the correct shift sequence and gear selection based on signals from the pulse generator and the throttle sensor (see Figure 10-22).

Convertor lock-up and overdrive are suppressed by the ECU when it receives a signal from the lock-up and overdrive inhibitor section (see Figure 10-23).

The ECU will not allow convertor lock-up to occur when the:

- engine coolant temperature is below 65 degrees Celsius;
- brake pedal is depressed;
- self-diagnosis section indicates a malfunction.

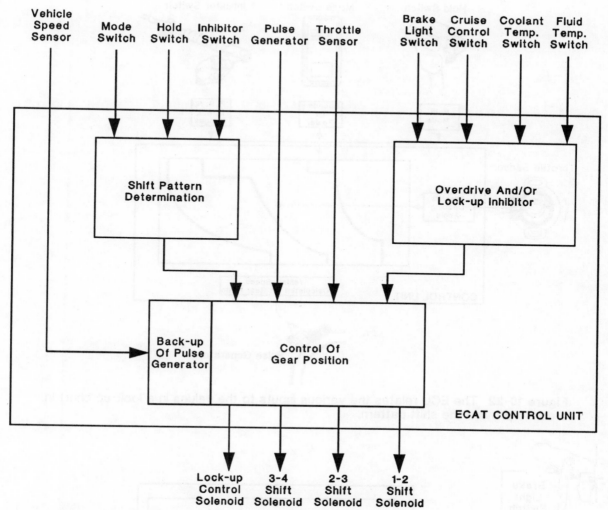

Figure 10-21 A schematic layout of the ECAT control unit.

Overdrive is inhibited by the ECU when the:
- set or resume switches are activated during cruise control operation;
- vehicle speed drops more than 3 km/h below the set speed.

SELF-DIAGNOSIS FUNCTION

The self-diagnosis section continually checks the operating condition of the four solenoid valves, the vehicle speed sensor, the pulse generator and the throttle sensor. The ECU, sensing a malfunction in one of these units, will through the self-diagnosis section flash the Hold indicator at one cycle every two seconds (see Figure 10-24). A code is also logged into the ECU's memory so that a service technician, through the code display function, can quickly determine the cause of the malfunction.

FAIL-SAFE FUNCTION

The fail-safe section permits the vehicle to be driven when a malfunction has occurred in the four solenoid valves, the vehicle speed sensor, the pulse generator and/or the throttle sensor.

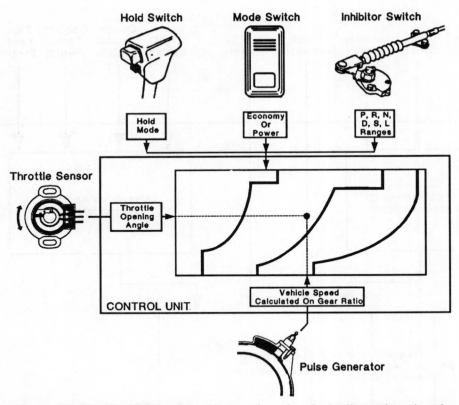

Figure 10-22 The ECU relates the various inputs to the values in a look-up chart in order to provide the shift pattern.

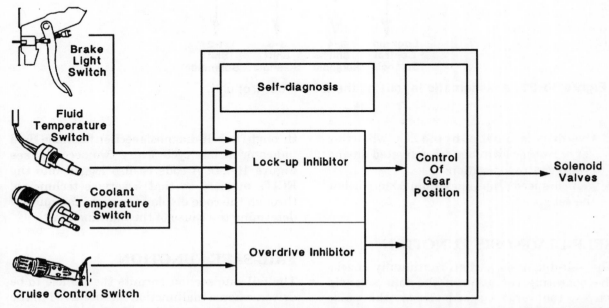

Figure 10-23 A schematic layout showing the inputs that are required by the ECU for it to prevent converter lock-up or overdrive selection.

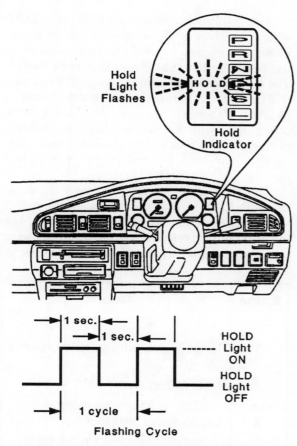

Figure 10-24 The HOLD indicator is used to warn the driver that a malfunction has occurred in the ECAT system.

The ECU, sensing a malfunction in one or several of these units, will configure the shift pattern according to the particular malfunction. In an event where all four solenoids are switched 'OFF', limited hydraulic control is available.

WARNING CODE DISPLAY FUNCTION

A set of service connectors, located in the wiring harness near the ECU, allows the service technician to read the warning codes that have been logged in the memory (see Figure 10-25). The grounding of one of the connectors causes the Hold light to flash the code provided the ignition switch is turned 'ON' after the

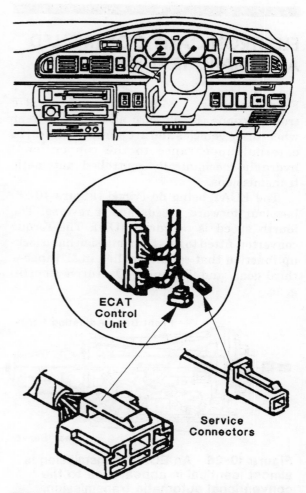

Figure 10-25 A typical location for the ECU and the service connectors is under the instrument panel.

connection is made. An example of a warning code is six short flashes which indicates a malfunction in the vehicle speed sensor.

Due to the vast number of codes that the manufacturers have devised, it is not practical to supply a list in this book. For further information related to warning or diagnostic codes consult the appropriate workshop manual.

ELECTRONICALLY CONTROLLED AUTOMATIC TRANSMISSION

Electronically controlled automatic transmissions (ECAT) are also fitted to rear wheel drive vehicles and are almost identical in external appearance to the conventional hydraulic-mechanically controlled automatic transmissions.

The ECAT being described (Figure 10-26) has four forward speeds and a reverse. The fourth speed is overdrive (OD). The torque convertor, fitted to this transmission has a lock-up function that can be applied in '3' range—third gear, and 'D' range—Overdrive (fourth) gear.

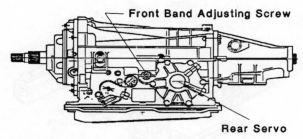

Figure 10-26   An ECAT transmission is almost identical in appearance to the conventional automatic transmission.

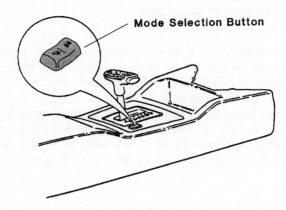

Figure 10-27 Two driving modes 'Power' and 'Economy' are available to the driver.

Two shift programs, available from the ECU, offer the driver a choice of economy or power mode (see Figure 10-27). The shift pattern in use is selected by the driver via a mode switch. The major difference between the shift programs will be evident to the driver at throttle openings between 40 and 60 per cent. The power mode offers optimum trailer towing and vehicle performance whereas the economy mode has earlier gear upshifts and offers greater fuel economy.

The shift selection pattern (see Figure 10-28) of P, R, N, D, 3, 2 and 1 provide the driver with greater transmission control than the earlier standard selection pattern. The following gears are available when the driver selects:

Range	Gear
1	1st gear only with engine braking.
2	1st and 2nd gear.
3	1st, 2nd and 3rd gear —convertor lock-up in 3rd gear.
D	1st, 2nd, 3rd and 4th gear —convertor lock-up in 4th gear.
N	No gears selected —engine can be started.
R	Reverse gear.
P	No gears selected —engine can be started.

The electronic controls of this transmission allow 'skip shifting', e.g. when the transmission is in 'D' range the following shifts are possible 1-2, 1-3, 1-4, 2-3, 3-4, 2-4, 4-3, 4-2, 4-1, 3-2, 3-1 and 2-1 as a function of vehicle speed, throttle position and rate of change of throttle position.

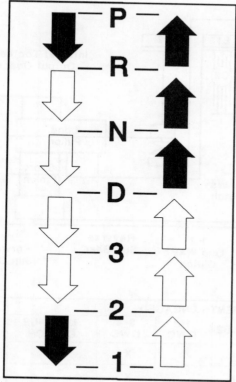

Figure 10-28 The shift selection pattern is more comprehensive than earlier transmissions. The black arrow indicates that movement in the direction of the arrow is inhibited by a detent mechanism.

COMPONENTS

An ECAT is a mechanical device, operated by hydraulics and controlled by electronics. The three main sections of an ECAT contain the:

- Mechanical components;
- Hydraulic components;
- Electronic components.

MECHANICAL AND HYDRAULIC COMPONENTS

The transmission power train uses the same type of components as a standard automatic transmission.

The major components of the power train (see Figure 10-29) are:

- torque convertor with lock-up clutch;
- hydraulic, fixed displacement, pump;
- four multi-plate friction clutch packs;
- two brake bands;
- two one-way clutches;
- Ravigneaux compound planetary gear set.

Many of the hydraulic valves have similar names to valves within a conventional automatic transmission. The electronic interface to many of these valves changes the way in which valve operation occurs. Therefore a limited explanation of hydraulic valve action will be included, where necessary, in conjunction with the electronic control.

It is not within the scope of this book to describe the construction and operation of the power train, pump and hydraulic components. Where the reader is not fully conversant with these components, it is recommended that the appropriate workshop manual be carefully studied.

ELECTRONIC COMPONENTS

The electronic components shown in Figure 10-30 can be grouped under three subsections.

- input devices;
- output devices;
- electronic control unit (ECU).

The nine input devices supply the ECU with information so that decisions can be made relative to the parameters in the ECU's memory. Once the action has been decided, the ECU controls the output devices to achieve the desired results.

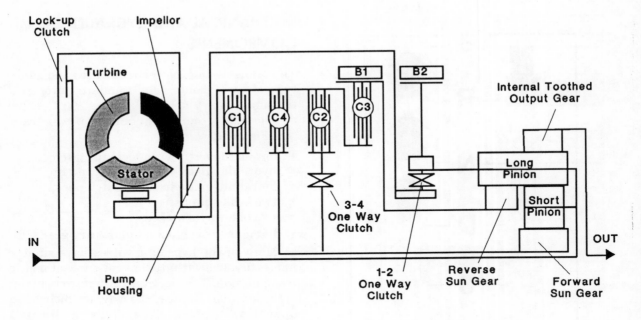

GEAR	GEAR RATIO	ELEMENTS ENGAGED								
		C1	C2	C3	C4	B1	B2	1-2 OWC	3-4 OWC	LOCK-UP CLUTCH
1st	2.393		X					X	X	
2nd	1.450		X		X	X			X	
3rd	1.000	X	X		X				X	X (MANUAL 3 ONLY)
4th	0.677	X	X			X				X
Reverse	2.094			X			X			
Manual 1st	2.393		X		X		X		X	

Figure 10-29 Major power train components.

INPUT DEVICES

The input devices consist of link wires, sensors and switches. These devices are located at various points on the vehicle and are connected to the input stages of the ECU by wires. The left-hand side of Figure 10-30 provides a typical list of these components:

- Battery voltage;
- Air conditioning compressor operation;
- Throttle position sensor;
- Engine speed;
- Vehicle speed sensor;
- Fluid temperature switch;
- Power/economy mode switch;
- Gear position and inhibitor switch;
- Self-test input.

BATTERY VOLTAGE

Battery voltage monitoring is not a physical device but is a wire connected between the ECU and the positive side of the battery. The monitoring action takes place within the ECU internal circuitry.

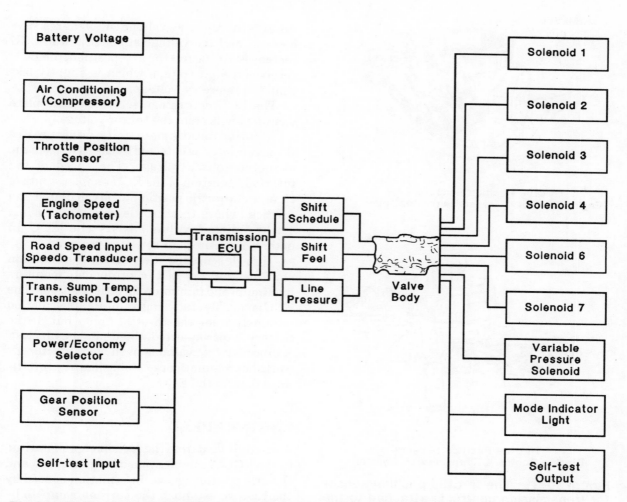

Figure 10-30 The electronic system can be grouped into inputs (left side), decision making (transmission ECU) and outputs (right side).

AIR CONDITIONER

Sensing of air conditioning compressor operation is provided by a wire connected between the ECU and the air conditioning compressor clutch circuit. The ECU can initiate a change in shift points in response to air conditioner operation. By delaying the upshift, compressor speed can be kept high and air conditioning performance maintained.

THROTTLE POSITION SENSOR

The throttle sensor is a variable resistor which is attached to the throttle body and connected to the throttle shaft (see Figure 10-31). Three wires connect the sensor to the ECU. The angle of throttle opening is determined by the ECU from the voltage signal that is received from the sensor.

Construction

A circular resistive element is positioned concentric to a wiper arm hub inside a small housing. The end of a wiper arm, attached to the

MPEFI

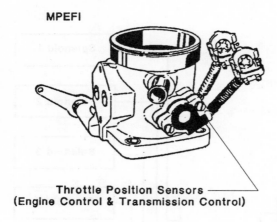

Throttle Position Sensors
(Engine Control & Transmission Control)

EFI

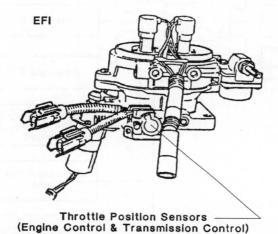

Throttle Position Sensors
(Engine Control & Transmission Control)

Figure 10-31 The throttle position sensor for transmission control is attached to the engine control throttle position sensor. Both sensors are operated by the throttle body shaft.

hub, is positioned on the resistive element. Each end of the circular resistive element and the wiper arm are connected by wires to terminals in a socket block.

Operation
A reference voltage is applied to one end of the resistive element and the other end is connected to ground at the ECU. When the throttle is closed, the wiper arm is positioned near the grounded end of the resistive element. In this position the voltage supplied to the ECU through the wiper arm connection is very low. As the throttle starts to open, the wiper arm moves

towards the reference voltage end of the resistive element and the voltage supplied to the ECU increases. At the full throttle position, the ECU receives a high voltage which is slightly less than the reference voltage.

The ECU can recognise the throttle plate opening angle from the sensor voltage.

A moving throttle plate, as the driver presses the accelerator, is detected by the ECU from the changing sensor voltage. The speed of the internal operation of the ECU enables a rate of change of throttle position to be calculated and predicts where the throttle will stop moving. The predicted value assists the ECU in determining the gear selection required.

Throttle sensor position is critical to maintain shift feel and the ECU continuously monitors maximum and minimum throttle positions. The ECU is capable of altering its internal calculation parameters for the throttle limits but, if the battery is disconnected, the limits may be lost. A procedure is available in the manufacturer's workshop manual for re-establishing the throttle limit data to the ECU.

ENGINE SPEED
A wire is linked from the tachometer signal line to the ECU A combination of engine speed and throttle position sensor information is used by the ECU to establish the current 'ramp' to be applied to the variable pressure solenoid. Further details of variable pressure solenoid operation will be provided later in this section.

ROAD SPEED
The vehicle to which this transmission is fitted, is equipped with an electronic speedometer. The speedometer is supplied with a square wave voltage signal from a transducer fitted to the transmission. The frequency of the wave form is related to transmission output shaft speed. The ECU shares the transducer output with the speedometer and is able to calculate road speed from this information. This information takes the place of the hydraulic governor pressure signal produced by non-electronic automatic transmissions.

TRANSMISSION FLUID TEMPERATURE SENSOR

The transmission fluid temperature sensor is an NTC type resistor. At low temperatures the resistor has high resistance (approx. 15 k ohm), but at high temperatures the resistance will decrease to approximately 80 ohms. The sensor is fitted to the internal wiring loom and is immersed in the fluid within the sump. At temperatures approaching 135 degrees C, the ECU detects the resistance value and initiates convertor clutch lock-up at lower than normal vehicle speeds. This action prevents convertor slippage. The convertor oil is diverted to the external cooler to reduce transmission heat load.

POWER/ECONOMY SELECTOR

A rocker button type switch is located on the dash panel to the left of the steering wheel for column shift vehicles. If the vehicle has a 'T' bar floor mounted console, the switch will be on the right of the selector panel. The switch is connected by a single wire to the ECU.

Construction

Two fixed contacts are located in the base of the switch where they are insulated from one another. A spring-loaded contact bridge is attached to an over-centre rocker mechanism but insulated from the body. The over-centre mechanism is connected to a button marked 'P' and 'E'.

Operation

If the button is in the economy position the contacts are not closed and the circuit from the ECU is open. When the driver pushes the side of the rocker button marked 'P', the over-centre mechanism moves the contact bridge against spring tension to connect the two contacts together. The mechanism locks and holds the switch in the 'P' position. The ECU wire is grounded to earth through the switch contacts. With the button in this position, the ECU will illuminate a 'Power' indicator jewel on the dash panel.

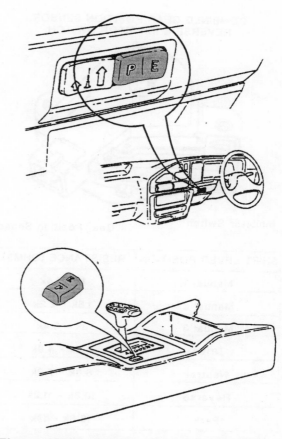

Figure 10-32 Power and economy mode switches are fitted to both steering column selector and T-bar equipped vehicles.

With the mode switch at 'P', the ECU configures its operation to suit the power mode.

With the mode switch at 'E', the ECU configures its operation to suit the economy mode.

GEAR POSITION SENSOR

A combined gear position sensor, reverse light switch and inhibitor switch is located on the transmission case and positioned on the gear selection mechanism (see Figure 10-33). The switch is a multipurpose type and has two output sockets.

COMBINED GEAR POSITION SENSOR, REVERSE & INHIBITOR SWITCH.

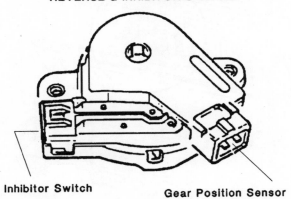

Inhibitor Switch

Gear Position Sensor

SHIFT LEVER POSITION	RESISTANCE (OHMS)
Manual 1	1k - 1.4k
Manual 2	1.8k - 2.2k
Manual 3	3k - 3.4k
Drive	4.5k - 4.9k
Neutral	6.8k - 7.2k
Reverse	10.8k - 11.2k
Park	18.6k - 19k

Figure 10-33 The position sensor output signal is varied by the resistor value present at each selector location.

Construction

A wiper arm is formed with a small hub centre and several wiper contacts at its outer end. The hub is attached to and moves with the gear selector shaft.

A series of fixed contacts and set resistors are positioned in a short arc around the centre of the hub. The switch housing is constructed of plastic and each contact is insulated from other contacts and earth. The fixed contacts are for the starter inhibitor and reverse lamp function. The series of set resistors are used as gear position sensors.

Operation

When the driver moves the gear selector lever, the switch hub turns with the gear selector shaft. Three different outputs are required from the switch depending on gear selector position.

1 Reverse selection rotates the wiper arm to form a bridge between two contacts, completing a circuit and allowing current flow to the reverse lamps.
2 Park or Neutral selection rotates the wiper arm to form a bridge completing the starter motor solenoid circuit. These two selector positions are the only positions that will allow starter cranking when the ignition key is moved to the start position.
3 The gear position sensing function operates in all selector positions when the engine is running. Two wires, one in and one out, form the link to the ECU Rotation of the wiper arm across the series of set resistors will alter the value of the output signal sent to the ECU. From this information the ECU can determine the position of the selector lever.

SELF-TEST INPUT

The self-test input (see Figure 10-34) is one of a series of diagnostic terminals fitted to vehicles of this type. The input is a normally open circuit

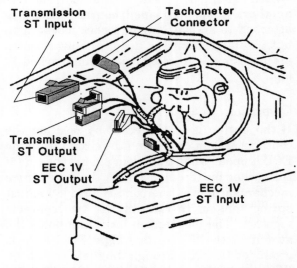

Transmission ST Input

Tachometer Connector

Transmission ST Output

EEC 1V ST Output

EEC 1V ST Input

Figure 10-34 Transmission fault codes can be accessed at the self-test connectors.

and, when grounded, activates the diagnostic output from the ECU. Grounding of this terminal should only be done in accordance with the manufacturer's instructions and in conjunction with the use of approved diagnostic equipment.

OUTPUT DEVICES

The output devices (Figure 10-30) consist of seven solenoids, a mode indicator light and a self-test output. Six of the solenoids are mounted to the valve body and the seventh is in the pump cover. Each solenoid is connected by a wire to the ECU.

The solenoids are of conventional electromagnetic construction and operate a plunger to open or close a hydraulic passage within the solenoid body. Six solenoids are of the on/off type and one is a variable pressure type. The on/off types can be further subdivided into 'normally open' or 'normally closed'.

NORMALLY OPEN

Normally open means that with no power applied the oil can flow through the solenoid valve to the specific circuit. Figure 10/35b. illustrates a solenoid with power off allowing oil to flow from the supply (line 70) to the outlet circuit (S1). The exhaust port is closed.

When power is applied, (Figure 10-35a.) the plunger moves and line 70 oil is blocked. The exhaust port is connected to S1 and oil in circuit S1 can be dumped through the open exhaust port.

NORMALLY CLOSED

Normally closed solenoids are the reverse of normally open types. With no power applied, oil cannot pass from a supply port to a specific circuit. The supply oil is usually dumped through an open exhaust port. When power is applied the solenoid plunger moves; the exhaust port is blocked and oil can flow from a supply passage across the plunger and out to the required circuit.

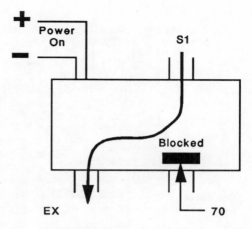

a) With power ON a normally open solenoid blocks oil flow to specific circuits.

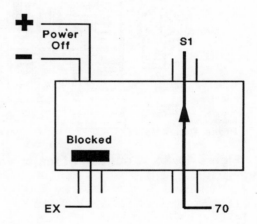

b) With power OFF a normally open solenoid blocks oil flow to exhaust and opens flow to specific circuits.

Figure 10-35 Schematic diagram of a normally open solenoid showing the flow conditions for both power ON & OFF.

SOLENOID APPLICATION

The total schematic (see Figure 10-36) of an automatic transmission can appear complex but by separating each section the task of understanding the role of the solenoids can be made easier. The transmission can be subdivided into the:

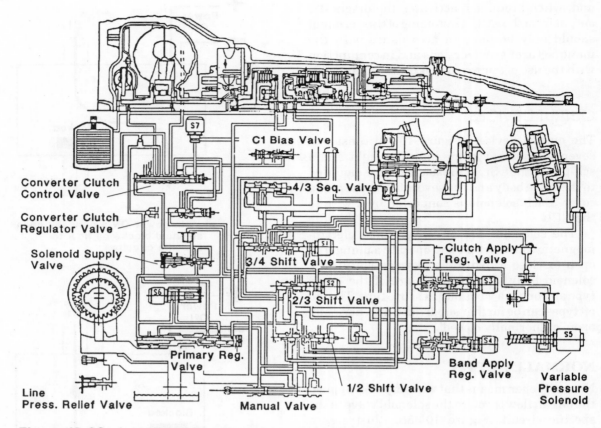

Figure 10-36 Location of major valves and solenoids.

- range selection;
- line and solenoid supply pressures;
- shift solenoids and valves;
- shift timing regulators and solenoids;
- convertor lock-up clutch controls and solenoid.

This explanation will deal with the electronic control of the transmission, and the reader's prior knowledge of pumps, hydraulic valves, bands, servos and clutch pack operation is assumed.

Range selection

No solenoids are present in the selector controls. Range selection is controlled by the manual valve (lower centre of Figure 10-36). The valve is mechanically linked to the shift lever operated by the driver. Oil pressure from the pump is

Manual Valve

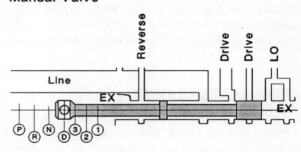

Figure 10-37 Range selection is controlled by the manual valve.

directed through the valve into different circuits, depending on where the valve has been positioned. This type of valve is common to most automatic transmissions.

Line and solenoid pressures

Several different hydraulic pressures are required within an automatic transmission. The base working pressure is known as the line pressure and it is from this pressure that all other pressures within the transmission are developed. The lower left area of Figure 10-36 contains the primary regulator valve, solenoid S6, which is fitted to the end of the line pressure boost valve and the solenoid supply valve (shown slightly above S6 in the illustration but is physically fitted to the end of the line pressure boost valve).

Primary regulator valve

This valve regulates the pump output pressure and in doing so, produces the transmission regulated line pressure. The basic operation of the valve is hydraulic; pump (line) pressure enters the valve and acts against spring force. The valve starts moving to the left as line pressure increases. The valve movement uncovers the convertor feed port and oil flows out to pressurise the torque convertor. Further movement of the valve to the left opens the pump suction port (see Figure 10-38). Fluid passes back to the pump inlet and the line pressure is held at the value required to compress the spring sufficiently to open the port. This condition is known as **high line pressure**.

Under light throttle cruise conditions the ECU (refer to the S6 solenoid for further details of this action) allows oil from the S6 line to enter the valve at the right-hand end. S6 oil assists the line pressure to act against the regulator spring and the suction port will open at a lower line pressure. This condition is known as **low line pressure**.

When reverse gear is selected, oil is routed from the manual valve to the left-hand end of the regulator. This reverse oil assists the regulator spring and the line pressure must rise to a higher value before the suction port is uncovered. This condition is known as **boosted pressure**.

Line pressure boost valve and solenoid supply valve

This valve is two valves in one. The right-hand end (solenoid supply) is hydraulic in action and regulates line pressure down to 483 kPa. This pressure is known as line 70 pressure (483 kPa = 70 psi) and is used as the supply pressure to all the solenoid valves in the transmission. The left-hand end (line pressure boost) contains an on/off, normally closed, solenoid. When the ECU detects the vehicle is under light load, current is allowed to flow through solenoid S6. (see Figure 10-39) S6 opens a flow path from line 70 oil to an outlet passage that connects to the primary regulator valve. The oil in this outlet passage is known as S6 pressure and has the effect of moving the primary regulator valve to the low line pressure condition. If the ECU detects load, e.g. heavy throttle application, S6 solenoid is switched off. Line 70 oil is blocked at the solenoid and the primary regulator valve moves to the high line pressure condition. High line pressure occurs when the S6 solenoid is switched off, therefore the default condition for any

Primary Regulator Valve

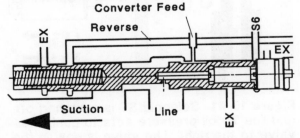

Figure 10-38 The primary regulator valve regulates the pump output pressure.

Line Pressure Boost Valve & Solenoid Supply Valve

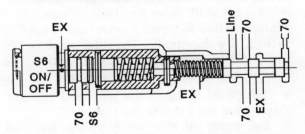

Figure 10-39 This valve supplies solenoid pressure and can also supply boost line pressure as required.

malfunction will be the high line pressure condition.

Before moving on to the other transmission valves a summary may be helpful.

The transmission is supplied with:

- either a high or low line pressure depending on whether S6 is 'OFF' or 'ON';
- line pressure in reverse range that is boosted above normal.;
- a pressure known as line 70 pressure which is routed from the solenoid supply valve to all transmission solenoid valves.

Shift solenoids and valves

The shift solenoids are in a group through the centre of Figure 10-36. There are three shift valves: 1-2, 2-3 and the 3-4. At the right hand end of the valves are the shift solenoids: S1 and S2. Both S1 and S2 are normally open on/off solenoids controlling the 3-4 and 2-3 shift valves respectively. The 1-2 shift valve does not have its own solenoid but can only move under the control of solenoids S1 and S2. The shift valves control the direction of oil flow to the various friction elements and operate under the control of the ECU. Readers familiar with hydraulic transmissions would expect to see these valves with throttle pressure at one end being opposed by governor pressure at the opposite end. In an electronic control system line 70 pressure is fed to the left hand end forcing the valve to the right. The solenoids S1 and S2 under instruction from the ECU control the admission of oil to the left hand end of the valve. Therefore valve movement and the resulting gear change occur from ECU solenoid control and not an inter-action of throttle and governor pressures.

The 2-3 shift valve has been chosen as an example of how electronic control can be applied to the hydraulic circuits. The valve, as shown in Figure 10-40, is in the second gear position. The conditions present in the circuits are:

- line 70 oil is fed in the left-hand end and pushing the valve to the right.;
- 2nd apply oil directed from the 1-2 shift valve is resting on the valve land but not passing into the valve.;

- solenoid S2 is currently switched on; line 70 oil is entering the solenoid but cannot pass into the valve area. Remember S2 is a normally open solenoid and because it is switched on, line 70 supply will be blocked from flowing to its specific outlet circuit.

The ECU determines a shift to third gear is required and switches solenoid S2 off. The actions which occur within the valve (see Figure 10-41) are:

- line 70 oil is routed through the solenoid S2 to fill the cavity to the left of the solenoid.

2/3 Shift Valve

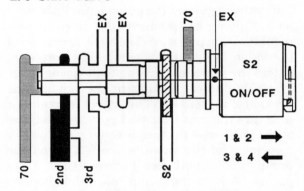

Figure 10-40 Solenoid S2 is switched on and the valve is held in the 1 & 2 position by line 70 oil pressure.

2/3 Shift Valve

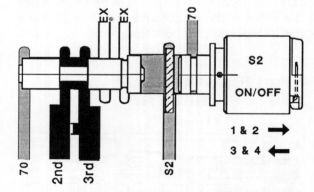

Figure 10-41 Solenoid S2 is switched off and line 70 oil pressure acts to push the valve to the right. The valve is now in the 3 & 4 position.

The pressure build-up in this cavity moves the valve to the left;

- line 70 oil can now flow across the valve and out the S2 outlet passage;
- the valve is now seated to the left of the housing and 2nd apply oil which was resting on the valve land can now connect to the 3rd apply passage and the transmission will shift to third gear.

The ECU does switch more transmission components than the 2-3 shift valve when a gear change is required but the oil to apply the friction components is routed through the shift valves.

By a combination of switching S1 and S2 'ON' and 'OFF' the ECU can move the transmission up or down through its gear range.

The solenoid logic for static gear states is:

Gear	Solenoid 1	Solenoid 2
1	'on'	'on'
2	'off'	'on'
3	'off'	'off'
4	'on'	'off'
REV	'on'	'on'

If an electrical failure occurred it can be seen from the logic chart that the transmission would move into third gear because the normal state for third gear is both solenoids 'OFF'.

Shift timing regulators and solenoids
The shift timing components under ECU control are the variable pressure solenoid (S5), clutch apply regulator (S3) and the band apply regulator (S4). These three solenoids and valves are located on the right of Figure 10-36 and are linked together hydraulically during gear shifts. The role of the solenoid regulator valves is to control the rate at which line pressure oil is applied to the friction elements during the gear shift. The combination of these valves and the convertor clutch control valve provide the control components to ensure smooth and precise shift action.

The variable pressure solenoid (VPS) is a regulator valve operating under ECU control (see Figure 10-42). Line 70 oil is passed through the valve and emerges as S5 control pressure. The pressure value of S5 oil varies depending on the current provided to the solenoid by the ECU.

Variable Pressure Solenoid

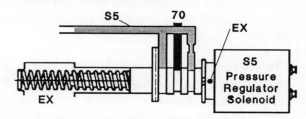

Figure 10-42 Line 70 pressure is regulated through S5 variable pressure solenoid to provide an output pressure (S5).

When the transmission is in a gear and no gear changes are imminent the current supplied to the solenoid will be approximately 200 milliamps. The solenoid valve is constructed so that maximum S5 pressure will be produced when the current is low. If current is increased from the 200 milli-amp value the output pressure will decrease. Two statements are possible from this pressure current relationship:

1 Increased current decreases S5 output pressure.
2 Decreased current increases S5 output pressure.

S5 output pressure is produced from line 70 oil, therefore it will always be less than line 70 pressure (480 kPa). At the 'standby' current of 200 milli-amps S5 pressure reaches its maximum value of approximately 450 kPa. During a gear shift, the ECU sets a current value to the solenoid; line 70 oil is regulated through the valve to produce an S5 pressure related to the current applied. The ECU will then set up a current 'ramp' through the solenoid. This means the current applied to the solenoid will be progressively increased or decreased from the set value. The effect on the S5 pressure will be the reverse of the current action. If the current is ramped up in value the S5 pressure will be ramped down. S5 controlled pressure is supplied to the clutch apply regulator, band apply regulator and the convertor clutch regulator valve.

Further explanation of how the varying S5 output pressure can be used in shift timing is included in a later part of the text. At this point it is sufficient to understand that S5 pressure can be ramped up or down in response to ECU signals.

Clutch apply regulator valve

The solenoid, S3, is a normally open ON/OFF type. When the transmission is in a gear and a gear shift is not in process, this solenoid will be turned 'OFF'. Figure 10-43 shows the solenoid in the 'OFF' position. Line 70 oil will flow through the solenoid into the cavity between the solenoid and the valve plunger. Because line 70 pressure is always greater than S5 pressure the plunger will move to the left. This movement will squeeze the S5 oil out from between the plunger and the valve. In this position 3rd apply oil can flow through the left-hand side of the valve and pass out to the clutch apply force (CAF) passage (see Figure 10-43).

When the ECU switches S3 solenoid 'ON', line 70 oil is blocked from entering the plunger

Clutch Apply Regulator Valve

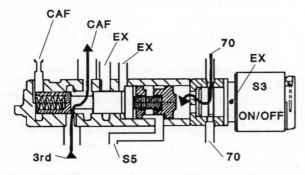

Figure 10-43 The solenoid is OFF and line 70 oil pushes the valve to the left. In this position, 3rd oil can pass through the valve to the C1 clutch (CAF passage).

cavity. S5 oil, at a pressure set by ECU action on the variable pressure solenoid (S5), pushes the plunger to the right until it is against the solenoid. The regulator valve spring will push the valve to the right attempting to close the 3rd oil passage. This movement will be opposed by the S5 pressure. As S5 pressure is 'ramped' up the valve will move to the left opening the passage to enable 3rd oil to pass through the valve, out the CAF passage and apply the clutch. The rate at which pressure can be applied to the clutch is controlled by the S5 pressure and the 'amplification' ratio of the valve. Note the difference in land areas of the valve at the points where CAF oil, S5 oil and line 70 oil can act. This difference in areas determines the amplification ratio of the valve. When line 70 oil is acting on the plunger the CAF pressure will be in excess of twice the line 70 value. When the solenoid is 'ON' and S5 oil is acting on the plunger the CAF pressure will be in excess of twice the S5 value.

To summarise the regulator action:

• When the transmission is in steady state (not shifting gear), S3 will be 'OFF' and pressure in the CAF line will be determined by the amplification ratio of the valve times line 70 pressure;
• During a gear shift requiring clutch apply force, S3 will be 'ON' and pressure in the CAF line will be ramped. This action controls the rate of change of the clutch. Shift quality is therefore a function of the ECU, VPS (S5) and the regulator (S3).

Band apply regulator valve

The band apply regulator valve operates in a similar manner to the clutch apply regulator valve. The solenoid, S4, is a normally open ON/OFF type. When the transmission is in a gear and a gear shift is not in process, this solenoid will be turned 'OFF'. Figure 10-44 shows the solenoid in the 'ON' position indicating a gear shift involving a band apply or release is in process. In the 'ON' position S5 oil has control of the valve and is regulating the flow between 2nd oil and the band apply force (BAF) passage.

The amplification ratio of this valve is less than S3 because the difference in land area of

Band Apply Regulator Valve

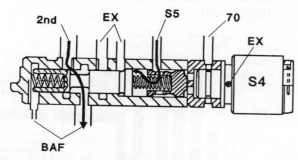

Figure 10-44 A gear change is in progress. Solenoid S4 is ON allowing line 70 oil to flow to exhaust. S5 oil is in control of the valve position and 2nd oil is being regulated as it flows into the BAF passage. (B1 band passage)

BAF, S5 and line 70 oil is not as great. To summarise the regulator action:

- When the transmission is in steady state (not shifting gear), S4 will be 'OFF' and pressure in the BAF line will be determined by the amplification ratio of the valve times line 70 pressure;
- During a gear shift requiring band movement, S4 will be 'ON' and pressure in the BAF line will be ramped. This action controls the rate of change of the band servo. Shift quality is therefore a function of the ECU, VPS (S5) and the regulator (S4).

A point to remember by the service mechanic is that this is an electronically controlled transmission with sophisticated controls to ensure shift quality. These controls rely on the movement of fluid over a timed period. If the service mechanic does not maintain the servo band adjustments at manufacturer's specification the shift quality will be degraded.

Convertor lock-up clutch controls and solenoid

The group of valves in the upper-left corner of Figure 10-36 control the torque converter clutch lock-up action. The C1 bias valve is hydraulic in action and supplies line 70 oil to solenoid S7 when C1 clutch oil pushes the valve open. Oil is only in the C1 clutch line when the transmission

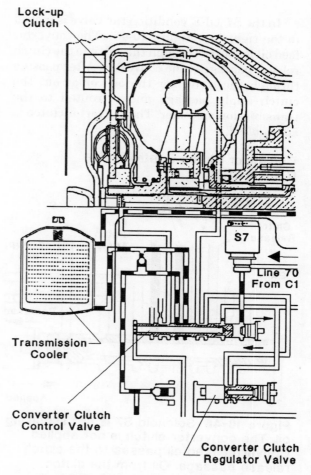

Figure 10-45 Torque converter lock-up clutch, associated control valves and solenoid.

is in third or fourth gear therefore convertor lock-up cannot occur in any other gear. The ECU controls solenoid S7 (see Figure 10-45), which is a normally closed ON/OFF solenoid. The ECU will supply current to switch S7 'ON' (and open) in '3 range third gear' and 'D range fourth gear' only. Lock-up of the convertor clutch is therefore available in two gears and no others. The other valves needed for convertor clutch operation are the convertor clutch control valve and the convertor clutch regulator valve.

Convertor clutch control valve

The convertor clutch control valve is a two position valve.

In the S7 'OFF' condition the valve is moved to the right (see Figure 10-46). The convertor feed oil is routed through the valve to the clutch release passage of the convertor. After passing through the convertor the oil comes out the clutch apply passage and is routed to the transmission oil cooler. The convertor clutch is *not applied*.

Converter Clutch Control Valve (Clutch Off)

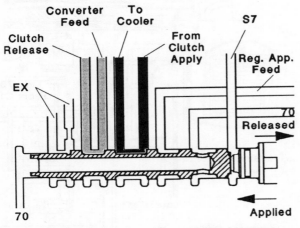

Figure 10-46 Solenoid S7 is not supplying oil. The converter clutch is not applied. Converter feed oil passes to the clutch release passage. Oil from the clutch apply passage passes out to the transmission cooler.
Note: Line 70 and regulated apply oil are not shown.

The ECU switches S7 'ON' and line 70 oil flows through the solenoid becoming S7 oil. The valve moves to the left. Oil from the regulator apply feed line is routed to the convertor clutch apply line. The position of the valve now routes convertor feed oil direct to the transmission cooler. Any oil in the clutch release line can now be dumped through the exhaust port (see Figure 10-47). The convertor clutch is *applied*.

Convertor clutch regulator valve

The convertor clutch regulator valve (see Figure 10-48) supplied the oil via the reg. app. feed passage to apply the convertor clutch (refer to

Converter Clutch Control Valve (Clutch On)

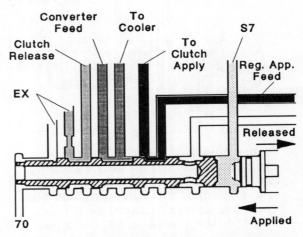

Figure 10-47 Solenoid S7 is supplying oil to push the valve to the left. Converter feed oil is sent to the cooler. Oil in the clutch release passage is dumped to exhaust. Oil is sent to the clutch apply passage from the apply feed line. The converter clutch is applied.
Note: Line 70 oil is not shown.

Converter Clutch Regulator Valve

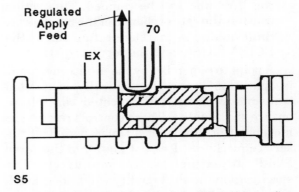

Figure 10-48 Oil is regulated by this valve and is sent to the converter clutch control valve to apply the converter clutch.

the previous valve). To ensure the apply and release of the convertor clutch is achieved smoothly the ECU controls the apply pressure. S5 oil from the VPS solenoid is applied to one end of the regulator valve. The ECU will cause the VPS to ramp S5 pressure whenever S7 is first switched 'ON' or 'OFF'. Therefore the pressure going to the convertor clutch is variable during a clutch movement and the engagement or disengagement is performed smoothly.

This completes the description of the valve operations which are effected by the ECU. There are several other valves within the transmission, e.g. B1R exhaust valve and the 4/3 sequence valve but their action is hydraulic and not within the scope of this book.

TRANSMISSION OPERATION

To link together the knowledge gained from the description of each individual valve it will be necessary to consider the transmission operation during an upshift and a downshift of gears.

Figure 10-50 is a schematic listing the friction elements that are on in each gear range.

The chart (Figure 10-49) shows the condition of each solenoid before, during and after the gear shift.

Figure 10-51 is an overall transmission schematic.

These three illustrations will assist you in following the descriptive text of the gear shifts.

The transmission gear shift selected as an example is first to second gear upshift in drive range.

'D RANGE' 1—2 UPSHIFT

The transmission is in first gear, therefore the C2 clutch is the only hydraulically applied component. The 1-2 and 3-4 one way clutches are locked but these types of clutches lock and unlock in response to reactions from the gear train and are not controlled directly by the ECU.

Second gear retains the C2 clutch but requires the C4 clutch and the B1 band to be applied. Therefore, to move the transmission from first to second gear, the ECU must send out information to apply the C4 clutch and the B1 band.

The information detailing which friction elements are required in any gear is listed in Figure 10-50.

The ECU, acting on information from the input sensors, determines the vehicle speed and load are correct for an upshift to second gear. While the vehicle is moving in first gear and before the gear shift is initiated, the solenoid condition will be:

3-4 shift valve	S1	'ON'
2-3 shift valve	S2	'ON'
clutch apply reg. valve	S3	'OFF'
band apply reg. valve	S4	'OFF'
variable pressure solenoid	S5	200milli-amps/ max.pressure

In an upshift, the ECU must switch the required shift valve/s and regulator valves simultaneously; the VPS current flow must also be moved to the start of the current ramp. Because this is a 1-2 upshift the ECU will initiate the shift by switching S1 'OFF' and S4 'ON'. In this specific shift the VPS (S5) is set at a current flow of 750 mA. and is ramped to 600 mA. during the shift. Information for the other gear shifts is available from Figure 10-49.

The ECU initiates the shift and S1 (controlling the 3-4 shift valve) turns 'OFF'. Line 70 oil pushes the valve to the left (2 & 3 position). Two oil flow actions occur:

- Line pressure flows through the 3-4 shift valve and applies the C4 clutch.
- Line 70 oil is redirected through the 3-4 shift valve and flows to the 1-2 shift valve placing it in the 2,3,4 position. Line pressure can then pass through the 1-2 shift valve and flow to the band apply regulator valve S4.

The ECU also turned S4 'ON' at the start of the shift and line pressure oil flows through S4 on its way to apply the B1 band. The VPS, operating under ECU current ramping control, can now affect the rate and pressure of the line pressure oil flowing through the band regulator valve. As the current ramps down, the S5 output pressure will increase and the band apply pressure will also increase. When the VPS stops

SOLENOID LOGIC FOR STATIC GEAR STATES

Gear	S1	S2
1st	ON	ON
2nd	OFF	ON
3rd	OFF	OFF
4th	ON	OFF
REV	ON	ON

SOLENOID OPERATION DURING GEARSHIFTS

Shift	To Initiate Shift	Typical S5 Current Ramp	To Complete Shift
1-2	S1 OFF S4 ON	750mA to 600 mA	S4 OFF
1-3	S1 OFF S2 OFF S3 ON S4 ON	850mA to 750mA	S3 OFF S4 OFF
1-4	S2 OFF S3 ON S4 ON	850mA to 750mA	S3 OFF S4 OFF
2-3	S2 OFF S3 ON S4 ON	650 mA	S3 OFF S4 OFF
3-4	S1 ON S4 ON	750mA to 600mA	S4 OFF
4-3	S4 ON	750mA to 900mA	S1 OFF S4 OFF
4-2	S3 ON	700mA to 950mA	S1 OFF S2 ON S3 OFF
4-1	S3 ON S4 ON	600mA to 1000mA	S2 ON S3 OFF S4 OFF
3-2 < 30 k.p.h.	S3 ON	900mA to 1000mA	S2 ON S3 OFF
3-2 > 30 k.p.h.	S2 ON S4 OFF	200mA at 30 k.p.h. 750mA at 100 k.p.h.	S4 OFF
3-1	S3 ON S4 ON	700mA to 950mA	S1 ON S2 ON S3 OFF S4 OFF
2-1	S4 ON	800mA to 950mA	S1 ON S4 OFF
CONV. CLUTCH ON OFF	S7 ON	700mA to 400mA 600mA to 1000mA	S7 OFF

Figure 10-49 Use this chart in conjunction with the transmission operation explanation.

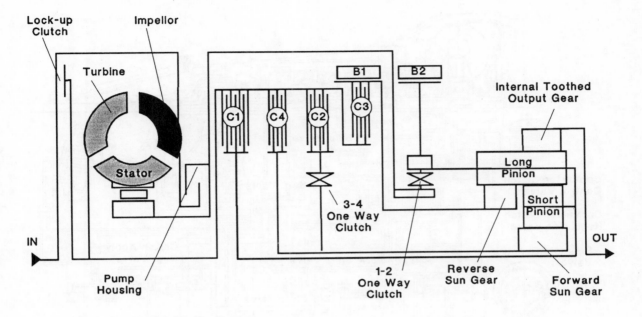

Figure 10-50 Application of the friction elements for each gear range.

GEAR	GEAR RATIO	ELEMENTS ENGAGED								
		C1	C2	C3	C4	B1	B2	1-2 OWC	3-4 OWC	LOCK-UP CLUTCH
1st	2.393		X					X	X	
2nd	1.450		X		X	X			X	
3rd	1.000	X	X		X				X	X (MANUAL 3 ONLY)
4th	0.677	X	X			X				X
Reverse	2.094			X			X			
Manual 1st	2.393		X		X		X		X	

its current ramp, the timed gear shift is over. The ECU will immediately switch S4 'OFF' and return the VPS S5 current to its standby value (200 milli-amps). The band apply force will increase rapidly as line 70 oil re-enters the band apply regulator.

Refer to the section detailing 'Band apply regulator operation' if further detail of the regulator action is required.

At the conclusion of the gear shift the transmission will be in second gear and the solenoid condition will be:

3-4 shift valve	S1	'OFF'
2-3 shift valve	S2	'ON'
clutch apply reg. valve	S3	'OFF'
band apply reg. valve	S4	'OFF'
variable pressure solenoid	S5	200 milli-amps/ max. pressure

This example has described how the fine control, offered by electronics within an ECU,

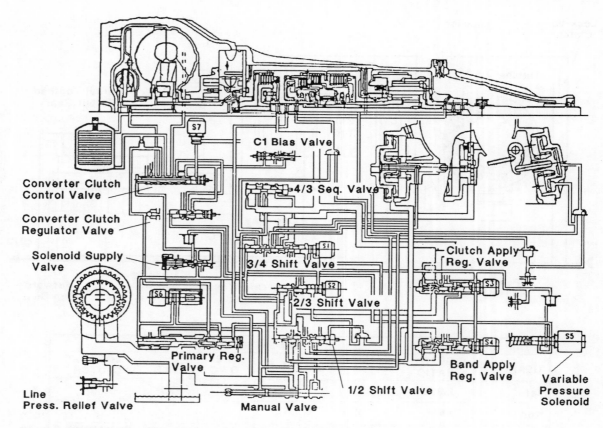

Figure 10-51 Location of major valves and solenoids.

enables gear shifts within an automatic transmission to be performed smoothly and precisely. The remaining upshifts occur in a similar way but a variation is when a down shift is required. The ECU detecting a down shift is required, switches the regulator valve/s 'ON', then sends the VPS through a current ramp. The current ramp is upwards, which means the VPS output pressure is ramped down with subsequent effect on the oil passing through the regulator valve/s. At the end of the downward pressure ramp, the ECU simultaneously switches the shift valve/s as required, the regulator/s to 'OFF' and the VPS to standby maximum pressure.

ELECTRONIC CONTROL UNIT

The electronic control unit (ECU) has all the features of the microprocessor described in

Chapter 7. This ECU has a number of functions. These functions are to provide:

- shift control and timing;
- memory retention during battery disconnected periods;
- self-diagnosis;
- limp home mode;
- diagnostic code output.

The ECU is located in the passenger compartment of the vehicle (see Figure 10-52). Its precise location varies with the manufacturer. Common positions include, behind the side kick panels, under the front seats or, as in this example, under the dash panel.

The shift control and timing features have been adequately described in the previous text, therefore they will not be repeated in this section.

- During service there are occasions when it may be necessary to disconnect the vehicle

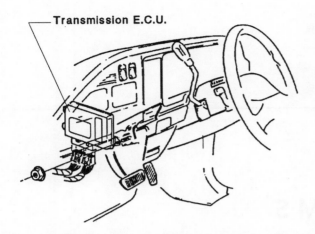

Figure 10-52 The E.C.U. may be attached to the brake pedal support bracket when an under-dash mounting location is specified.

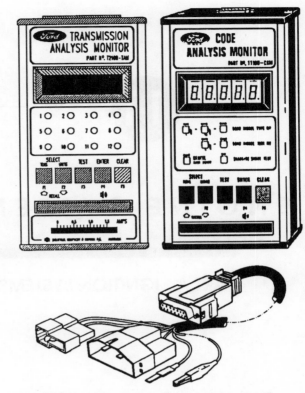

Figure 10-53 Test equipment for both diagnostic code read-out and road testing is available.

battery. The throttle input calibration constants and the diagnostics information are stored in a separate RAM supported by its own capacitor. The memory will be retained during a power shut-down for a period up to 10 minutes at temperatures of approximately 25 degrees C.

- The ECU has the ability to store a significant number of fault codes covering many transmission faults.

- The limp home mode consists of allowing the transmission limited availability of gear ranges, e.g. third gear and reverse gear. In addition to the limp capability the transmission has a number of default conditions which are staged to allow most transmission functions to continue. The actual restriction placed on transmission operation will depend on the fault detected, e.g. firm shifts caused by an engine speed input fault.

- Modern transmissions have available a comprehensive range of test equipment to read the fault codes stored within the ECU memory. These test units are usually connected by a harness to a series of plugs located at a convenient service point on the vehicle. Do not use non-approved methods of extracting self-test codes, as high current draw devices could damage the ECU.

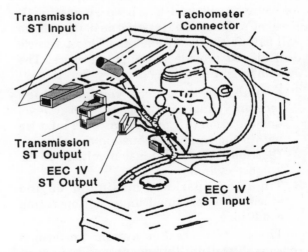

Figure 10-54 Transmission fault codes can be accessed by connecting the test units to the appropriate connection plugs on the wiring harness.

11

ENGINE SYSTEMS

PART A   IGNITION SYSTEMS

The illustration (see Figure 11A-1) is of a typical, contact breaker triggered, ignition system. This type of system has been in service for many years and has been successful. If periodic maintenance is performed, the system will satisfy the requirements of most vehicle owners and provide reasonable service. The system must meet three primary objectives:

1 Produce, from battery voltage, a high tension voltage of sufficient energy level and duration to ignite the air-fuel mixture within the engine combustion chamber.
2 Distribute the high tension voltage to the 'correct' cylinder.
3 Time the spark so maximum engine efficiency is obtained under all operating conditions.

The contact breaker system performed reasonably well until stringent conditions were imposed on engine operations. The need to change from contact type systems resulted from a combination of events. Engine development,

regulatory controls on emission standards and the need for long periods between services challenged the ability of the ignition system to perform at:

• higher engine r.p.m.;
• increased cylinder compression ratios;
• leaner air-fuel mixtures;
• higher spark firing voltages;
• extended intervals without maintenance;
• precise timing tolerances.

Most readers would be aware that contact breaker systems have been almost discarded in favour of 'electronic' ignition systems. However, before moving on to the modern systems, it is useful to consider why the change-over became necessary. Any ignition system, contact triggered or electronic, must meet the three objectives previously listed. Describing the areas, in which contact triggered units failed to continue meeting these objectives, provides an insight to the need for continuing system development.

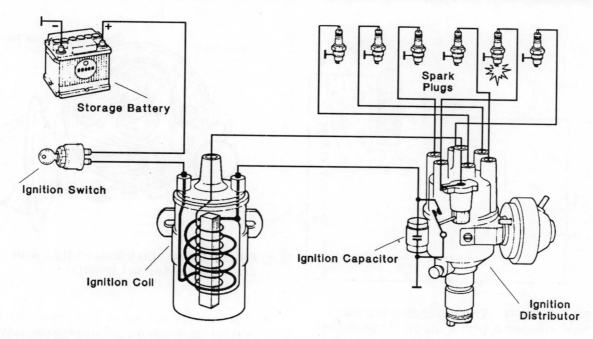

Figure 11A-1 Conventional coil ignition system (six cylinder engine).

OBJECTIVE 1

Produce, from battery voltage, a high tension voltage of sufficient energy level and duration to ignite the air-fuel mixture within the engine combustion chamber. Attaining this objective at air-fuel mixtures of 17:1 and leaner became difficult, particularly in engines which retained high compression ratios. To 'fire' the mixture, these engines require greater HT voltages from the ignition coil. This problem, combined with an increased engine r.p.m. (which shortens the 'build-up' time available to the coil) and extended maintenance periods, ended the use of contact breaker systems.

A brief review of ignition circuit operation will help the reader to understand why it was necessary to change specific ignition components.

When the contact points are closed (see Figure 11A-2), a current of approximately 4 amperes flows through the primary winding of the coil to earth. This action causes a magnetic field to form around the primary winding.

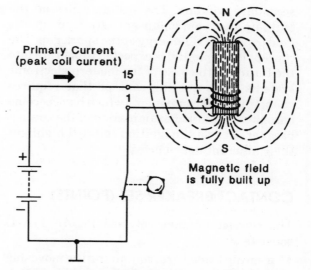

Figure 11A-2 Magnetic field created by current flowing through primary winding.

Continued rotation of the distributor cam causes the contact points to open, interrupting the current flow through the primary winding. The magnetic field 'collapses' (see Figure 11A-

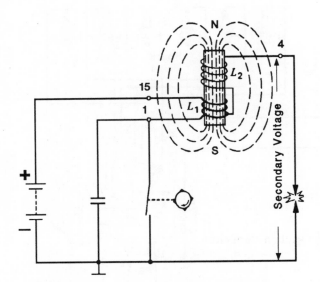

Figure 11A-3 The collapsing magnetic field induces a high voltage surge in the secondary circuit.

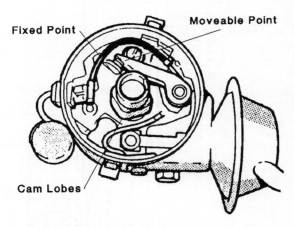

Figure 11A-4 Distributor fitted with contact breaker set (points).

3) and an HT voltage is induced into the secondary winding. The system relies on the build-up and rapid collapse of a strong magnetic field to provide HT energy at the spark gap. The concept is simple and is expected to operate into the future as the base of the modern electronic systems. It is not the concept that requires change; it is the components which have become obsolete. A detailed examination of the contact breaker set and the ignition coil will highlight their inherent deficiencies.

CONTACT BREAKER SET (POINTS)

The contact breaker set (see Figure 11A-4) consists of:

• a spring loaded arm containing the moveable point. The arm is moved by a rubbing block which contacts the lobes of the rotating distributor cam.
• a non-moving contact point which is rigidly attached to the distributor plate.

The contact set contains a number of features which, even though improved by redesign, limit the switching ability available to the primary circuit current.

• The contact set has a maximum current carrying capacity of approximately 5 amperes.
• The contact faces are subject to contact erosion, burning and deterioration in service. The condenser cannot completely eliminate contact sparking due to switching voltage.
• The rubbing block wears against the cam lobes causing a change in dwell angle and ignition point.
• The spring-loaded arm is subject to 'contact bounce' at high engine r.p.m.
• The maximum dwell period ('on' time) available to the coil primary circuit is restricted by the mechanical limitations of the moving arm. Time is lost in the period where the arm swings open then closed.

These limitations mean the maximum coil energy is restricted, because current flow cannot be increased beyond the capacity of the contact set. The small dwell period available (particularly on 8 cylinder engines) is insufficient to allow the ignition coil to reach magnetic saturation at high engine r.p.m. If an ignition coil does not attain magnetic saturation, the electrical energy available to the spark plugs is not at full potential. Finally, the wear rate of the contact faces and the rubbing block prevent the engine operating for extended service intervals.

A significant step forward in system design occurred with the introduction of a solid state

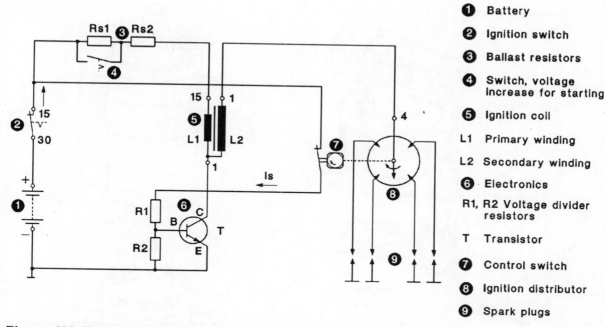

Figure 11A-5 Simplified block diagram of a breaker-triggered inductive semiconductor ignition system (TCI-k).

switching device (see Figure 11A-5) for the coil primary circuit. The contact breaker was used as a trigger to the solid state switching circuit. The current passing through the contact faces of the points is reduced to a level just sufficient to allow self-cleaning of the contact surfaces. It is theoretically possible to reduce the triggering current to very small values but the points may mist with oil and fail to operate correctly.

Current in the primary winding is passed through the electronic switch to earth and is not required to pass through the contact points. The points are only required to switch at battery voltage which ensures contact sparking is minimal. With suitable redesign of the ignition coil, a current of approximately 9 amperes can be passed through the primary winding.

These types of ignition systems are grouped under a broad heading of transistor assisted contact breaker systems. Several systems appeared on the market including a few which contained a form of electronic dwell extension. Collectively, these systems provided an answer to many of the deficiencies of the standard contact breaker system. The most significant problem remaining was the need for periodic service. The current reduction through the points did increase contact life but the service period was still limited by rubbing block wear.

Replacement of the contact set with a solid state triggering device eliminated a major obstacle to ignition system performance. Modern systems use a triggering device, which can be based on inductive pulse, Hall effect (see Figure 11A-6) or optical (light based) concepts. The trigger signal from these devices is fed to an electronic control module which switches the coil primary current 'on' or 'off' as required. As there are no moving parts in contact, the maintenance requirements of electronic systems are minimal and extended service periods are possible.

In addition to minimal maintenance requirements, these systems can switch up to 10 amperes through the primary winding of suitable ignition coils. The electronic control module may also provide an increased dwell period for the ignition coil.

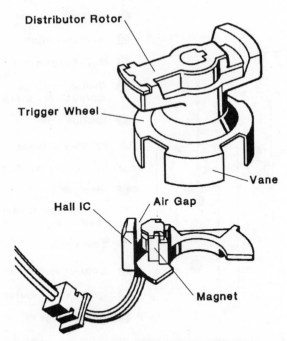

Distributor Rotor

Trigger Wheel

Vane

Hall IC   Air Gap

Magnet

Figure 11A-6 Hall generator used as a solid state triggering device.

To assist in achieving the high energy required from the secondary circuit, ignition coils have been redesigned. The conventional coil was replaced by a high energy coil, similar in external appearance but, with internal modifications. Due to the low primary current flow, the secondary voltage potential of the conventional coils is 25 000 volts. The high energy coil has a potential of 40 000 volts.

The graph (see Figure 11A-8) compares the primary current and rise time of a conventional ignition coil and a low inductance ignition coil. The combination of high primary current flow and adequate dwell period provide a HT energy level to satisfy the requirements of most modern engines.

Development of the ignition coil is continuing with the introduction of moulded ignition coils. The type shown in Figure 11A-9 is for a twin system. The power transistors for primary circuit switching are attached to the mounting bracket of each coil.

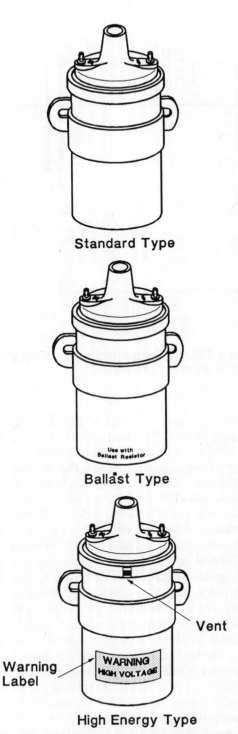

Standard Type

Use with
Ballast Resistor

Ballast Type

Vent

Warning
Label

WARNING
HIGH VOLTAGE

High Energy Type

Figure 11A-7 Ignition coils may be similar in appearance, but have significantly different operating characteristics and must not be inter-changed.

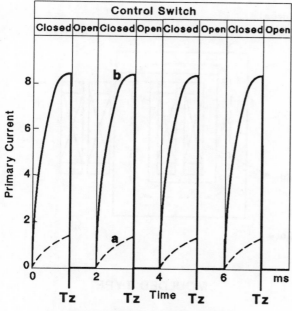

a) Conventional Ignition coil

b) Very low inductance ignition coil

TZ = Point of ignition

Figure 11A-8 Comparison of primary current rise time at a spark rate of 7500 sparks per minute (8 cylinder engine).

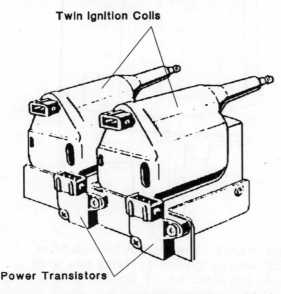

Figure 11A-9 Twin, lightweight moulded ignition coils.

MOULDED TYPE IGNITION COIL

This type of construction avoids the magnetic flux loss associated with conventional type coils as the flux passes through the air (see Figure 11A-10).

The iron core former around which the coil windings are fitted concentrates the magnetic flux, improving coil performance and reducing the current required in the primary circuit.

OBJECTIVE 2

Distribute the high tension voltage to the correct cylinder

The familiar rotor button and distributor cap (see Figure 11A-11) has continued for many years. A few systems which have solid state triggering devices have retained caps which differ only slightly in appearance from the earlier models.

The need for change becomes apparent as air-fuel mixtures become leaner and engine firing voltages increase. The possibility of secondary system HT voltage leakage or distributor cap cross-fire increases as the HT voltage rises. The problem can be solved by several methods.

1. Upgrade existing systems by:
   a increasing the size of the distributor cap allowing more space between the HT towers (see Figure 11A-12);
   b improving the socketing and insulation shrouds of all HT wires;
   c increasing the insulation surrounding the core of the HT wires.
2. Delete parts of the HT system by design changes.
   a fit multiple coils 'fired' by a module which is triggered from a crankshaft sensor. The coils are direct HT wired to individual spark plugs and a distributor is not fitted to the engine (see Figure 11A-13);
   b fit an ignition coil directly attached to each spark plug eliminating all HT wires and the distributor.

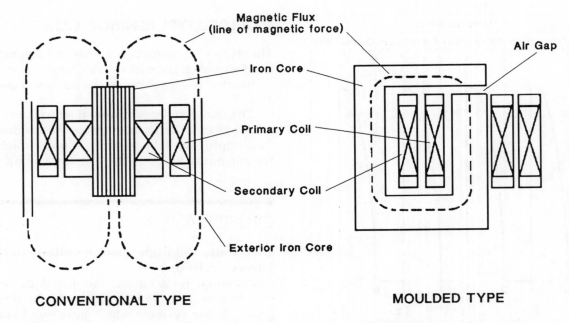

CONVENTIONAL TYPE

MOULDED TYPE

Figure 11A-10 Conventional type coils have a loss in magnetic flux as it passes through the air. Performance of a moulded type ignition coil is improved as all of the magnetic flux passes through the iron core.

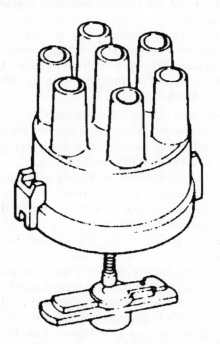

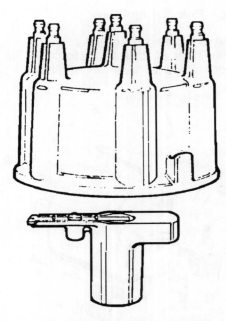

Figure 11A-11 Conventional distributor cap and rotor. Note the relatively close spacing of the HT towers.

Figure 11A-12 The distributor cap fitted to high energy ignition systems has wide spacing between the HT towers.

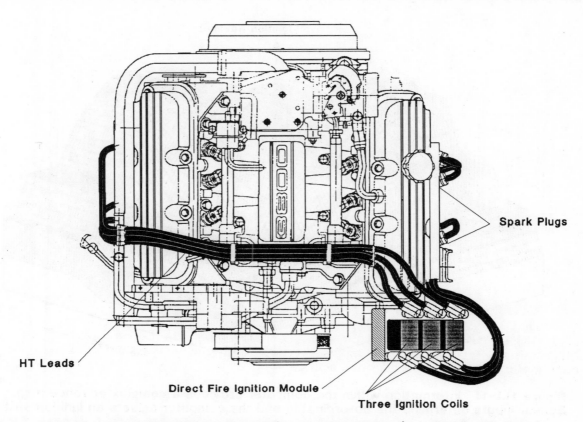

Figure 11A-13 V6 engine fitted with DFI (Direct Fire Ignition). A distributor is not required with this type of ignition system.

OBJECTIVE 3

Time the spark so the maximum engine efficiency is obtained under all operating conditions.

The dual-advance (mechanical and vacuum advance) distributor continues to find service in a few current engines. The distributor is installed to the engine and rotated until a base timing point is located. This operation is usually referred to as 'timing' the engine and is accomplished with the aid of a flashing timing light to strobe the crankshaft timing marks.

The term 'dual-advance', when used as part of the distributor description, refers to the two spark advance mechanisms housed in the distributor.

• Mechanical advance—the distributor cam lobes, or the solid state trigger, are mounted on a movable upper section of the distributor shaft (items 1 and 12—Figure 11A-16). As distributor speed increases, the upper section advances slightly in the direction of rotation. The advance movement is caused by a pair of pivoted weights which move outward in response to the increased r.p.m., but the rate of movement is controlled by a pair of calibrated springs.

• Vacuum advance—a ported vacuum signal is sent to a vacuum diaphragm control unit, which is linked to a movable stator/plate (items 13 and 20—Figure 11A-16). The strength of the vacuum signal is related to throttle position and engine load. The diaphragm will move in response to vacuum signals and will retract a link arm attached to the stator. The stator will move around

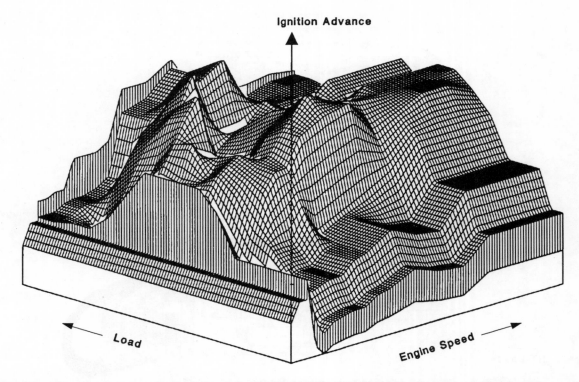

Figure 11A-14 Information within the computer provides a complex advance map. Sensor inputs determine the co-ordinates and the computer selects an ignition setting.

against the direction of distributor shaft rotation advancing the point of ignition.

The point of spark 'firing' for a system fitted with a dual advance distributor is therefore based on:

- initial positioning of the distributor;
- mechanical advance as a result of engine r.p.m.;
- vacuum advance as a result of engine load condition.

For many years engine load and speed have been accepted as the two basic inputs from which to develop the ignition point. The dual advance distributor would appear to suit this requirement with a combination of vacuum and r.p.m. related advance mechanisms. Unfortunately, this type of advance system is incapable of determining timing variations required by changes in engine temperature, octane rating of the fuel or air-fuel mixture ratio.

For efficient operation, a spark timing system must be designed to ignite the air-fuel mixture so peak combustion pressure is present in the cylinder approximately 10 degrees ATDC. If the pressure rise is too early, the pistons, as they approach TDC, will have to 'fight', against the rising pressure. Alternatively if the pressure rise is too late, the pistons will be moving rapidly downwards, ahead of the expanding gases, and effective work on the piston will be lost.

To position the ignition timing accurately requires the assistance of a computer and sensor inputs from the:

- intake manifold (MAP sensor);
- throttle position (TPS switch);
- engine coolant (engine coolant temperature sensor);
- engine speed (engine speed sensor).

These sensors all feed information into the system computer, where, from tables and calculations, the most suitable ignition advance point can be calculated. Figure 11A-14 illustrates the type of spark advance map possible from a modern system. Systems of this type can be broadly described as electronic spark timing (EST). A number of systems within this group are also capable of altering the ignition point to suit low-octane fuels and the load placed on the engine by accessory units, such as air conditioning.

IGNITION SYSTEM APPLICATIONS

The five ignition systems to be reviewed in this chapter were selected on the basis of either widespread application or significant features which will be of interest to the reader. The earliest design selected for study will be based on solid state triggering of the primary circuit and it is assumed the reader is familiar with contact triggered systems with, or without, ballast resistors.

System 1   Dual advance distributor, induct-

ive triggering, miniature ignition module.

System 2   EST ignition computer, inductive triggering from flywheel.

System 3   EST integrated computer, Hall effect triggering in distributor, ignition module.

System 4   EST integrated computer, LED/ photo-diode triggering in distributor, ignition module.

System 5   EST integrated computer, Hall effect triggering from crankshaft sensors, DFI module, no distributor.

SYSTEM 1: DUAL ADVANCE DISTRIBUTOR, INDUCTIVE TRIGGERING, MINIATURE IGNITION MODULE

The system (see Figure 11A-15) has: a dual advance distributor fitted with inductive triggering, a miniature electronic control unit and a low inductance ignition coil.

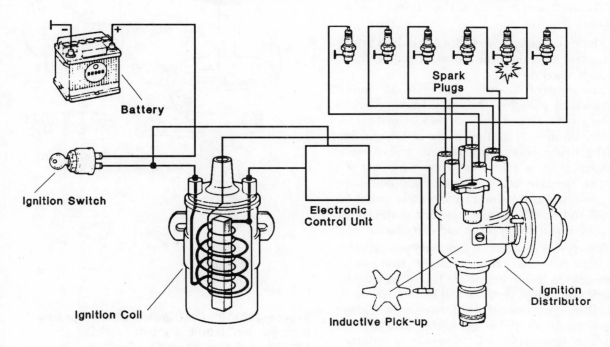

Figure 11A-15 Breakerless high energy ignition system with miniature electronic control unit.

The distributor (see Figure 11A-16) is similar in appearance to the earlier contact breaker types. Ignition advance is supplied by the distributor shaft and flyweight assembly (engine speed related) and the vacuum control unit (engine load related). The significant changes within the distributor are the pulse generator (inductive pick-up) and the electronic control unit (module).

Figure 11A-17 illustrates the wiring connections required for a breakerless ignition system. The ignition switch, battery, high energy ignition coil, distributor cap rotor and spark plug operate in a similar method to most other ignition systems therefore our description of the system will focus on the pulse generator and the control module.

The pulse generator, located within the distributor, is linked to terminals 3 and 7 of the control module by two wires. The ignition coil primary winding negative terminal is connected to terminal 16. A 12 volt supply from the ignition coil supply is linked to terminal 15. The control module is earthed through the attaching screws and lower mounting face.

The system operation can be summarised as follows:

- The pulse generator emits AC pulses to the control module. One pulse is required for each spark firing.
- The control module, for each AC pulse received, switches the coil primary current on, sets the dwell period, limits the maximum current flow in the coil, identifies the firing point from the AC pulse wave form, then switches the primary current off.
- The ignition coil magnetic field collapses when the coil primary current is switched off and an HT voltage is routed to the spark plug through the rotor and distributor.

The system operation summary provided an insight into the role each of the major components has in providing HT voltage to the spark plug. Most readers will be familiar with ignition coil operation however a closer examination of pulse generator and control module operation will provide a greater understanding of how each component fulfils its task.

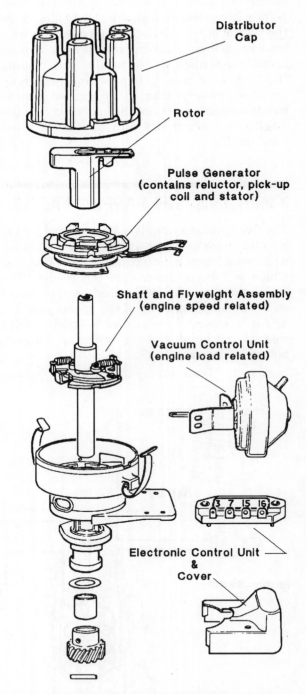

Figure 11A-16 Distributor with inductive pick-up, electronic control module, mechanical and vacuum advance mechanisms.

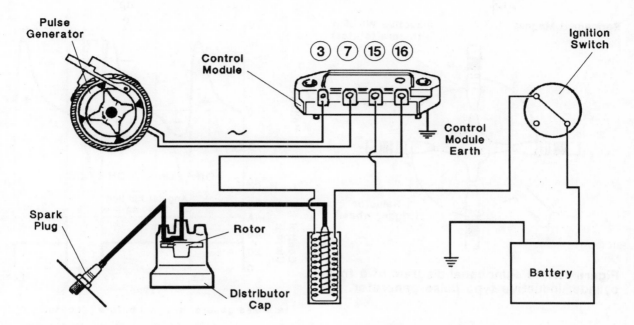

Figure 11A-17 Wiring circuit for a typical breakerless ignition system.

PULSE GENERATOR

The pulse generator (see Figure 11A-18) consists of an inductive winding fitted within a multi-toothed magnetic stator. A reluctor (multi-toothed starwheel) is mounted onto the distributor shaft and rotates with it.

The number of teeth on the stator and the reluctor are the same and correspond to the number of cylinders in the engine, e.g. 4 teeth– 4 cylinder engine. As the distributor shaft and reluctor rotate, the reluctor teeth will approach, momentarily align, then move away from the stator teeth (see Figure 11A-19). At the point where the teeth momentarily align, the gap between the stator and reluctor teeth will be approximately 0.5 mm. The movement of the reluctor teeth towards and then away from the stator teeth causes a change in the magnetic field of the stator. The changing magnetic field induces a small voltage in the pick-up coil winding. This voltage rise in the pick-up coil can be loosely compared to the point action of the early contact breaker system, i.e. a four cylinder distributor has four lobes to open the points four times each 360 degrees—the pulse generator has four pairs of teeth which cause four induced

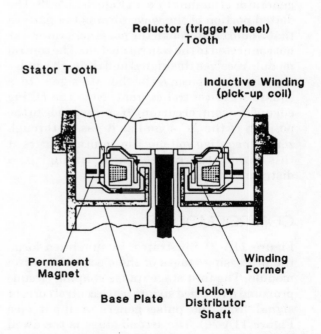

Figure 11A-18 Induction-type multitooth pulse generator. Arrows show magnetic circuit.

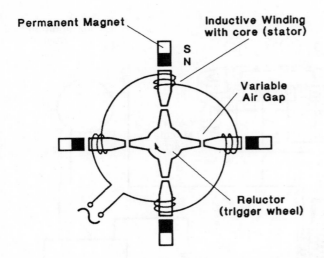

Figure 11A-19 Functional diagram of a four cylinder inductive type pulse generator.

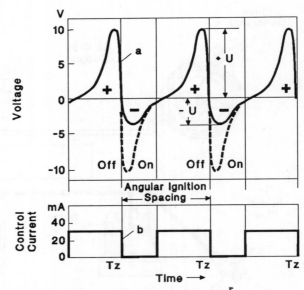

a. Pulse generator signal before processing

b. Rectified square wave control current

Tz = Point of Ignition

Figure 11A-20 The control module changes the pulse generator AC wave form into a square wave control current.

voltages in the pick-up coil each 360 degrees. Further details of pulse generator operation are included in Chapter 7.

The wave form emitted from the pulse generator is similar to 'a' in Figure 11A-20. The dotted portion of the wave form is the pattern that would be present if the pulse generator was not connected to the control module. The control module receives this signal and through a pulse shaping circuit converts the AC signal to a square wave control current. Note the falling edge of the control current 'b' corresponds to the position of the AC signal as it passes through zero to negative voltage. The ignition occurs at this point, therefore 'spark timing' is a distributor function.

CONTROL MODULE

Figure 11A-21 illustrates, in simplified form, the processing stages of the electronic control module. The first stage (pulse shaping circuit) produced a rectified square wave control current signal from the pulse generator input (see Figure 11A-20). The second stage is the dwell angle and current control functions.

The ignition coil fitted to this system is of the low inductance type, therefore current control of the primary circuit is essential.

A comparison of primary current and rise time for a conventional and a low inductance ignition coil is shown in Figure 11A-22. The maximum primary circuit current which can occur is in excess of the system requirement, therefore the current must be regulated. In this system the current limit is set at 7 amperes. Unfortunately the regulation of primary circuit current produces significant heat dissipation within the electronic control module. To minimise the possibility of heat damage to the control module, the period during which current regulation occurs must be the shortest period possible consistent with adequate ignition performance. Theoretically it is possible to close the primary circuit and commence ignition coil build-up period (dwell) immediately after the HT spark firing is finished. Maximum dwell period would be available to the ignition coil, however this arrangement is not suitable for use with low inductance, high energy ignition coils. The fast rise time of the coil would require

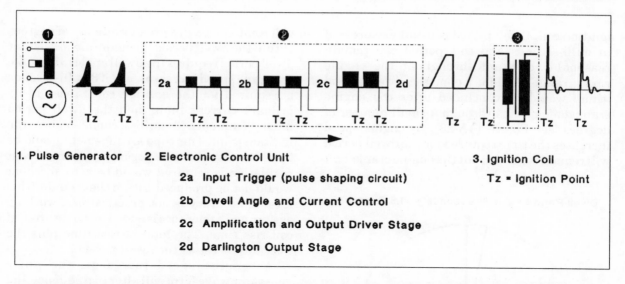

1. Pulse Generator
2. Electronic Control Unit
   2a Input Trigger (pulse shaping circuit)
   2b Dwell Angle and Current Control
   2c Amplification and Output Driver Stage
   2d Darlington Output Stage

3. Ignition Coil
   Tz = Ignition Point

Figure 11A-21 A simplified schematic of how the pulse generator signal is processed through the control module to switch the primary circuit of the ignition coil.

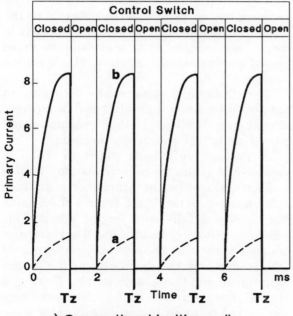

a) Conventional ignition coil

b) Very low inductance ignition coil

TZ = Point of ignition

Figure 11A-22 Comparison of primary current rise time at a spark rate of 7500 sparks per minute (8 cylinder engine).

significant periods of current regulation with consequent heat problems. A dwell angle matched to engine speed would solve this problem, therefore additional electronic circuitry is contained within the control module to provide variable dwell angle.

The change in dwell angle is directed to stage 2c and 2d of the control module (see Figure 11A-21) where the wave form is amplified to act as a driver for the output transistors. The output transistors must be capable of switching the coil primary current on and sustaining the significant amperage required by the coil during the dwell period. The start of primary circuit current flow occurs the exact number of crankshaft degrees (as set in stage 2b of the module) required for the coil to reach maximum energy slightly before ignition occurs. The concept of a relatively constant dwell period (time) and a variable dwell angle is the opposite of the earlier contact breaker systems. The following paragraphs provide an explanation of the relationship between dwell angle, dwell period and a simplified explanation of dwell angle control within the module.

The terms 'dwell period' and 'dwell angle' refer to the time period in which current is flowing through the primary winding of the ignition coil. Dwell period is usually expressed in milli-seconds and this period can remain reasonably constant throughout the engine r.p.m. range. At idle r.p.m. the dwell period (time) would be concluded after the engine crankshaft rotates through a small number of degrees (see Figure 11A-23). As engine speed increases the time required for the dwell period will remain constant, but the engine crankshaft

Dwell Period = 3 milli-seconds = 10.8 Degrees

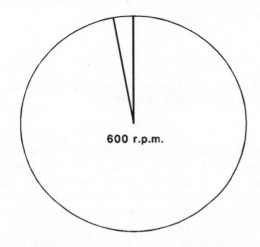

Dwell Period = 3 milli-seconds = 54 Degrees

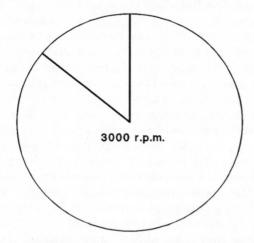

Figure 11A-23 An engine crankshaft will turn through 10.8 degrees in 3 milliseconds at 600 rpm. For the same time period, at 3000 rpm the crankshaft will turn through 54 degrees.

will rotate through a greater number of degrees within the dwell period. To maintain a constant dwell period requires the dwell angle to increase as engine speed increases, e.g. if 3 milli-seconds is required for the ignition coil to rise to its primary current limit then the dwell period must commence 3 milli-seconds before the ignition point. This time arrangement would be ideal because the time period in which current regulation is required would be zero and heat would not be produced within the module. This level of accuracy is not practical in a working system and a tolerance is added to the theoretical coil rise time. The total of rise time plus the tolerance is set as the dwell period.

As engine speed increases, the pulse generator wave form will alter in frequency and amplitude (see Figure 11A-24). The point on the wave form, at which the coil primary current is turned 'off', must remain constant to ensure that ignition timing does not alter. The difference between the 'on' point and the 'off' point on the wave form is the dwell period. The objective of dwell angle control is to maintain the dwell period approximately constant at all engine speeds.

The following paragraphs explain, in simplified form, how the control module provides a changing dwell angle (but constant dwell period) suitable for each engine speed.

A voltage reference point (V ref.) is established within the control module circuitry by a section known as the tachometer and voltage reference generator. 'V ref' is not a fixed value but rises and falls with engine speed. This reference voltage causes the dwell angle control components to alter the input trigger threshold (the point where the module turns the output stage on/off and starts/stops current flowing in the coil primary circuit). As the reference voltage varies, the point on the pulse generator wave form where the input trigger threshold will act also varies.

Part A of Figure 11A-24 illustrates the pulse generator wave form when the engine is idling. The reference voltage is below the trigger threshold. If the engine stops, the pulse generator output will be zero and the control module output stage will be turned 'off'. Current

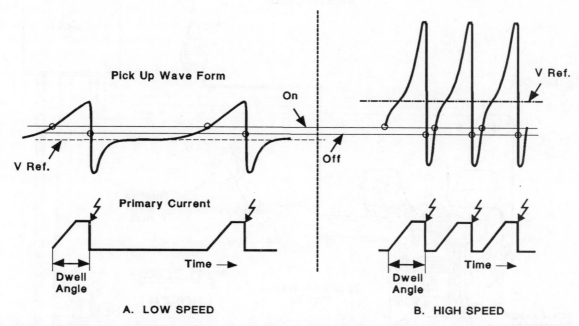

Figure 11A-24 The dwell angle increases with speed enabling a constant dwell period to be maintained.

will not flow through the ignition coil even though the ignition key is in the 'run' position. The lower part of the illustration depicts the coil primary current. The line starts with current increase as the coil current rises to its control limit. The small horizontal section of the line is where regulation of current occurs, followed by a vertical line to represent the drop in current as the output stage turns off and the ignition coil 'fires'. The long horizontal line before the next angled rise indicates the output stage remains off and current is not flowing in the ignition coil primary circuit. By using this method, sufficient time is provided for the coil to reach its maximum energy level (rise time) but the current regulation period is kept to a minimum.

Part B of Figure 11A-24 shows the change to the pulse generator wave form at high engine speed. The voltage reference point is now higher but its effect on the internal circuitry has been to cause the input trigger threshold to act on the lower portion of the wave form. This action has maintained the dwell period (time) at a constant value even though the number of degrees within the dwell period will have increased. The time in which the coil is turned 'off' is considerably reduced as the ignition 'firings' become closer together. The time between one ignition point and the coil primary switching 'on' again is controlled to allow adequate HT spark duration at the spark plug. If this was not done, the optimum dwell period for an 8 cylinder engine, at high engine revolutions, might require the coil to be switched 'on' before the spark duration from the preceding cylinder was concluded.

SYSTEM 2: EST IGNITION COMPUTER, INDUCTIVE TRIGGERING FROM FLYWHEEL

This type of EST system can be fitted to fuel injected engines (see Figure 11A-25) or to carburettor equipped engines. The number, type, construction and appearance of the components used for both systems is almost

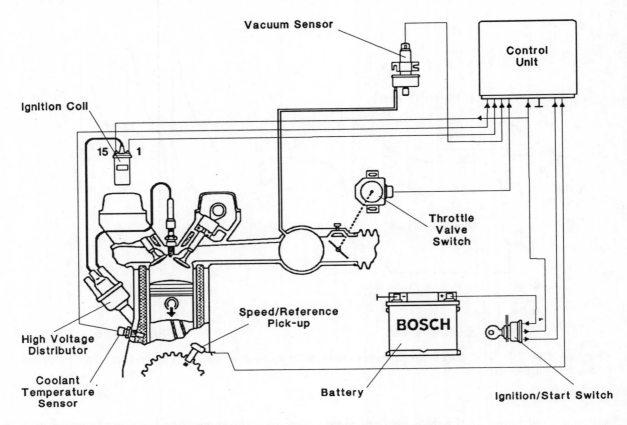

Figure 11A-25 Electronic spark timing systems representation.

identical. The EST systems described later in this chapter are all fitted to EFI engines. Therefore, to show the wide applicability of EST, a carburettor equipped vehicle using an inductive type sensor, triggering from the flywheel, has been selected for this system description.

As shown in Figure 11A-26, the EST computer receives information from the:

- engine speed sensor;
- engine coolant temperature sensor;
- throttle position switch;
- engine vacuum sensor;
- air conditioning switch;
- starter motor.

Based on input from these sensors and its internal programming the EST computer controls the 'earth' switching of the ignition coil. The ignition, or firing point, occurs when the ignition coil primary circuit is 'opened' from

earth. Therefore the EST computer has full control of ignition timing and neither mechanical nor vacuum advance mechanisms are fitted to the distributor. The ignition coil used with this system is a standard high energy type common to many other applications.

DISTRIBUTOR

The distributor fitted to complete the system (see Figure 11A-27), is used to distribute secondary system HT voltages only and is not connected to the primary ignition triggering device. Examination of the distributor shows that contact points, base plate or a condenser are not fitted and the distributor is virtually a hollow shell.

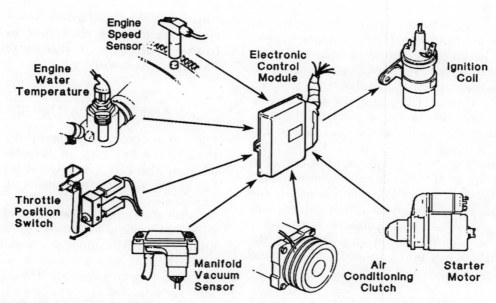

Figure 11A-26 Input sensors are used by the module to control coil primary current.

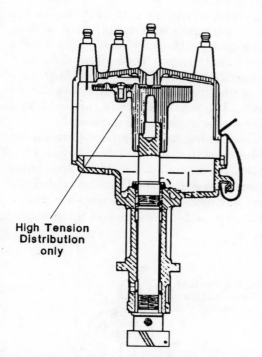

Figure 11A-27 The EST distributor is not fitted with contact points or any other primary circuit triggering device.

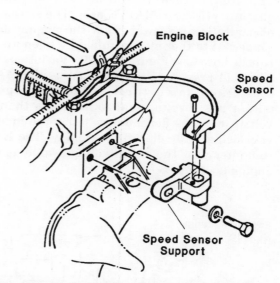

Figure 11A-28 The engine speed sensor is mounted to the rear of the engine block.

SPEED SENSOR

The engine speed sensor (see Figure 11A-28) is mounted at the rear of the engine block and faced down towards the flywheel. The flywheel (or torque convertor), on 6 cylinder engines, is fitted with three steel pins equally spaced around its circumference and adjacent to the starter ring gear. The engine pistons, in an in-line 6 cylinder engine, operate in pairs. When piston No. 1 is at TDC, piston No. 6 will also be at TDC. After 120 degrees of crankshaft rotation , pistons 2 and 5 will be at TDC followed 120 degrees later by pistons 3 and 4. The flywheel steel pins are 120 degrees apart but arranged so they are under the sensor 6 degrees BTDC for each cylinder pair. As each pin passes the inductive speed sensor, an output pulse is sent to the computer. The computer is therefore aware of engine position and by timing the interval between any two pins the engine speed can be calculated.

VACUUM SENSOR

The engine vacuum sensor is a manifold absolute pressure (MAP) sensor. The sensor is normally located away from the engine and connected to the inlet manifold by a length of rubber hose (see Figure 11A-29). Variations in manifold pressure are converted by the MAP sensor into an electrical signal. The sensor output has a voltage range from less than 1 volt up to 5 volts, depending whether the engine is cruising or accelerating strongly. The EST computer uses the signal voltage value as an engine load input.

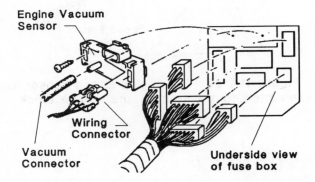

Figure 11A-29 The engine vacuum sensor is located away from the engine.

Caution: The operation voltage of this sensor is 5 volts, therefore never apply higher voltages or permanent sensor damage will occur.

COOLANT TEMPERATURE SENSOR

The engine coolant temperature sensor (see Figure 11A-30) has an important role in advising the EST computer of both high and low engine temperature during idle and drive conditions.

After start-up, a cold engine has substantial frictional loads to overcome. To maintain idle quality, the EST computer acts on information from the temperature sensor and provides additional ignition advance. The quantity of the advance supplied is variable and depends on the temperature at start-up and the time lapsed after start-up. As the engine warms up, the amount of advance is progressively reduced.

If the vehicle is driven with a cold engine the EST computer improves drivability of the vehicle by supplementing the calculated advance point with additional advance. The additional advance is progressively reduced as engine temperature rises and the time since the engine was started increases. The maximum time the additional advance can be present is 8 minutes.

Engine cooling can be improved during hot idle conditions if radiator cooling fan speed is increased. The sensor advises the EST computer of a potential over-heat condition and additional ignition advance, causing an increase in idle speed to be provided until coolant temperature is reduced.

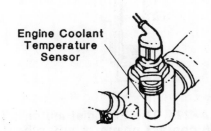

Figure 11A-30 The engine coolant temperature sensor is mounted on top of the engine thermostat housing.

THROTTLE POSITION SWITCH

The throttle position switch is mounted to the carburettor (see Figure 11A-31) and advises the EST computer when the throttle butterfly is closed or open. If the switch is out of adjustment and does not close when the throttle plate shuts, the EST computer may not allow the engine r.p.m. to fall to the correct idle speed.

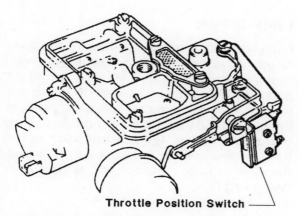

Throttle Position Switch

Figure 11A-31 The throttle position switch is mounted to the carburettor.

CONTROL MODULE (EST COMPUTER)

The 'computer' is a microprocessor-based control module (see Figure 11A-32) which:

- stores the operating program;
- contains data tables for spark advance as a function of speed and load, temperature, throttle switch position and lapsed time after start;
- executes the program stored in memory;
- converts the spark timing signals developed within the microprocessor into signals for controlling coil primary current switching;
- controls coil dwell time, including compensation for varying battery voltage levels.

For every combination of the two basic inputs, engine speed and engine load, there is a unique spark advance value specified. The spark advance information is stored in a table within the module memory. Because the engine must operate efficiently for both highway and urban use, there are two spark information tables within the module memory. Effectively one table (optimum advance table) is for highway use and the second table contains a modified advance to ensure low emission output.

The microprocessor uses the advance tables for every ignition pulse calculation.

Unlike the dual advance distributor ignition module, the EST computer intervenes to control ignition advance at different levels to suit the operating mode of the engine. These modes are defined as:

CRANKING

The starter motor solenoid signals the engine is being cranked. For cranking speed less than 400 r.p.m. the ignition will 'fire' when the flywheel pin causes a pulse from the sensor. The pins are positioned 6 degrees BTDC therefore cranking advance is set at 6 degrees BTDC.

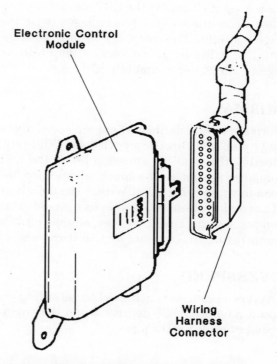

Electronic Control Module

Wiring Harness Connector

Figure 11A-32 EST control module.

IDLE

Idle is defined as any engine speed between 400 r.p.m. and 900 r.p.m. with the throttle position switch closed. The base idle advance is set by the computer at 6 degrees BTDC. This advance setting has been selected as the most suitable value for effective emission control. However, to maintain idle quality and speed, during hot and cold temperatures and loads imposed by power steering, air conditioning or electrical units, requires the use of several compensation values for idle advance.

The effects of hot and cold operation were described in the paragraph related to the engine coolant temperature sensor. Additional advance will be applied if the engine slows due to alternator or power steering load. This compensation is progressively applied if engine speed drops below a preset minimum. The amount of advance applied is limited but should be sufficient to bring the engine idle up to an acceptable speed.

The use of an air conditioner causes additional heat loads as well as increased engine load. A signal is sent to the EST computer when the air conditioner clutch is engaged (with the engine at idle). The computer will increase the advance value by 12 degrees to maintain an engine idle of approximately 850 r.p.m.

NORMAL

Normal mode is defined as engine speed above 900 r.p.m. and throttle switch open. While still meeting the required emission levels the EST computer optimises the spark advance for fuel economy and drivability. Within the definition of normal mode it is possible to have advance compensation caused by low engine coolant temperatures (see coolant temperature sensor).

OVERSPEED

Overspeed control is achieved by retarding the spark advance by 20 degrees when the engine speed exceeds 5800 r.p.m.

DECEL

Decel mode occurs when the accelerator is released closing the throttle (the throttle switch closes) yet the engine r.p.m. is greater than 900 r.p.m. The engine speed and throttle switch position signals advise the computer the vehicle is decelerating . The computer selects a spark advance value which will minimise emission levels during this mode.

LIMP HOME

Limp home is a term used to describe the mode of operation adopted if a mechanical mal-function prevents the system operating normally. If the computer enters the limp home mode a warning lamp on the dash panel is illuminated and the ignition advance drops back to the base setting. Ignition will occur as the flywheel pins pass the inductive pick-up sensor therefore ignition advance will remain at 6 degrees BTDC.

DIAGNOSTICS

A diagnostic link is provided to assist the service technician to isolate any system problems. If a fault develops in the system the EST lamp will illuminate. Under these conditions the engine should be left idling and the diagnostic link (refer to wire to pin 18 on Figure 11A-33) connected to earth by a jumper lead. The link is usually taped to the harness at the rear of the engine. This action will place the computer into a diagnostic mode and the EST warning lamp will flash a coded signal. The lamp will flash at a rate of 1,2 or 3 pulses every 8 seconds indicating a fault/s with the engine coolant temperature sensor, throttle position sensor or vacuum sensor. If the lamp is on continuously in normal mode and does not flash a recognisable code in diagnostic mode it is possible the EST module has sustained damage. The appropriate workshop manual should be referenced for exact determination of the fault as time after start, engine r.p.m. and throttle position can influence the final diagnosis.

To reset the diagnostic lamp for normal

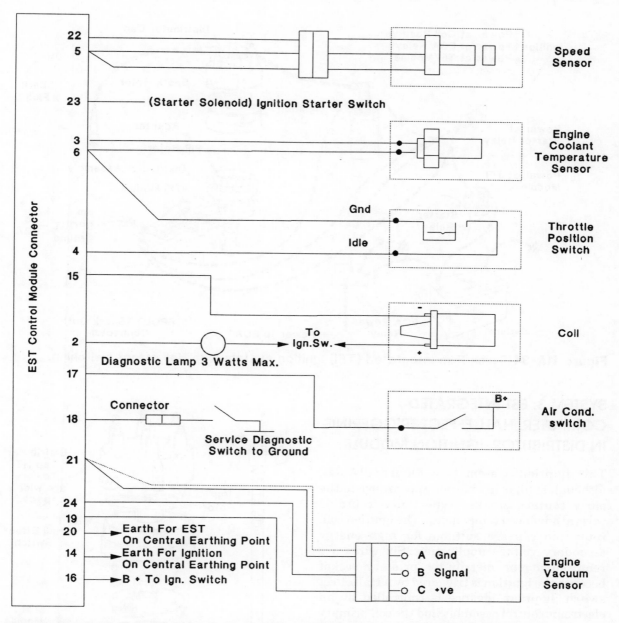

Figure 11A-33 The EST wiring diagram includes a service diagnostic link to ground. When this link is connected to ground, the diagnostic lamp will flash in a coded sequence.

operation, turn off the ignition and remove the jumper lead. Restart the vehicle and if the fault is still present the lamp will illuminate when the engine condition, which caused the original problem, occurs.

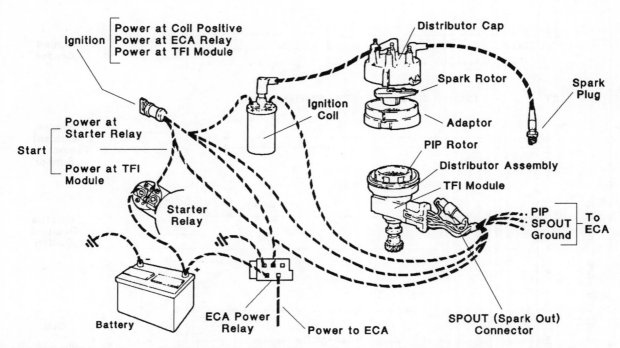

Figure 11A-34 Thick Film Integrated (TFI) ignition system with Hall effect switching.

SYSTEM 3: EST INTEGRATED COMPUTER, HALL EFFECT TRIGGERING IN DISTRIBUTOR, IGNITION MODULE

This ignition system (see Figure 11A-34), although similar in exterior appearance to the early contact breaker type units, contains several advanced components. The ignition coil, rotor and cap are suitable for high energy secondary voltage application. Six wires are required at the distributor primary socket because the ignition is triggered by a Hall effect switch, ignition advance is controlled by an electronic control assembly and the coil primary current is turned 'on' and 'off' by a thick film integrated ignition module.

The distributor, depicted in Figure 11A-35, does not contain mechanical or vacuum advance mechanisms. The distributor is positioned into the engine and set at an ignition timing point. This initial setting is important as all ignition advance values derived from the ECA assume the distributor is correctly installed. The distributor cap and rotor button are designed, to suit high secondary voltages, with wide tower

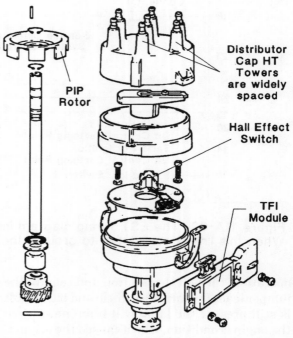

Figure 11A-35 This distributor does not contain mechanical or vacuum controlled ignition advance mechanisms.

spacing and protruding terminals for improved HT lead socketing.

Ignition triggering is based on a Hall effect switch and magnet which are securely mounted to a baseplate assembly. The Hall effect switch is an integrated circuit (IC) which contains very complex circuitry. The Hall voltage produced as a result of the magnetic flux acting on the Hall layer must be amplified, pulse shaped, modified by voltage stabilisation and temperature compensation circuits, before emerging from the IC as the signal voltage. The profile ignition pick-up (PIP) rotor is mounted on, and turns with the distributor drive shaft. The rotor, IC and magnet are often referred to as a Hall generator assembly. (See Chapter 7 for further details of the Hall effect device.)

As the distributor shaft rotates, the PIP rotor blades and windows pass through the Hall effect switch. When a window is at the switch position (see Figure 11A-36 upper) the magnetic flux path from the magnet, through the Hall effect switch, and back to the magnet, is completed. Further shaft rotation places a rotor blade into the gap between the Hall effect switch and the magnet (see Figure 11A-36 lower). The blade provides a shunt for the magnetic flux. The flux lines leave the magnet, pass on to the blade and return to the magnet.

Continuous rotation of the distributor rotor creates an on/off effect of the magnet on the Hall layer of the IC. The Hall IC produces a rising and falling output signal in response to the voltage changes across the Hall layer.

The output voltage signal pulses, which are the signals of interest to the servicing technician, can be viewed very simply for this particular application.

The Hall effect sensor (switch) has three wires attached to it. Figure 11A-37 depicts these wires: power, signal and ground in schematic form. When the rotor blade enters the switch air gap, the signal line voltage rises to the same value as the power line voltage. As the window of the rotor enters the switch; the voltage on the signal line drops to zero.

The on/off voltage signal, known as a PIP signal, produced from the Hall effect switch is shown in Figure 11A-38. The rising edge of the signal is caused by the leading edge of the rotor

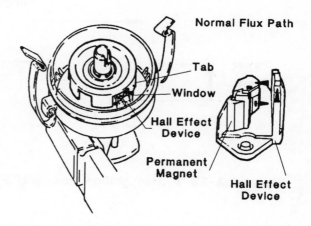

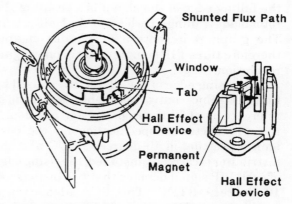

Figure 11A-36 The movement of blades (tabs) and windows between the magnet and the Hall switch cause the switch to turn on and off.

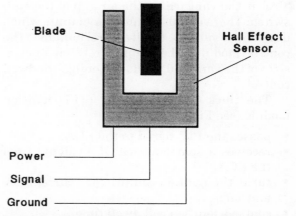

Figure 11A-37 Three wires are attached to the Hall effect sensor.

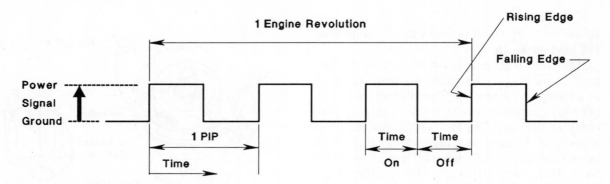

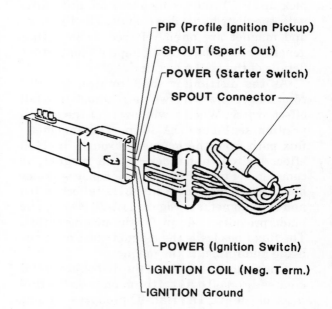

Figure 11A-38 Profile ignition pick-up (PIP) wave form.

blade as it enters the Hall switch. Conversely the falling edge of the signal is a result of the rotor blade trailing edge leaving the Hall switch. The graph represents a 6 cylinder engine, therefore three PIP signals occur in one engine revolution.

The rising edge of the PIP signal is used by the electronic control assembly (ECA) as a reference point for the engine piston position. If the manufacturers specification for the base ignition setting is 10 degrees BTDC and the distributor is correctly installed, the rising edge of each PIP signal denotes that two pistons are 10 degrees BTDC. The ECA also uses a combination of the rising and falling edges of the PIP signal to calculate engine r.p.m.

Note: When checking other types of Hall effect triggered systems, the service technician may find the ignition module will switch the primary circuit 'off' and cause the ignition coil to 'fire' as the vane *leaves* the air gap of the Hall switch. There are also a number of units which do not have variable dwell control built in to the control module. These units rely on uneven sizes of the vane and window openings for dwell control.

The thick film integrated (TFI) ignition module (see Figure 11A-39):

- passes the PIP signal to the ECA;
- receives a spark out (SPOUT) signal from the ECA;
- turns the ignition coil primary current 'on' and 'off';
- controls ignition coil dwell time.

A PIP signal is passed through the TFI

Figure 11A-39 TFI module and connector.

module to the ECA. The ECA, from the rising edge of this signal, is aware two pistons are at 10 degrees BTDC and from the rising and falling edges can calculate engine r.p.m. This information is combined with sensor inputs advising the ECA of the quantity and temperature of the air entering the engine and the engine coolant temperature. The ECA determines the most suitable advance point for the next cylinder 'firing'. At the precise ignition point required, a SPOUT signal is sent to the TFI module. The TFI module opens the coil primary circuit, current flow ceases, the magnetic field in the ignition coil collapses and an HT voltage is routed to the spark plug via the distributor cap and rotor.

If the ECA malfunctions and fails to send a SPOUT signal to the TFI module, the engine can continue to operate. Circuitry within the TFI module can sense the absence of SPOUT signals when it is passing PIP signals to the ECA. If this condition occurs, the TFI module will switch coil primary current 'off' on receipt of a PIP signal from the Hall effect switch. Because the leading edge of the PIP signal is set at 10 degrees BTDC (in this application) the ignition firing point will remain at 10 degrees BTDC until repairs are effected.

For further details of how this type of distributor/ignition system can be linked with electronic fuel injection, see Chapter 11, Part C.

SYSTEM 4: EST INTEGRATED COMPUTER, LED/PHOTO-DIODE TRIGGERING IN DISTRIBUTOR, IGNITION MODULE

It is possible to 'trigger' an ignition system by building a light sensing device into the location normally occupied by the contact points. A rotor plate (see Figure 11A-40), containing a series of slits, is rotated past a light emitting diode (LED). The light from the LED is aimed at a photo-diode. Rotation of the rotor plate will allow light from the LED to reach the photo-diode through a slit before being blocked off as the slit moves away from the LED. An on/off signal is produced at the photo-diode which is then fed to a wave forming circuit. The wave output can be used to 'trigger' an ignition coil. A number of 'after market' ignition systems use this method as an alternative to contact breaker points and the LED assembly is fitted within a conventional distributor. This specific system has two LED photo-diode pairs which produce reference signals for both cylinder position and engine speed. The wave forming circuit sends these signals to an electronic spark timing computer.

The EST computer sends two signals back to a power transistor (see Figure 11A-41) fitted to the side of the distributor. One signal turns the transistor 'on', allowing current to flow in the ignition coil primary circuit. The second signal turns the transistor 'off' stopping primary current flow, the coil magnetic field collapses and an HT spark appears at the ignition coil tower. The ignition coil is a moulded type and is connected to the distributor cap by an HT wire in a conventional manner.

The power transistor has a socket connection for two terminals (see Figure 11A-42). The base 'B' terminal receives the on/off signals from the computer. The 'C' terminal is in series with the

Figure 11A-40 A rotor plate containing a series of slits is rotated between a light emitting diode and a photo diode.

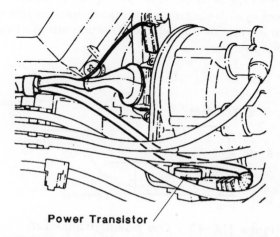

Power Transistor

Figure 11A-41 A power transistor mounted to the distributor is used to turn primary current ON and OFF.

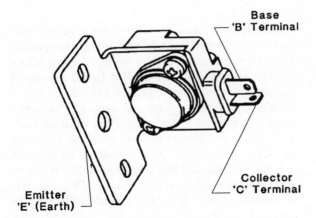

Figure 11A-42 The power transistor has three connection points; B (base), C (collector) and E (emitter - earth through the mounting bracket).

negative side of the ignition coil primary winding. The emitter 'E' connection is to earth through the mounting bracket of the assembly.

The power transistor can be represented schematically as a single ordinary transistor (see Figure 11A-43). When signal current is applied to the base 'B', the coil primary circuit current will flow from the collector 'C' to earth through the emitter 'E'. The internal construction of this transistor is not a single unit, therefore consult the manufacturer's service manual for the correct test method.

Further details of system operation and how this type of distributor/ignition system can be linked with electronic fuel injection are included in the engine management section later in this chapter.

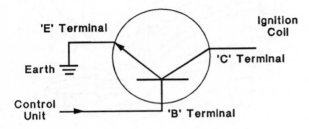

Figure 11A-43 The power transistor, depicted as a single transistor, is more complex and must be tested in accordance with manufacturers procedures.

SYSTEM 5: EST INTEGRATED COMPUTER, HALL EFFECT TRIGGERING FROM CRANKSHAFT SENSORS, DFI MODULE, NO DISTRIBUTOR

All the previous ignition systems had one item in common: they contained a distributor. A number of systems are now available that do not use a distributor for either crankshaft position sensing or HT distribution. From your study of previous systems you will be aware that an ignition system requires a:

- primary circuit 'trigger';
- primary circuit switch for on/off;
- timing reference point on which to base the ignition advance;
- HT distribution system.

The following paragraphs will explain the basics of the system; for a more detailed system explanation see Chapter 11, Part C.

A crankshaft sensor (see Figure 11A-44) containing two Hall effect sensors is mounted onto a bracket.

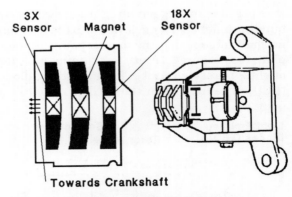

Figure 11A-44 Crankshaft sensor.

The sensor assembly is mounted behind the crankshaft balancer (see Figure 11A-45). Two interrupter rings are fitted to the balancer and are positioned so the vanes pass between the Hall sensors and the magnet. Rotation of the interrupter rings through the Hall sensors cause two signals to be sent to the direct fire ignition. (DFI) module. The DFI module is linked to an electronic spark timing function within the

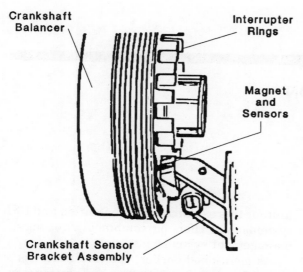

Figure 11A-45 The crankshaft sensor is mounted behind the crankshaft balancer.

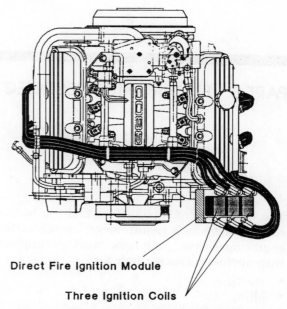

Figure 11A-46 Three 'double-ended' ignition coils are mounted onto the DFI (Direct Fire Ignition) module.

engine management computer. From the signals sent via the DFI module (and other sensor inputs), the computer calculates the required ignition advance and sends a primary circuit switching signal back to the DFI module.

Three ignition coils are mounted on top of the DFI module (see Figure 11A-46). These coils are different to a conventional ignition coil. The secondary winding is not connected at one end to earth. Each end of the secondary winding terminates in an HT outlet socket. A single ignition coil has two HT leads, each of which is connected to a spark plug in different cylinders. Three coils, each with two outlets, supply HT voltage to six spark plugs. When the computer sends a signal to switch 'off' a primary circuit, the DFI module remembers which of the three coils is the correct unit to use for the current cylinder positions. The coil sequencing order was provided from one of the Hall sensor signals.

When the coil 'fires', the HT voltage is sent to a forward firing spark plug, where it passes across the spark gap to the engine block. At the second, or reverse firing, spark plug the HT voltage appears at the earth electrode, 'jumps' the gap back to the centre electrode, passes through the HT lead and enters the ignition coil at the opposite end to the original spark. The HT voltage path can be compared to a series circuit with current flow always flowing in the same direction.

Only one engine cylinder will be at a suitable position for ignition when one coil discharges to two spark plugs, therefore one spark plug firing is wasted. All six spark plugs will fire each engine revolution but only three will be effective in igniting a combustible mixture.

Further details of system operation and how this type of distributorless ignition system can be linked with electronic fuel injection are included in the engine management section, Chapter 11, Part C.

PART B ELECTRONIC FUEL INJECTION

The basic aim of mixing fuel and air in the correct ratios is a requirement for successful engine operation. An engine must be supplied with air/fuel ratios suitable for:

* starting;
* idling;
* part throttle load;
* full throttle load;
* cruising;
* deceleration.

The conditions listed above will occur during cold start, warm-up and hot engine operations. Therefore the air/fuel ratio must be varied for both the operating condition and the engine temperature.

Improvements in carburettor design, aided by 'add on' emission control devices, extended the use of carburettors into the 1980s. During this period of gradual tightening of permissible exhaust emission levels, electronic fuel injection systems became more compact in size and found acceptance in mass production vehicles. The updating of EPA emission regulations reduced the emission levels permitted from vehicle exhaust systems. To comply with the regulations, very accurate metering of air/fuel ratios, beyond the capabilities of most carburettors, is required.

Electronic fuel injection is only part of the answer to efficient engine operation and permissible exhaust emission levels. The EFI is normally coupled into a total engine management control system when fitted to current production vehicles. However, to avoid the complexity inherent in the total management system, the EFI will be treated as a separate unit for this explanation. The ignition and EFI systems will then be combined into engine management systems.

As mentioned earlier, the carburettor is a device for metering fuel. The carburettor uses several different systems to achieve its aims and can deliver fuel from different supply outlets, e.g. idle port, accelerator pump discharge nozzle and the main jet discharge. In simple terms, as the fuel requirement altered, the number and source of the fuel ports supplying fuel could be altered. This not the case in an EFI system; all fuel is supplied from the nozzle opening of the injector. The size of the nozzle opening is fixed and the fuel supply pressure is maintained at a fixed value above manifold pressure. Given these restraints on nozzle opening size and fuel delivery pressure, the only variables available to alter the amount of fuel delivered are the amount of time the injector is held open and the frequency (how many degrees of engine crankshaft rotation between openings) at which the injector is opened.

As you progress through the various parts of the system, and learn about the components, it is worth remembering that the processing unit and sensors are only fitted to provide information that will determine:

* how long the injector will remain open to inject fuel;
* the frequency at which the injector will operate.

The EFI system will have basic components (e.g. pumps, ducts and lines) to provide fuel and air to the engine and electronic components to monitor and control the system.

For ease of understanding, these components are grouped into three sub-systems:

- fuel system;
- air supply system; and
- electronic control system.

The system chosen to explain EFI operation is a model made up of components from each of the L-Jetronic, LE-Jetronic and Nippondenso EFI systems.

FUEL SYSTEM

The components of this subsystem (see Figure 11B-1) include the following:

- injector;
- cold start valve;
- fuel supply rail;
- pressure regulator;
- fuel filter;
- fuel pulsation damper;
- fuel pump;
- fuel lines;
- fuel tank.

INJECTOR (FUEL-INJECTION VALVE)

FUNCTION

To spray fuel into the inlet manifold of the engine.

LOCATION

The injector/s are located in different positions in the inlet manifold, depending on the type of EFI in use.

Throttle body injection (TBI) or centre point injection (CPI) systems have the injectors housed

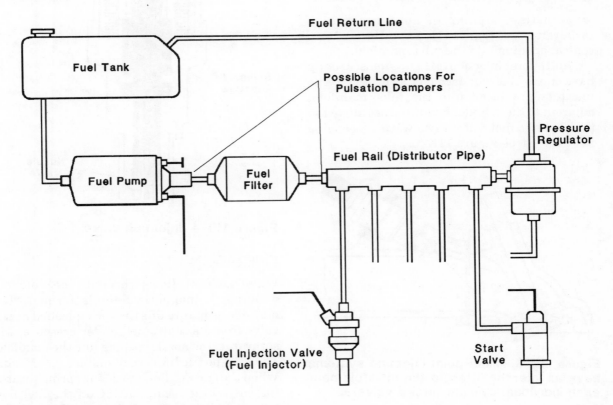

Figure 11B-1 Block diagram of the fuel system.

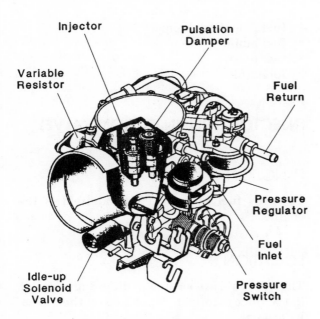

Figure 11B-2 Injection mixer of a type used with central fuel injection.

in a casting, similar in appearance to a carburettor (see Figure 11B-2), and located on a central opening into the inlet manifold.

Multi-point injection (MPI) systems, usually have one injector for each engine cylinder. The injectors are fitted into the inlet manifold, adjacent to the cylinder head, and are angled so that injected fuel will spray towards the engine inlet valve (see Figure 11B-3).

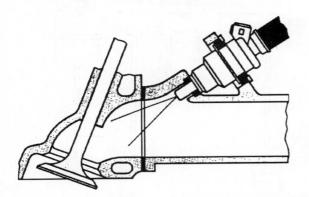

Figure 11B-3 Multi-point injection systems have an injector fitted to the manifold near each individual cylinder intake valve/s.

CONSTRUCTION

The injector (see Figure 11B-4) consists of a valve body: the lower section contains a fuel opening, nozzle valve, valve sealing seat and nozzle valve guide. The upper section houses; the solenoid winding with its moving armature attached to the nozzle valve, the nozzle valve seating spring, and external points for fuel and electrical supply connections.

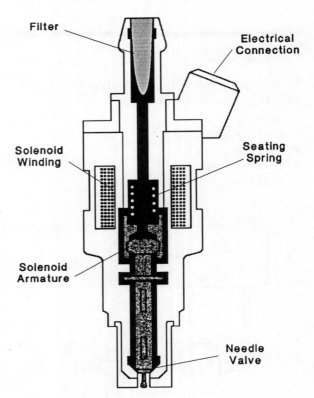

Figure 11B-4 Injection valve.

OPERATION

When current flows through the solenoid winding, a magnetic field is formed. The movable armature lifts the spring-loaded nozzle valve from its seat. Fuel, under pressure, will spray from the nozzle opening into the manifold. The total nozzle lift is very small, e.g. 0.15 mm. When the current flow stops, the spring returns the valve to its seat, cutting off the fuel flow from the injector. The timing and the period of injection is controlled by the electronic control unit (ECU) switching the current on and off in the injector circuit.

Note: Two versions of current control to the injector are in use. To achieve fast mechanical response, the injector solenoid coils are relatively few in number and have low inductance. Current flow must be accurately controlled to prevent 'burning out' of the solenoid coil. To prevent this problem occurring, one method is to fit a 'dropping' resistor in series with each injector coil (see Figure 11B-5). The combination of a resistor and an inductive coil will prevent excess current flow.

Caution: 12 volts must never be connected directly to any injector in this type of system, as damage will occur to the injector.

The second method of current control is obtained through additional components within the ECU and does not require the use of 'dropping' resistors. The injector valves switch on with high current but immediately the valve opens, the current is reduced to a 'holding' current by the ECU current regulation process. The substantial reduction in current enables the use of an ECU which can operate more injectors from a single final stage.

COLD START VALVE

FUNCTION

To enrich the intake air/fuel mixture during cold engine cranking.

LOCATION

This valve is not needed in all countries, therefore it is possible the valve has been deleted after suitable redesign of the control unit. When fitted, it is located after the throttle body, at a central point in the inlet manifold.

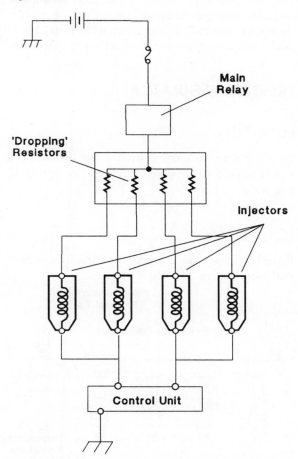

Figure 11B-5 Fitting a solenoid 'dropping' resistor in series with the solenoid coil will prevent excess current flow.

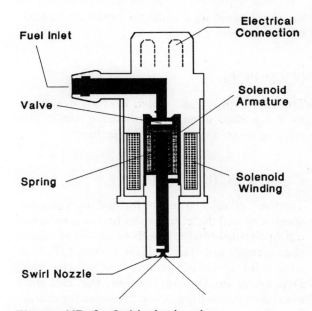

Figure 11B-6 Cold start valve.

CONSTRUCTION

The valve (see Figure 11B-6) consists of a body containing a solenoid winding, moving armature and valve, spring, valve seat, swirl nozzle and connections for fuel and electrical supply.

OPERATION

When current flows through the solenoid winding a magnetic field is formed. The armature moves the spring-loaded valve from its seat. Fuel, under pressure, passes down the sides of the armature to the nozzle. The nozzle design causes the fuel to swirl, thus ensuring the fuel leaves the nozzle in a finely atomised form. When the current flow stops, the spring returns the valve to its seat cutting off the fuel flow from the injector. The injection time is controlled by the thermo-time switch.

FUEL SUPPLY RAIL (DISTRIBUTOR PIPE)

FUNCTION

To supply fuel at a uniform pressure to each injector.

LOCATION

The fuel supply rail, which is often described as a manifold or pipe, is mounted across the inlet connections of the fuel injectors.

CONSTRUCTION

The fuel rail may be cast or formed from tubular steel. The rail distributes the fuel to a series of outlet fittings (one for each cylinder) to which the injectors are attached (see Figure 11B-7). A fuel inlet is fitted at one end of the rail. Depending on the application, the rail may also have connections for a pressure regulator, fuel pressure test point and a cold start valve fuel supply.

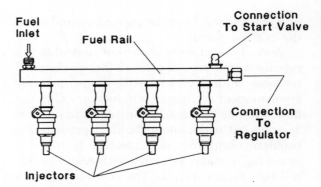

Figure 11B-7 Fuel supply rail.

OPERATION

The fuel pressure at individual injectors must remain uniform during an injection cycle, therefore the rail storage capacity must be of adequate size to prevent variation in pressure.

PRESSURE REGULATOR

FUNCTION

To control the pressure within the fuel supply rail to a fixed value above engine inlet manifold pressure.

LOCATION

Usually fitted to a connection on the fuel supply rail (see Figure 11B-8).

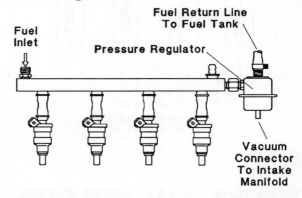

Figure 11B-8 Fuel supply rail fitted with a pressure regulator.

CONSTRUCTION

The regulator consists of a pressed metal housing divided into two compartments by a flexible diaphragm (see Figure 11B-9).

One compartment contains a pipe connection and a coil spring which exerts a force on the diaphragm. This compartment is connected to the engine inlet manifold by a rubberised hose. The other compartment has two connection points, one from the fuel rail (inlet) and the other to the excess fuel return line (outlet). The fuel outlet fitting extends into the regulator to form a seat for the valve plate and holder which are attached to the diaphragm.

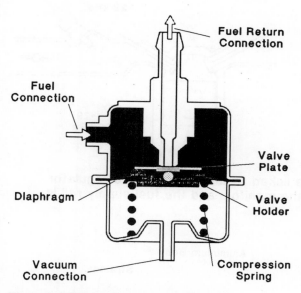

Figure 11B-9 Pressure regulator.

OPERATION

The regulator operates on an excess pressure overflow concept. When the fuel rail pressure overcomes the diaphragm spring tension, the diaphragm moves, allowing fuel to flow across the valve plate into the excess fuel return line. Returning fuel to the tank in this way ensures the system will be continually flushed and cool fuel will be supplied to the injectors.

Unfortunately this simple process would keep the fuel pressure at a set value above atmospheric pressure. If the injection system operated in this way the quantity of fuel sprayed from the injector, for the same opening time, can be slightly different depending on whether the manifold vacuum is high or low.

To solve this problem the spring pressure side of the pressure regulator diaphragm is connected by a tube to the intake manifold (see Figure 11B-10). As the intake manifold vacuum changes, the fuel pressure set by the regulator will also change.

Figure 11B-11 shows a pressure difference of 250 kPa is maintained between the injector fuel pressure and the inlet manifold absolute pressure. Therefore regulated pressure is not a constant but will vary to ensure that the fuel quantity delivered from the injector is proportional to the time the injector is open.

Pressure regulator operation, during hot engine restart conditions, has been modified on a number of vehicles. As the ignition key is turned, fuel pressure in the fuel rail is raised by allowing atmospheric pressure to enter the vacuum line between the inlet manifold and the pressure regulator. The increased fuel pressure reduces vaporisation at the injectors; engine operation after restart is improved. After a short time, pressure regulator operation is returned to normal.

FUEL FILTER

FUNCTION

To filter impurities from the fuel.

LOCATION

Fitted to the fuel supply line between the fuel pump and the fuel supply rail. The actual location on the vehicle can vary from adjacent to the fuel tank up to the engine compartment.

CONSTRUCTION

Porous paper is pleated, held by a support plate and housed in a cylindrical metal (aluminium) body (see Figure 11B-12). The filter

Fuel Pressure Pre-set For Atmospheric Pressure

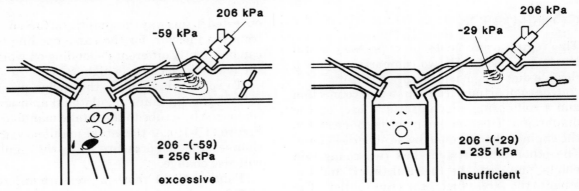

206 kPa
-59 kPa

206 -(-59)
= 256 kPa

excessive

206 kPa
-29 kPa

206 -(-29)
= 235 kPa

insufficient

Fuel Pressure Regulated By Manifold Vacuum

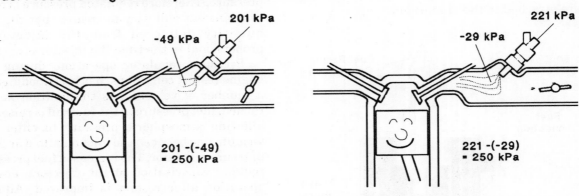

201 kPa
-49 kPa

201 -(-49)
= 250 kPa

221 kPa
-29 kPa

221 -(-29)
= 250 kPa

Figure 11B-10 The manifold vacuum must be linked to the fuel pressure regulator to ensure the pressure difference between the manifold and the fuel supply rail is maintained at a constant.

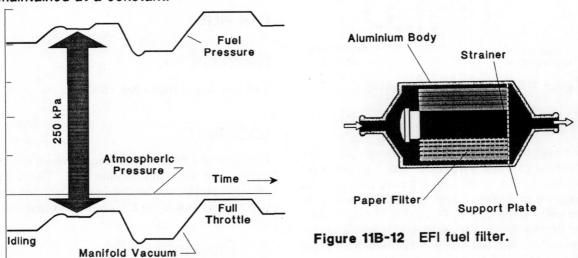

Fuel Pressure

250 kPa

Atmospheric Pressure

Time →

Full Throttle

Idling

Manifold Vacuum

Figure 11B-11 Fuel pressure is maintained at 250 kPa above intake manifold absolute pressure (usually referred to as manifold vacuum by mechanics).

Aluminium Body

Strainer

Paper Filter

Support Plate

Figure 11B-12 EFI fuel filter.

also contains a strainer at the outlet end of the filter material. Inlet and outlet pipe connections are located on the ends of the housing.

OPERATION

The filter must be installed with the indicator arrow, stamped on the casing, pointing in the direction of fuel flow. Fuel flows into the area surrounding the porous paper, passes through the paper and exits through the outlet connection. A strainer is fitted across the exit end of the filter material to trap any particles of the porous paper that may become dislodged in service.

TECHNICAL DATA

- Filter paper pore size—approximately 10 micro metre.
- Service life—varies between 30 000–80 000 km.

FUEL PULSATION DAMPER

FUNCTION

Reduce fuel pulsations in the fuel supply lines.

LOCATION

The damper/s can be located at the fuel pump outlet, on the end of the fuel supply rail, attached to, or in some applications be part of, the fuel supply line between the pump and fuel rail.

CONSTRUCTION

A metal housing containing two chambers (see Figure 11B-13). The air chamber contains a compression spring which forces against a flexible diaphragm. The diaphragm acts as a separator between the air chamber and the fuel chamber.

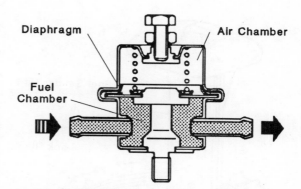

Figure 11B-13   Fuel pulsation damper.

OPERATION

Fuel pressure variations can be caused by the opening and closing of the fuel injectors and fuel pump action. As the pressure wave (pulsation) enters the damper, the damper diaphragm is forced against the compression spring. The small increase in the volume of the damper fuel chamber reduces the pulsation and pressure variation is kept to a minimum.

Fuel injection systems, which do not have separate pulsation dampers, normally contain a section of flexible fuel line. This line expands and contracts (changing internal line volume) to provide a dampening action for fuel pressure variations.

FUEL PUMP

FUNCTION

To supply, at the required working pressure, the quantity of fuel needed by the engine under all operating conditions.

LOCATION

The pump location can vary depending on the system type chosen by the manufacturer.

1 High pressure pump mounted to the vehicle body near the fuel tank.
2 High pressure pump mounted inside the fuel tank (see Figure 11B-14).

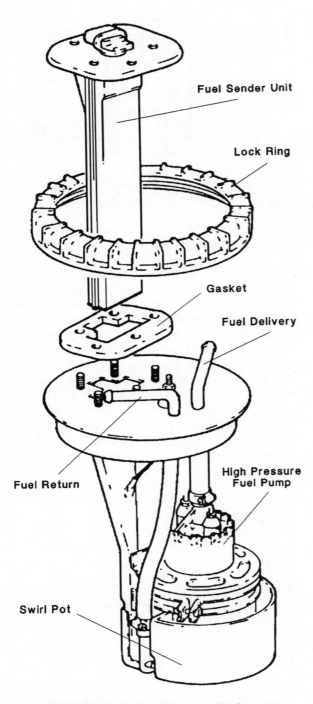

Fuel Sender Unit

Lock Ring

Gasket

Fuel Delivery

Fuel Return

High Pressure Fuel Pump

Swirl Pot

Figure 11B-14 High pressure fuel pump mounted inside the fuel tank.

3 High pressure pump mounted external to the fuel tank but supplied with fuel by a low pressure pump mounted inside the fuel tank (see Figure 11B-15).

CONSTRUCTION

The pump is normally a permanent magnet, electric motor type in which the rotating armature, brushes and pumping chamber are permanently submerged in fuel (see Figure 11B-16). The fuel ensures the pump is cooled and lubricated. The pump is always full of fuel, therefore an ignitable mixture is never present and there is no danger of explosion. The pump also contains a pressure limiter, non-return check valve and a pumping chamber within a common housing. The pumping chamber will be one of two basic types—roller cell or fluted impeller.

Roller cell: A set of rollers are contained in a series of recesses cut into a rotor (see Figure 11B-17). The rotor is fitted into a hollow cylindrical chamber which contains both the inlet and outlet recesses and ports. The mounting spindle for the rotor is not positioned exactly in the centre of the housing therefore the rotor is closer to one side of the chamber, i.e. eccentrically fitted within the cylindrical chamber.

Impeller: A fluted rotor is centrally mounted within a hollow cylindrical chamber which contains the inlet and outlet ports and recesses (see Figure 11B-18). A close tolerance is maintained between the rotor tips and the chamber wall.

OPERATION

A low pressure area is formed at the pump intake by the pumping elements. Fuel passes through the pump chamber into the motor section of the pump, across the armature and exits the pump through a non-return check valve fitted in the outlet connection (see Figure 11B-19).

The check-valve (residual line pressure valve) helps to prevent fuel vaporisation during hot engine shutdown periods by maintaining

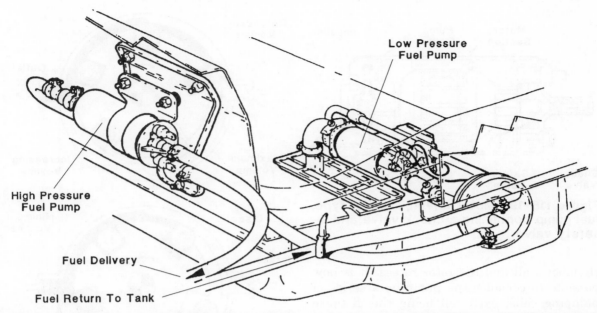

Figure 11B-15 A combination of a low pressure (inside tank) pump and a high pressure (externally mounted) fuel pump.

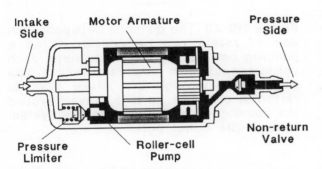

Figure 11B-16 Roller-cell type fuel pump.

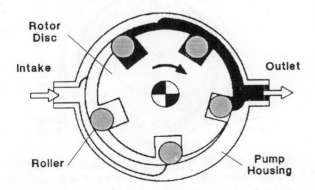

Figure 11B-17 The rotor is not mounted centrally in the pump housing.

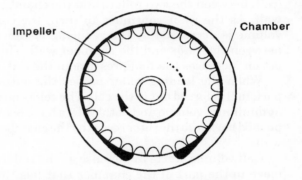

Figure 11B-18 The pumping element consists of a fluted impeller (rotor) spinning inside a chamber.

pressure in the fuel line. The safety valve (internal pressure limiter) prevents excessive fuel pressure in the event of a fuel flow restriction, e.g. accident damage which has crushed the return line to the fuel tank.

The operation described in the preceding paragraph applies to most pumps in common use, however, the pumping element may be of either roller cell or fluted impeller design.

Roller cell: When the rotor spins, the rollers are forced outward against the wall of the chamber. Each roller forms a seal between the

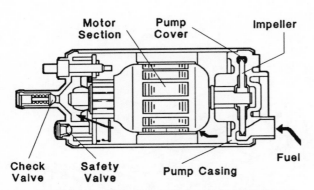

Figure 11B-19 Motor driven, impeller type, fuel pump containing a check valve and a safety valve.

chamber wall and the rotor recess. It is now possible to consider the pump as a series of pumping cells, each cell being the distance between one roller and the following roller.

From the illustration (see Figure 11B-20) it can be seen that inlet fuel is trapped in the gap (cell) between the two rollers and the chamber wall. As the rotor continues to turn, the cell volume is decreased by the reducing distance between the rotor and the chamber wall. This action increases the fuel pressure in the cell.

When the leading roller passes the outlet port, fuel is forced out of the cell. The cell is now at minimum volume and both rollers have been pushed back into the rotor recess by the chamber wall.

Cell volume starts to increase as the rollers move to the part of the chamber that has the greatest gap between the chamber wall and the rotor. As the cell volume increases, the pressure drops to less than atmospheric pressure. When the leading roller passes the inlet port, fuel from the tank will enter the cell and the action is repeated.

This action is occurring in each cell in rotation and fuel is delivered in rapid pulses to the main body of the pump.

Fluted impeller: The fluted impeller runs with close tolerance within a cylindrical pumping chamber. As the individual flutes pass the intake port, a low pressure area is formed. Fuel enters the flute cavities and is carried around the pumping chamber by the impeller (see Figure 11B-21). Fuel on the pumping chamber wall

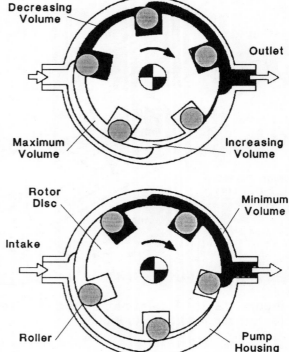

Figure 11B-20 The distance between each pair of rollers can be considered as a pumping cell. As the rotor disc turns each cell increases, then reduces in volume. The fuel enters through the intake port, is raised in pressure as it is moved around the housing, then delivered to the system through the outlet port.

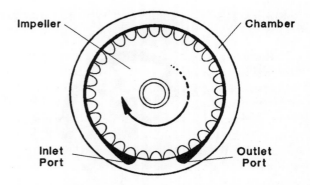

Figure 11B-21 The impeller operates with close tolerance to the chamber walls. Fuel enters at the inlet port and is carried by the impeller round the chamber to the outlet port.

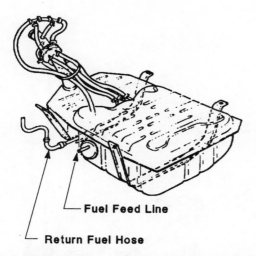

Figure 11B-22 Fuel tank with return line fitting for EFI system.

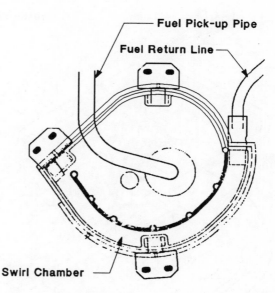

Figure 11B-23 A swirl pot, fitted inside the fuel tank, provides a reservoir of fuel to prevent the fuel pick-up pipe becoming uncovered during cornering.

causes a frictional drag (opposing force) on the impeller and the fuel in the flute cavities. The force provided by the rotational speed of the impeller raises the pressure of the fuel. When the flutes pass the outlet port, fuel is discharged, under pressure, into the main body of the pump.

FUEL TANK

The fuel tank is of conventional metal or plastic design (see Figure 11B-22). The changes required for fuel injection applications are minimal and relate to the pick-up and return of fuel.

A number of manufacturers fit a swirl pot to the fuel pick-up pipe (see Figure 11B-23). This will prevent the fuel pick-up becoming uncovered during cornering or when the fuel level is low. Fuel bypassed at the pressure regulator can flow via a return pipe into the fuel tank. The returning fuel is routed into the swirl pot where the swirling action reduces the possibility of aeration of the fuel.

AIR SUPPLY SYSTEM

The air supply system (see Figure 11B-24) contains an:

- air cleaner;
- air flow meter (or similar device);
- air ducting;
- throttle body;
- air bypass valve (auxiliary air/BAC/etc.);
- surge tank;
- intake manifold.

AIR CLEANER

FUNCTION

To filter inlet air.

LOCATION

The air cleaner is located in the engine compartment, usually close to the front of the

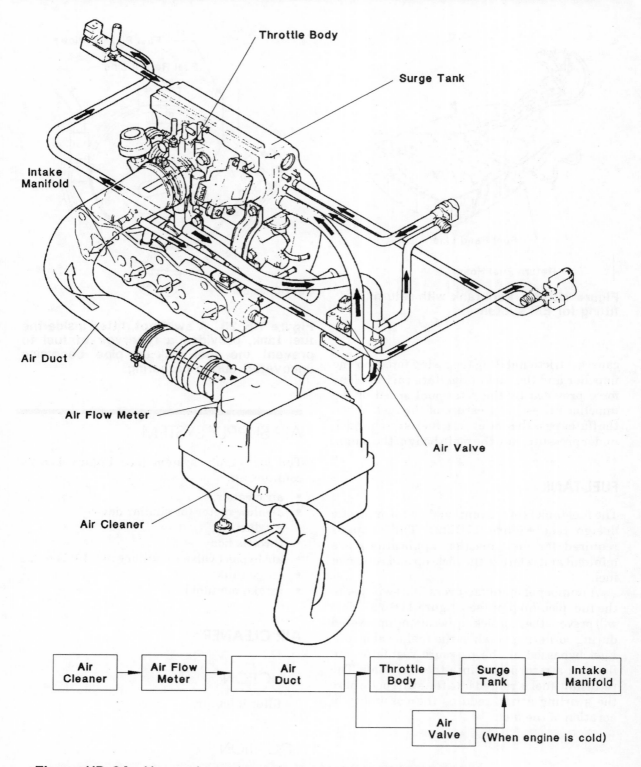

Figure 11B-24 Air supply system of a typical fuel injected engine.

vehicle with the air intake fitted behind the headlamp area. This is not always possible because of engine size or component location, however ducting is often used to route air from the front of the vehicle to the air filter body.

CONSTRUCTION

The industry has tended to favour rectangular box designs manufactured from a plastic material (see Figure 11B-25). The 'box' consists of an entry and an exit chamber separated by a filter. The filtering material is made from pleated paper, packaged as a 'throw away' element.

OPERATION

As air is routed to the air cleaner entry chamber, it passes through ducting designed to reduce induction noise. The air cleaner filtering element extracts dust particles from the air. The clean air then flows out the exit chamber to the air vane meter (or similar device).

AIR FLOW METER (AIR FLOW SENSOR)

Several different types of sensors are in common use. Each sensor type provides the computer with similar information (the quantity of air entering the engine) but their construction and

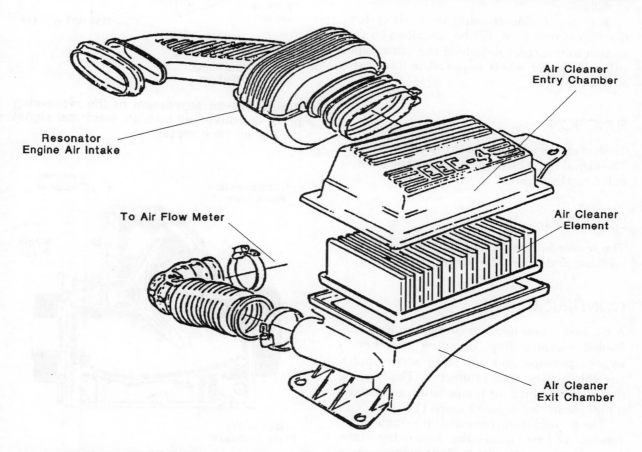

Figure 11B-25 Rectangular air cleaner assembly with inlet and outlet ducting.

operating principle vary considerably. A number of systems have deleted the air flow sensor in favour of a different device called a manifold absolute pressure (MAP) sensor. MAP sensors are linked to the inlet manifold and provide the computer with signals derived from manifold vacuum.

Intake air flow signals, sent to the computer, are obtained from:

- flap movement (vane air meter).;
- current flow through a wire (LH System).;
- air turbulence (Karman vortex type where air vortices are detected by ultrasonic waves).;
- frequency changes derived from a MAP sensor.

For ease of understanding the basic system, the flap (vane) type will be described in this section and further details of the other types will be included, where required, in later parts of this chapter.

FUNCTION

To send an electrical signal to the EFI computer. The signal must be proportional to the air flow entering the engine.

LOCATION

The sensor is fitted in the inlet duct after the air filter and before the throttle body.

CONSTRUCTION

A die cast metal housing containing a spring-loaded movable flap, damping chamber, a bypass passage and adjusting screw and a variable resistor (potentiometer). The housing may also contain a air temperature sensor and a fuel pump switch (see Figure 11B-26).

The movable flap, mounted on a single pivot, consists of two plates—the measuring plate, located in the air stream, and the compensation plate, which is positioned in a damping chamber (see Figure 11B-27). The wiper arm of the potentiometer is rigidly attached to the flap pivot.

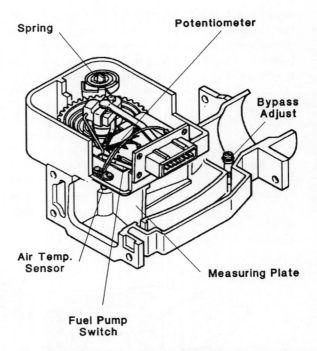

Figure 11B-26 Movement of the measuring plate is converted into an electrical signal by the air flow meter.

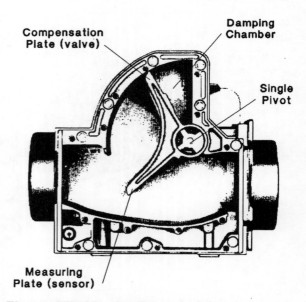

Figure 11B-27 Air-flow sensor.

OPERATION

Air passing into the engine causes both flaps to deflect in response to the air flow. Sudden movement of the accelerator will cause a rapid change in the air flow and the air flow meter could over-respond. Under these conditions, the compensation plate moves in the damping chamber. The air in the damping chamber acts as a buffer and the measuring plate moves smoothly in response to air flow changes (see Figure 11B-28). The computer is advised of a change in air flow by a varying voltage signal from the potentiometer.

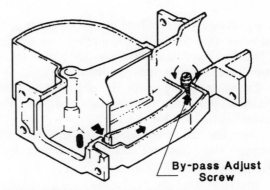

Figure 11B-29 The air-fuel ratio can be changed by altering the position of the bypass adjust screw.

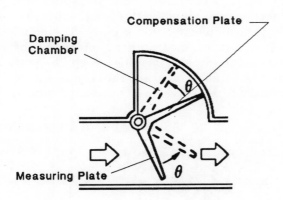

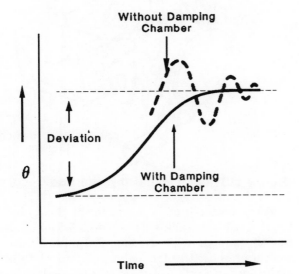

Figure 11B-28 Over-response of the measuring plate is prevented by movement of the compensation plate into the damping chamber.

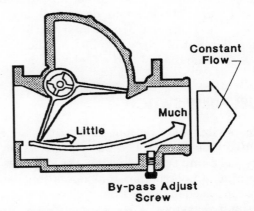

Figure 11B-30 Rotating the screw out of the passage increases bypass air and will lean the mixture.

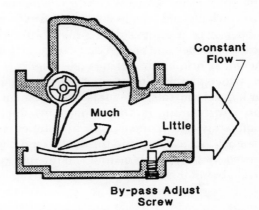

Figure 11B-31 Rotating the screw into the passage decreases bypass air and will richen the mixture.

Air flowing through the bypass passage is not sensed by the measuring plate (see Figure 11B-29). This allows the air-fuel ratio to be adjusted for engine idle conditions.

Turning the adjusting screw counter-clockwise will open the passage and 'lean' the air-fuel ratio (see Figure 11B-30).

Turning the screw clockwise restricts the passage and 'enriches' the air-fuel ratio (see Figure 11B-31).

AIR TEMPERATURE SENSOR

This unit is contained within the air flow sensor (see Figure 11B-26) and detects changes in inlet air temperature. Construction and operation details are contained in the section on electronic control of the fuel system.

FUEL PUMP SWITCH

This switch may be mounted in the air flow sensor (see Figure 11B-26) and will prevent the fuel pump operating when the engine is not running. Construction and operation details are contained in the section on electronic control of the fuel system.

AIR DUCTING

The air ducting connects the air cleaner to the throttle body and may be constructed of formed rubber or be a composite of a metal tube with flexible rubber end connections. The duct may include a number of fittings (see Figure 11B-24) which enable filtered air to be obtained for air bypass controls. The major requirement for the serviceman is to ensure the duct is completely sealed and outside air cannot enter the duct. Air leaking into the duct would not be sensed by the computer and the air-fuel ratio would be weakened.

THROTTLE BODY

The basic throttle body is constructed of aluminium alloy and contains a throttle valve controlled by the driver through a cable or linkage from the accelerator pedal. The operation of the throttle body can be compared to the lower section of a carburettor. The quantity of air entering the engine is controlled by the angle to which the throttle valve is opened.

The external appearance of the throttle body (see Figure 11B-32) makes the unit appear complex, however the major functions are as previously explained. Many pipes and connections are routed to this location because it is a convenient mounting point. The small pipes are vacuum lines to other components, the dashpot is simply a cushioning device when the throttle is rapidly closed and the large pipes are connected to an air valve or bypass air control valve.

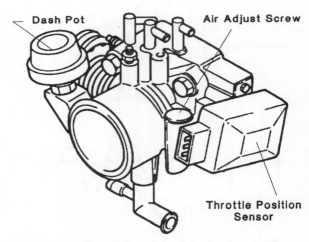

Figure 11B-32 Typical throttle body.

A complex switch is connected to the end of the throttle valve shaft. This switch is known as the throttle position sensor and advises the computer of the throttle valve position. Details of the operation of the sensor are included in the section on electronic control of the fuel system.

An air bypass passage containing a flow control screw (see Figure 11B-33) allows air to pass into the engine when the throttle valve is

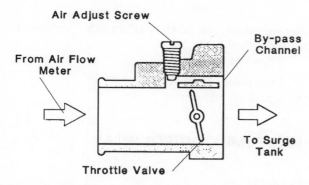

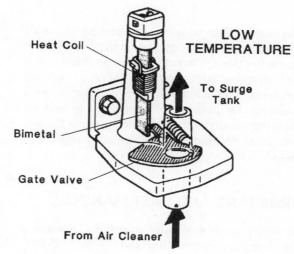

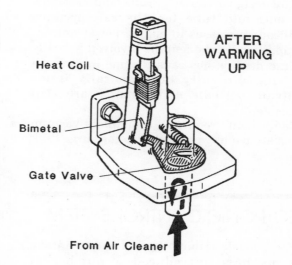

Figure 11B-33 Engine idle speed can be altered by rotating the air adjust screw. The screw position controls the quantity of air bypassing the closed throttle valve.

closed. The idle speed of the engine can be controlled by allowing more or less air through the bypass passage.

AIR VALVE (OR BYPASS AIR CONTROL VALVE)

The air valve controls the quantity of air flowing through a pipe which bypasses the throttle valve. When the air valve is fully open, the quantity of air bypassing the throttle valve is large and engine idle r.p.m. will increase. The r.p.m. increase compensates for engine friction during warm-up.

The valve body is cast from aluminium alloy and contains a disk valve, bi-metal strip, a heating element and coolant passages (see Figure 11B-34). During cold engine operation the valve is open and engine idle speed is increased. As the engine warms up, the bi-metal strip (receiving heat from both engine coolant and the electric heating element) deflects and the disk valve progressively closes the pipe opening. The engine idle speed will return to normal when the engine reaches operating temperature.

This type of valve is incapable of providing increased engine idle speed to compensate for the engine load from air conditioning, power steering, automatic transmission or electrical unit operation. To overcome the problem, many systems used separate bypass valves for each

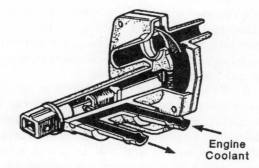

Figure 11B-34 The air valve increases engine idle speed, during warm-up, by opening a bypass air passage around the throttle plate.

component. This causes a congestion of small pipes on the throttle body as each bypass requires two connections. Current design systems use a BAC (bypass air control) valve receiving electrical signals from the computer and adjusting the quantity of bypass air to suit all operating conditions. On these systems the air valve and individual idle speed valves are deleted.

SURGE TANK AND INLET MANIFOLD

The surge tank and inlet manifold tubes are cast from aluminium alloy (see Figure 11B-24). Careful design of the surge tank volume, shape and manifold tube lengths can produce significant increases in engine power output. The additional tube lengths involved in many of these designs has caused some increase in servicing difficulty but generally items requiring periodic maintenance are still accessible.

Components requiring a manifold vacuum source are usually connected to the surge tank area.

ELECTRONIC CONTROL SYSTEM

The sections describing the fuel and air intake systems have introduced a number of components which were either sensors or actuators within the EFI system. In this section a few new components will be described but, if you have understood the previous text, most of the hard work has been done.

An EFI control system consists of an ECU (electronic control unit), usually known as the 'computer', and a number of sensors, actuators and relays. To complete the control system, a 12 volt supply source and the necessary wiring looms to link the components are required.

The sensors 'tell' the ECU what the engine is doing; the ECU makes a decision based on the sensor information and sends electrical signals to 'work' the actuators.

A number of components within the electrical control system do not rely on computer capability

for operation, nor do they affect the duration of injection.

They are:

- Fuel pump control relay;
- Auxiliary air device.

FUEL PUMP CONTROL RELAY

FUNCTION

To supply electrical power to the fuel pump. The relay must also contain a safety circuit to prevent pump operation if the engine is stopped with the ignition 'on'.

LOCATION

Usually within the engine compartment in a bank of relays.

OPERATION

Electrical power to operate the fuel pump is supplied via a fuel pump or control relay. This would appear to be a very simple arrangement but for safety reasons it is essential that the fuel pump ceases to supply fuel should the vehicle be involved in a collision. The easiest approach to monitoring the fuel system is to design the system so that if the engine is not operating the relay will open, switching off the fuel pump. It is also necessary to provide a means of bypassing the safety arrangements when the engine is cranking or fuel would not be available to start the engine. Two different methods of achieving safety control of fuel delivery are described. These are not the only methods available but they are widely used on many different vehicle makes.

CONTROL RELAY—IGNITION PULSE CONTROLLED

The control relay (see Figure 11B-35) provides power to the fuel pump and the EFI system. The main system and the fuel pump both switch off when the relay points open. When the

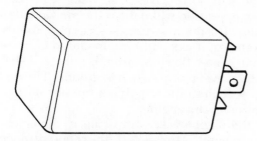

Figure 11B-35 Control relay.

ignition switch is turned 'on', the relay points will close provided one of the following two conditions is present.

a The engine is cranking.

b The ignition coil pulses being over the equivalent of 150 r.p.m. (the engine is running).

During engine cranking a voltage is present at the control relay terminals from both the ignition switch and the starter motor solenoid. Therefore the relay points will close and the fuel pump (and the EFI system) will operate (see Figure 11B-36).

After the engine starts, the ignition pulses from the ignition coil provide an engine running signal at the control relay terminal. This signal and the ignition switch 'on' voltage combine to keep the relay points closed (see Figure 11B-37).

If the engine rotational speed drops below 150 r.p.m (stalls or vehicle collision) or the ignition switch is turned 'off', the control relay points will open. The current supply to the fuel pump and the EFI control system is disconnected.

CONTROL RELAY—CONTROLLED BY SWITCH LOCATED IN AIR VANE METER

The circuit opening relay used to supply current to the fuel pump may also be used to power other system components (air bypass valve etc.). The internal construction of the relay requires the ignition switch to be 'on' and either the engine must be drawing in air or the starter

Engine Cranking

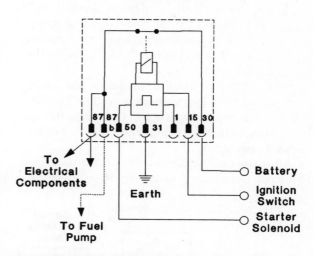

Figure 11B-36 The control relay contacts are closed when both ignition switch and starter solenoid voltages are supplied to the relay.

Engine Running

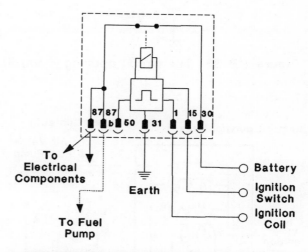

Figure 11B-37 The control relay contacts remain closed after the engine starts provided both the ignition switch 'ON' signal and ignition coil pulses continue to be supplied to the relay.

motor must be operating before current will be supplied to the fuel pump (see Figure 11B-38).

When the ignition switch is turned 'on', voltage is present at relay coil (L1) but current will not flow because the engine is not running and the vane flap has not deflected to close the pump switch.

By turning the ignition switch to the 'start' or cranking position, voltage is applied to relay coil (L2), the relay contacts will close and current is supplied to the fuel pump allowing the engine to start operating.

After the engine starts, and the ignition key is moved from 'start' to the 'ignition' position,

voltage will be removed from relay coil (L2). Engine operation will deflect the air vane flap and close the pump switch (see Figure 11B-40). The relay contacts will remain closed because current is now flowing through relay coil (L1) and the closed pump switch to ground. The fuel pump will continue supplying fuel for normal operation of the engine.

If the engine ceases drawing in air or the ignition switch is turned 'off', the relay will open the pump circuit and fuel supply will stop.

Under conditions of rapid deceleration it is possible that the air vane flap may close, opening the fuel pump switch. To prevent the relay

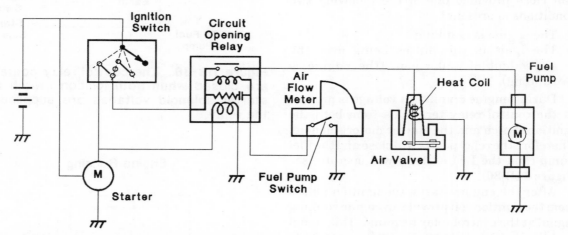

Figure 11B-38 The circuit opening (control) relay may control several components.

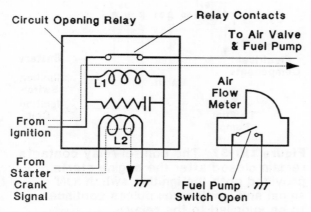

Figure 11B-39 During starting, current flows from the starter crank signal, passes through L2 and closes the relay contacts.

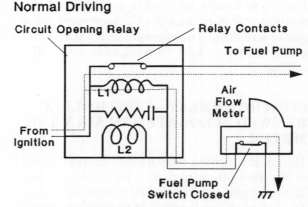

Normal Driving

Figure 11B-40 During normal driving, current flows through L1 and the fuel pump switch to close the relay contacts.

When Decelerating Rapidly

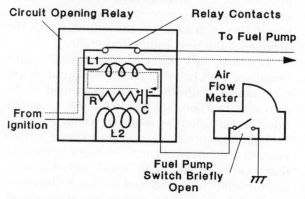

Figure 11B-41 During rapid deceleration, the capacitor (C) discharges through L1 keeping the relay contacts closed.

opening the circuit, a resistor and a capacitor (resistive/capacitive circuit) are fitted in parallel with relay coil (L1). Current is discharged through relay coil (L1) and, even though the fuel pump switch may be momentarily open, the relay contacts will remain closed (see Figure 11B-41).

AUXILIARY AIR DEVICE (AIR VALVE, BAC, ETC.)

FUNCTION

The auxiliary air device increases the engine idle speed during engine warm-up.

OPERATION

Electrical control of the air valve is very simple and in many applications is provided by the fuel pump control relay (see Figure 11B-42). When the engine is running, current will flow through the valve heating coil. For a detailed description of the valve operation and construction refer to the air supply section.

Further refinements to engine idle speed can be provided by idle up solenoid valves. These valves are controlled by the load source, e.g. when the air conditioner is switched on,

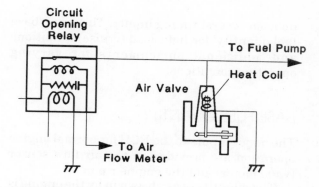

Figure 11B-42 Electrical current for the auxiliary air valve is usually supplied from the control relay.

voltage is applied to the air conditioner idle up valve, opening the air passage and raising the engine idle speed (see Figure 11B-43). A number of these valves may be linked through the computer or a separate idle up control unit to provide greater control.

Bypass air control (BAC) devices are more complex and details of the control methods will be included in the EFI system description to which they are fitted.

INJECTION PULSE TIME CONTROL

The injector opening time (duration) determines the amount of fuel sprayed into the inlet manifold. The injector opening duration is built

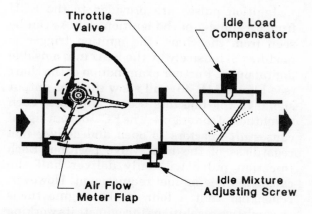

Figure 11B-43 Idle load compensation (idle-up) valve fitted to the throttle body.

up from several timing inputs. These are 'base fuel quantity' for light load (cruise) conditions and additional enrichments for varying operating conditions.

BASE FUEL QUANTITY

The major inputs to the ECU for normal engine operation are provided by the air flow sensor (vane meter) and the engine r.p.m.

The quantity of air drawn in by the engine is measured from the angular position of the air vane meter sensor flap. This position is transformed into a voltage. From the voltage signal the computer can calculate the air quantity being consumed by the engine (see Figure 11B-44).

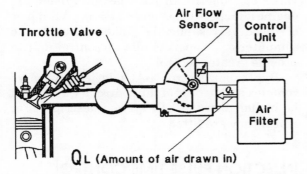

Figure 11B-44 The air-flow sensor in the intake system measures the quantity of air drawn into the engine.

Ignition pulses are provided to the ECU from terminal 1 of the ignition coil. As can be seen from the chart, the ignition trigger is modified extensively by the ECU into a usable digital pulse. Further examination of the chart (see Figure 11B-45) will show that the output signal to the injectors occurs only once each revolution of the engine. The ECU output signal causes *all* injectors to open and deliver fuel regardless of the inlet valve position. Each injection pulse will only deliver half the calculated engine fuel requirement. However, remember that a four stroke engine turns through two revolutions to complete its working cycle, so the other half of the required fuel is injected on the second revolution. Any fuel

injected at the 'wrong' time (when the inlet valve is closed) is temporarily stored in the inlet manifold and drawn into the cylinder with the air during the next induction stroke.

The base fuel quantity varies with engine speed and load, however the sensor inputs (vane air meter and ignition pulse) described in the previous text enable the ECU to keep the injectors open for the required time.

The concept of all injectors being pulsed simultaneously is not true of every system. A few systems use sequential pulsing or a combination of sequential and simultaneous pulsing, depending on engine operating range. However all systems do require the ECU to provide a base fuel quantity.

ENRICHMENTS

The base fuel quantity provides a fuel-air ratio that would enable a warm engine to operate within narrow limits of load, temperature range and throttle position. Carburettor equipped engines require idle, power, cold start, warm-up and deceleration circuits to alter the air-fuel ratio for varying conditions. Similarly, the EFI system, must allow for these conditions but is restricted to altering the frequency and duration of injector opening instead of introducing additional circuits. The ECU accepts voltage signals from a range of sensors and adds to the base fuel injection quantity to suit the engine operating condition (see Figure 11B-46).

COLD START AND WARM-UP ENRICHMENT

When an engine is required to start from cold the fuel requirement, depending on engine temperature, can be as high as two to three times the amount required for a warm engine. Immediately after the engine 'fires', a reduction in air-fuel ratio occurs. Progressive leaning of the mixture continues until the engine is at operating temperature. During this period of engine operation the ECU computes the frequency and duration of injection related to

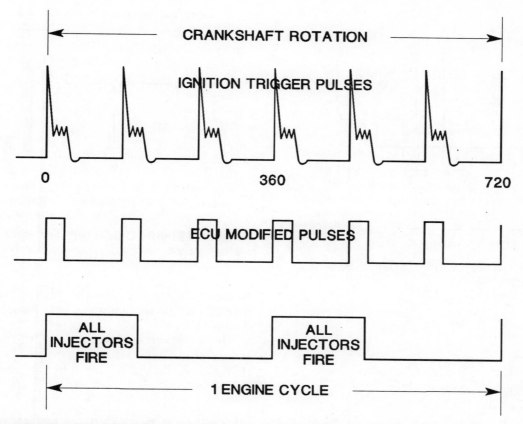

Figure 11B-45 The chart shown is for a six cylinder four-stroke engine. Ignition pulses are modified by the ECU into a usable form. At each third pulse, the earth circuit is completed and all injectors spray fuel simultaneously.

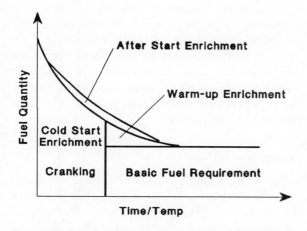

Figure 11B-46 The basic fuel requirement is supplemented by several enrichment factors to suit different operating conditions.

the time after starting and the coolant temperature.

COLD START ENRICHMENT

For systems using start control enrichment (not fitted with a cold start valve) and a coolant temperature sensor (see Figure 11B-47).

The ECU is aware of engine temperature (coolant temp. sensor), and when the ignition key is turned to the 'start' position, the control unit will deliver additional pulses to the injectors for each revolution of the engine. Up to three additional injector pulses per revolution are used to deliver the additional fuel required. This enrichment period is timed (maximum 10 seconds) so continued engine cranking will not cause a 'flooded' condition (see Figure 11B-48).

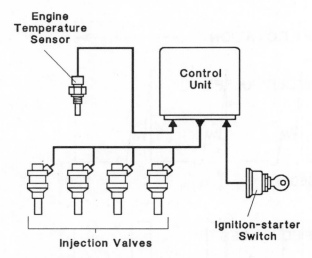

Figure 11B-47 Cold-start enrichment by start control.

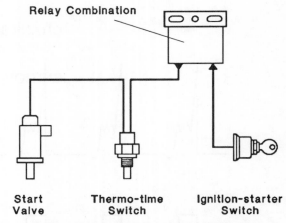

Figure 11B-49 Cold-start enrichment by start valve.

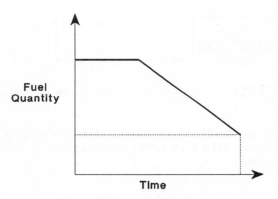

Figure 11B-48 Additional fuel injected during cold engine cranking is reduced over time (max 10 sec) to prevent flooding.

For systems fitted with a cold start valve, additional injector pulses will not be delivered during cold engine cranking. To control the start valve, the engine must be fitted with a thermo-time switch (see Figure 11B-49). The cold start valve electrical supply circuit is completed through the thermo-time switch contacts to ground. During cold engine cranking, the electrical winding will rapidly heat the bi-metallic strip, the contacts will open, and the cold start valve will cease fuel delivery to the engine. The electrical 'self-heating' is essential to prevent the engine being over-enriched during starting. The length of time the valve will operate depends on the temperature. (e.g. 8 sec. at -20 degrees C)

During warm engine cranking, the coolant temperature has caused the points to open, therefore no fuel is delivered by the cold start valve.

AFTER COLD START ENRICHMENT

After the engine 'fires', the abrupt reduction in fuel quantity, caused by the control unit no longer receiving the 'start' signal and ceasing to provide the additional injector pulses, could cause the engine to stall. To overcome this problem the control unit calculates an additional fuel supply to be added to the base fuel quantity. Some systems 'time' this period; other systems wait for engine r.p.m. to rise to an acceptable running speed before deleting the enrichment ratio.

WARM—UP

When the engine has passed through the cold start and after-start periods it is still necessary to provide additional fuel until the engine reaches operating temperature. The additional fuel required is only slightly greater than the base fuel quantity and the reference signal required can be supplied by the coolant temperature sensor. The control unit will use

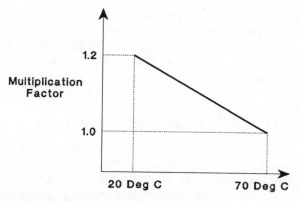

Figure 11B-50 Enrichment will be progressively reduced as the engine warms up.

this signal to progressively reduce the additional fuel as the sensor indicates a rise in coolant temperature (see Figure 11B-50).

LOAD ENRICHMENT

To provide load enrichment, the ECU requires information from the throttle body, the air flow sensor and the coolant temperature sensor. The throttle body switch provides information related to the throttle position, the air flow sensor advises of increased air flow and the engine temperature is sensed by the coolant temperature sensor.

ACCELERATION—WARM ENGINE

When an engine is running at a constant speed and the throttle position is suddenly changed, an increased air quantity must flow through the air flow sensor. The rapid change in air flow will cause the sensor flap to momentarily 'over swing' past the normal wide open position. The voltage signal produced at the air flow sensor is sent to the ECU and the injector pulse duration is increased. Because the air must pass through the air flow sensor, the enrichment signal can be processed before the air reaches the engine cylinders and excellent response to throttle opening is obtained.

ACCELERATION ENRICHMENT— COLD ENGINE

Generally the engine is considered to be in need of additional fuel, during acceleration, if the coolant temperature is less than 70 degrees Celsius. The opening speed of the air sensor flap can also be detected by the ECU and this factor is included when injector opening duration is calculated. The control unit calculates the increased fuel requirement and delivers an increased injector opening duration. This increase in fuel delivery occurs for approximately two seconds (see Figure 11B-51).

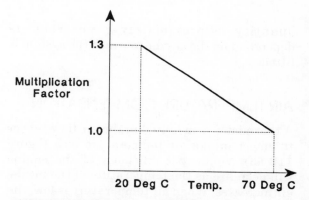

Figure 11B-51 Up to 2 seconds of enrichment will occur, under acceleration, if the engine coolant temperature is less than 70 degrees C.

FULL LOAD—WIDE THROTTLE OPENINGS

An engine must provide its greatest output when the accelerator pedal is pressed down and the vehicle is under high load. The throttle valve (body) switch has a set of contacts which close at wide throttle settings (see Figure 11B-52). The ECU receives this signal and increases the injector duration providing more fuel to the engine. The percentage increase of fuel is programmed into the ECU. The additional fuel

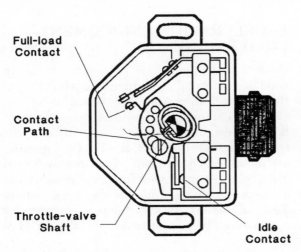

Figure 11B-52 Throttle-valve switch.

quantity, approximately 8 per cent, is dependent on the engine to which the system is fitted.

AIR TEMPERATURE COMPENSATION

A thermistor is mounted in the air flow sensor to detect intake air temperature (see Figure 11B-53). Volumetric efficiency of the engine will decrease as the temperature of the intake air increases. When air temperature is low, the oxygen content of the air will be high, therefore the quantity of fuel injected should be increased.

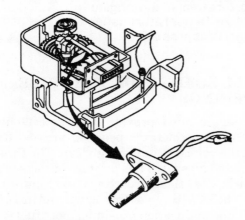

Figure 11B-53 Air temperature compensation is provided by a sensor (thermistor) mounted in the air flow meter.

The change in resistance value of the air temperature sensor (thermistor) causes a change in voltage signal to the ECU and the injector opening duration is adjusted accordingly.

FEEDBACK COMPENSATION

The use of catalytic convertor exhaust systems require the engine combustion process to be maintained at the theoretically correct air-fuel ratio of 14.7 to 1. An oxygen sensor is installed in the lower part of the exhaust manifold and generates a small voltage relative to the oxygen content detected in the exhaust gas. Over-rich (low content of oxygen) mixtures cause the generated voltage to rise and lean mixtures (high content of oxygen) cause low voltage output. Therefore a low voltage reading received at the ECU indicates the injector opening duration should be increased. If over-correction occurs, the mixture will be rich, the sensor voltage output will increase and the ECU will decrease the injector opening duration.

DECELERATION COMPENSATION

Continued injection of fuel during deceleration is wasteful, therefore many EFI systems are designed to shut off the injector pulse signal under deceleration conditions. Two inputs advise the ECU that the vehicle is decelerating. The throttle switch has a set of contacts which close when the throttle returns to the idle position. The ECU receives this signal, but also is aware from the engine r.p.m. signal that the engine is not idling. Provided the engine r.p.m. is not below a programmed figure, the pulse signal is stopped (see Figure 11B-54). The fuel injector pulse signals are restored when the engine r.p.m. drops below the program setting or the throttle position is moved from fully closed (idle contacts open). The resumption of fuel supply is progressive (up to the base fuel quantity) to prevent surging.

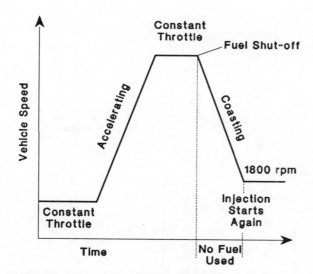

Figure 11B-54 Fuel is not injected during deceleration. This feature provides greater economy and improved pollution control.

ENGINE R.P.M. LIMIT

The system designer may include, within the ECU, a preprogrammed engine r.p.m. limit. When the engine r.p.m. exceeds this limit the injector pulses are discontinued until engine speed is safely reduced.

VOLTAGE CORRECTION

If the system voltage is low, the injector opening may be slightly slower than normal. The ECU monitors the system voltage. When low voltage conditions are detected, the injector duration is increased to compensate for the delay in opening time.

ELECTRONIC CONTROL UNIT (ECU)

The internal process stages of the ECU are beyond the scope of this book, however knowledge of the inputs and outputs is essential to gain an understanding of system operation. To consolidate the information you have gained, examine the basic ECU schematic (see Figure 11B-55) closely and relate back to the detailed explanation of each component to clarify any

problems. The ECU shown does not feature 'closed loop' feedback compensation and is of a type used with vehicles operating on leaded fuel.

LIMITED OPERATIONAL STRATEGY

Despite the exceptional reliability which EFI systems have displayed, modern vehicles are fitted with ECUs that can operate with sensors that are malfunctioning. ECU failures are rare and a system that can compensate for electronic failures, external to the ECU, is unlikely to cause a roadside breakdown. The components most likely to cause problems are connection points and sensors. The ECU is designed to substitute warm or hot engine control values for components such as air flow meters, intake air temperature sensors, coolant temperature sensors and oxygen sensors whenever a signal from this type of component is missing. The effect on vehicle drivability varies depending on which sensor fails. When the ECU detects the lack of input signal and the back-up control value is substituted, the vehicle is described as being in 'limp home' mode. The vehicle can now be driven to a service area for repair.

SELF-DIAGNOSTIC CAPABILITY

The inclusion of a self-diagnostic function into the modern EFI control unit has simplified the service technicians role. Many ECU's can emit either a number code or a coded flash sequence which, when decoded, directs the service technician to the fault area. This diagnostic area is expanding rapidly. The industry has moved from units that have no inbuilt diagnostic or 'limp home' capability to modern units that can accurately pin-point the system problem and in many cases the driver is only aware of the problem because a warning light is illuminated. Further details of diagnostic capabilities are included in the section on engine management systems.

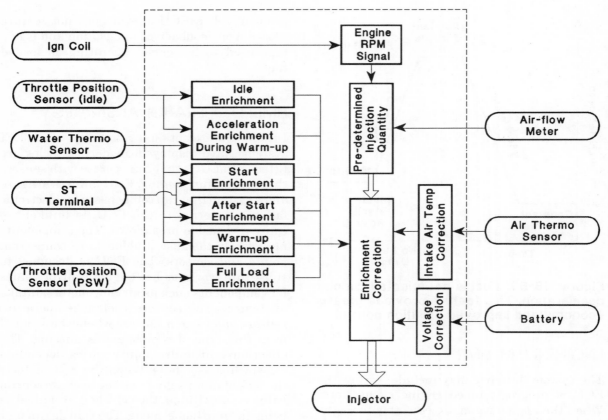

Figure 11B-55 The injector opening duration is set by the ECU after all the inputs have been evaluated.

TYPICAL VEHICLE APPLICATION OF AN EFI SYSTEM

Not all of the EFI component variations, previously described, would appear within a single system. The following system, fitted to a popular vehicle, has been chosen to illustrate the number and type of components which are required to form an operating system.

Figure 11B-56 shows the location of the EFI components which are attached to the engine. To supply, monitor and control these components requires the fuel, air and electronic systems described in detail in the previous section of this chapter.

Fuel supply pressure is provided by a 'wet', roller-type electric fuel pump mounted within the fuel tank. The wagon versions of this model use two pumps; a low pressure pump is fitted in the fuel tank and supplies an externally mounted high pressure pump.

The fuel line is fitted with an in-line paper filter contained within an aluminium housing. The filter is non-serviceable and must be renewed at intervals as specified by the manufacturer. (nominal service life—30 000 km.).

Fuel rail pressure is controlled by a pressure regulator with a nominal setting of 248 kPa. The regulator operation is affected by intake manifold pressure, which can vary between 69 kPa below atmospheric pressure up to atmospheric pressure. Therefore fuel pressure measurements, at the fuel rail, can be in a range from 179 kPa (idle) to 248 kPa (full throttle).

The fuel injectors (6) are mounted on a common fuel rail and are angled into the intake

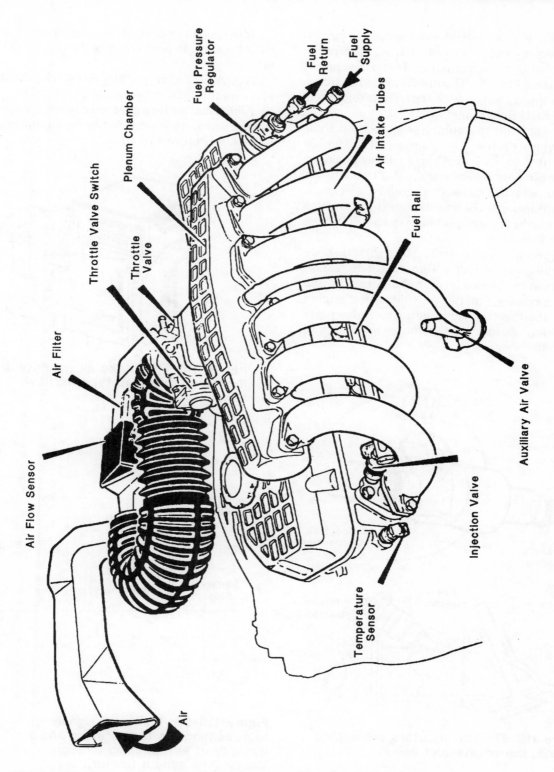

Figure 11B-56 Location of EFI components on a typical engine.

manifold. Rubber mouldings are used to mount the injectors and these reduce heat absorption and vibration being transferred from the engine (see Figure 11B-57). The injectors are not fitted with dropping resistors and current is regulated by the ECU final stage.

Air entering the engine is measured by a 'flap' type air sensor (see Figure 11B-58) which is mounted to the end of the air filter housing.

The air flow sensor casing also contains an air temperature sensor, therefore electrical signals relating to the quantity and temperature of the intake air are provided to the ECU (see Figure 11B-59).

Engine speed is controlled by a throttle body containing a butterfly flap (throttle valve) connected to the accelerator pedal. In addition to the primary function of controlling engine speed, the throttle body casting contains ports to provide vacuum for distributor advance and EGR valve control.

Idle speed is adjusted by a screw fitted in a throttle valve bypass passage (see Figure 11B-60).

A contact-point type throttle switch provides idle and full load information to the ECU.

Additional air for increased engine idle speed during warm-up is provided by an auxiliary air valve (see Figure 11B-61).

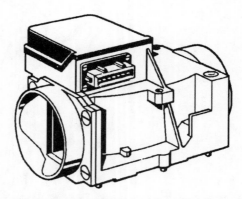

Figure 11B-58 Intake air quantity is measured by a flap type air flow sensor.

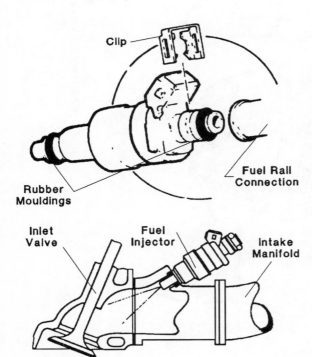

Figure 11B-57 The Injectors are angled towards the engine inlet valve.

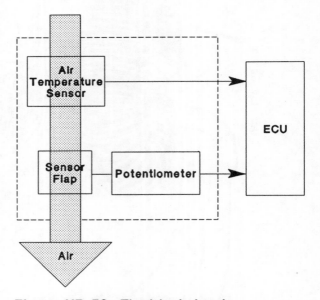

Figure 11B-59 Electrical signals representing the temperature and quantity of air, passing through the air flow sensor (meter), are sent to the ECU.

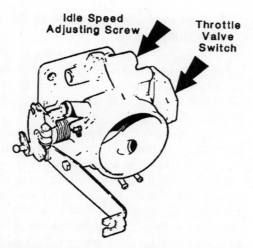

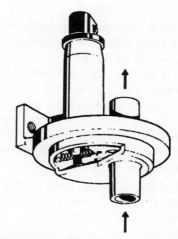

Figure 11B-60 Engine speed is controlled by a throttle valve mounted within the throttle body.

Figure 11B-61 Idle speed, during warm-up, is increased by supplying additional air through the auxiliary air valve.

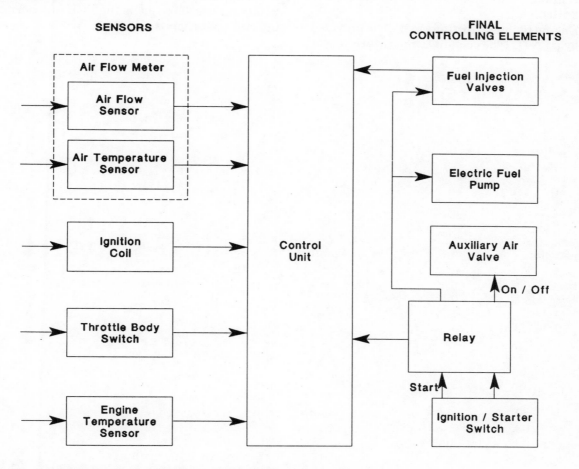

Figure 11B-62 Electronic fuel injection block diagram.

Electrical supply to the system is via an ignition switch controlled relay which provides current to the ECU, electric fuel pump and auxiliary air valve. Voltage is also provided to the injectors but current flow will only occur when determined by the ECU (see Figure 11B-62).

Essential input information to the ECU is provided by a small number of sensors (see Figure 11B-62):

- Air flow quantity and temperature is sensed by the air flow sensor.
- Rotational speed of the engine is provided by the ECU detecting the ignition pulses present at the negative terminal of the ignition coil.
- Engine throttle position is provided by the contacts contained in the throttle body switch.
- Engine temperature information is provided by an NTC type coolant temperature sensor located in the water jacket at the front of the cylinder head.

The engine is not fitted with a cold start valve and uses start control enrichment. To supply the fuel for cold engine starting, additional injector pulses are provided during engine cranking.

When the vehicle is decelerating under closed throttle conditions, continued fuel injection would result in fuel wastage and high emission output. Under these conditions the ECU can detect the closed throttle position from the throttle switch contacts and the high engine revolutions from the ignition signal, then takes action to shut off the injectors. This action only occurs at engine rotational speeds in excess of 2100 r.p.m. Movement of the throttle valve off the idle position or engine speed dropping below 2100 r.p.m. is detected by the ECU and normal injector operation is resumed.

ELECTRONIC ENGINE MANAGEMENT

The name sounds impressive and the mechanics who have just bridged the technology gap encountered with earlier systems, may consider the 'new technology' beyond their capabilities. In fact, the reverse is the case. New systems contain 'on board' diagnostics to assist the servicing mechanic and the majority of components are 'old technology' linked together into an integrated system.

Control devices for crankcase ventilation, exhaust gas recirculation, evaporative emissions, fuel and ignition timing have been fitted to vehicles for many years. These devices have become reliable and efficient, allowing the driver to have a vehicle which is both pleasant to drive and reasonably fuel efficient. The major problem presented by these vehicles, is not related to economy or performance but is the level of pollutants emitted from the exhaust pipe. Current emission control regulations are stringent and demand extremely close control of the engine operating parameters. This requirement has hastened the demise of individual systems such as ignition and fuel operating independently of each other.

Engine management is the linking together of earlier subsystems into a coordinated approach to pollution control. The new generation management systems collect information in similar ways to the earlier systems but additionally operate in a 'closed loop' or 'feedback' process (see Figure 11C-1). Expressed simply, 'closed loop' means the system samples the results (exhaust gas) of the engine

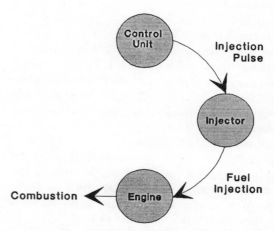

Open Loop: The output has no effect on, and does not modify the input.

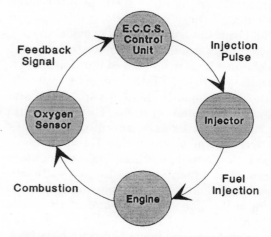

Closed Loop: The output has an effect on, and modifies the input.

Figure 11C-1 Open and closed loop engine systems.

settings. The sample causes a 'feedback' signal to the computer and the engine settings are altered until the exhaust sample confirms the optimum engine operation is being achieved. This function is particularly important to ensure the exhaust system catalytic convertor operates efficiently. The different system functions have been collected into a single computer which is capable of 'multi-tasking', i.e. controlling several systems and tasks from a single integrated unit. A modern computer (ECU) can provide system support for:

- ignition timing;
- ignition function;
- fuel injection;
- canister purge—evaporative emissions;
- exhaust gas recirculation (EGR);
- engine idle speed—unloaded and loaded (air conditioner, electrical load, transmission engagement, power steering, temperature change)
- self-diagnostic function;
- 'limp home' capability—substitution of set values when a sensor signal is outside the operating range (malfunctioning);
- compensation for engine age—is only featured on a few systems at this time.

As previously indicated, electronic engine management systems have developed from separate EFI and ignition systems. A thorough knowledge of these earlier electronic ignition and fuel injection systems will provide an easy path to understanding the 'new technology'. This chapter has encompassed ignition theory, ignition systems, EFI theory, EFI systems, ancillary components (EGR, Evap., BAC) and engine management systems. Gifted students may have 'jumped' sections of the chapter until they arrive at 'new' knowledge. If this course of action was adopted and the student encounters a component which appears to lack sufficient information, it is highly probable the component was described, in detail, in one of the basic systems.

Four engine management systems have been selected as representative of the many types available on vehicles. Each system has at least one different control or measuring method to the other three systems yet all provide satisfactory exhaust emission and performance levels.

System 1 A central fuel injection system using a vortex theory air flow sensor and fitted to a turbocharged engine.

System 2 Multi-point fuel injection system utilising a MAP sensor, Hall effect ignition triggering and EST.

System 3 Multi-point fuel injection system containing a hot wire type air flow sensor, optical ignition triggering and EST.

System 4 Multi-point fuel injection system linked with direct fire ignition and EST.

SYSTEM 1 CENTRAL FUEL INJECTION ON A TURBO-CHARGED ENGINE

Engine management systems are not only for family sedans but are also suitable for high performance versions of these vehicles. The system (see Figure 11C-2) chosen as a typical example is based on a vehicle fitted with a turbocharger, knock control, central fuel injection and an air flow sensor utilising Karmans vortex theory.

INTAKE AIR SYSTEM

The intake air system (see Figure 11C-3) consists of an:

- air cleaner;
- air flow sensor;
- injection mixer;
- turbocharger and ducting.

The air cleaner, turbocharger and ducting are of conventional design, however the airflow sensor is of specific interest in this system and a detailed explanation is provided. Air entering the engine is controlled by the throttle valve in the injection mixer. The injection mixer also houses the injectors and construction details are included in the fuel system description.

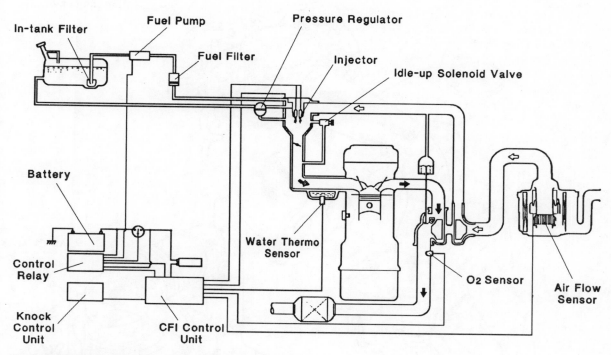

Figure 11C-2 Central fuel injection system.

AIR FLOW SENSOR

The air flow sensor is mounted inside the air cleaner to prevent contamination from dust or water. An air intake temperature sensor (see Figure 11C-4) is fitted to the air flow sensor to monitor temperature of the incoming air. The fuel injection quantity will be altered by the computer to suit the inlet air temperature.

The air flow sensor (see Figure 11C-5) consists of three major sections:

* flow regulator;
* vortex generator;
* vortices detection.

Filtered air from the main air cleaner is passed through the air flow regulator to suppress any turbulence that may have been present in the air stream.

The air then flows past a centrally mounted vortex generating rod or pole which creates vortices in the air stream.

The third section consists of a transmitter

and a receiver mounted opposite each other. The transmitter emits constant frequency ultrasonic waves which are detected by the receiver. If vortices are not present in the air stream, the ultrasonic wave form will arrive at the receiver at constant time intervals. However, if a vortice passes between the receiver and the transmitter, the wave form will be sped up by the front, then slowed by the tail of the vortice. Low air flows will only generate a low number of vortices, therefore the number of vortices detected will also be low. Conversely, high volume air flow will generate a high number of vortices. The receiving unit senses the change in air flow by detecting the varying time intervals at which the ultrasonic wave is arriving. These varying time intervals are changed into a rectangular wave form which is sent to the central fuel injection (CFI) control unit. The control unit uses this information to set an injector frequency suitable for the volume of intake air.

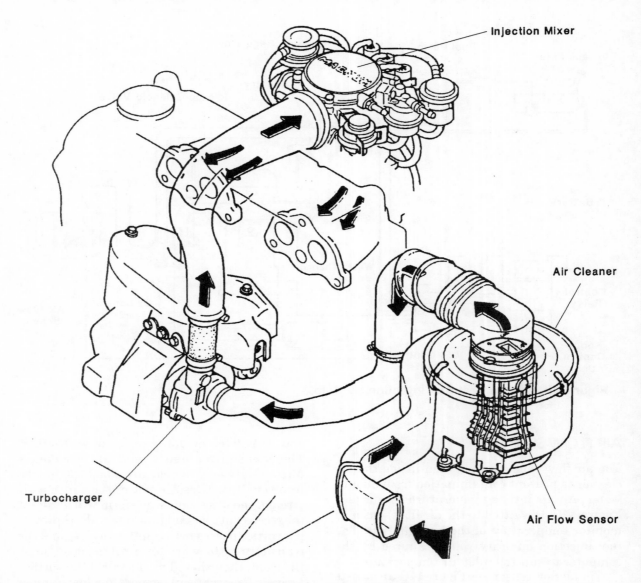

Figure 11C-3 Intake air system for a turbo-charged engine.

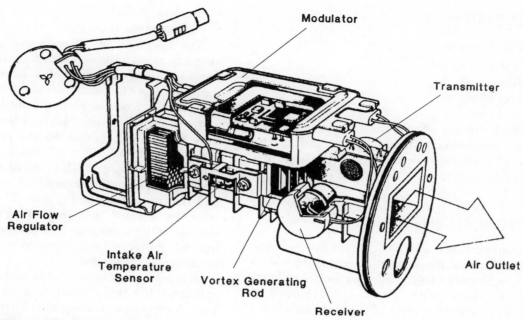

Figure 11C-4   Air flow sensor.

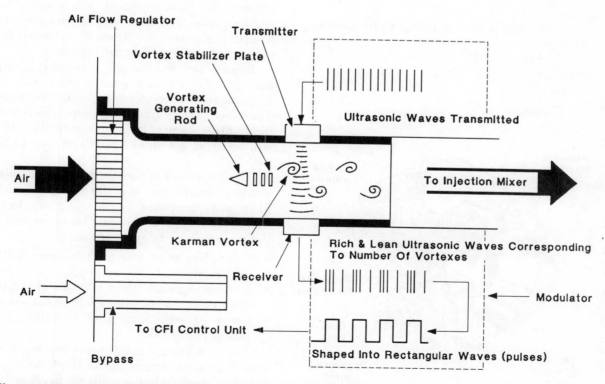

Figure 11C-5   Intake air metering system (Karman vortex theory).

FUEL SYSTEM

The fuel system contains an:

- injection mixer;
- filters;
- fuel pump;
- fuel tank.

With the exception of the injection mixer, all other components should be familiar to the reader and details are available in the basic fuel injection section.

INJECTION MIXER

The major functions of the mixer are to house the electromagnetic injectors and to provide a mixing chamber for the injected fuel (see Figure 11C-6). The mixer is also a suitable point at which to attach several of the sensors needed for system operation, therefore the explanation of the mixer will include:

- injectors;
- full throttle inhibitor and fast idle mechanism;
- variable resistor;
- throttle sensor;
- idle switch;

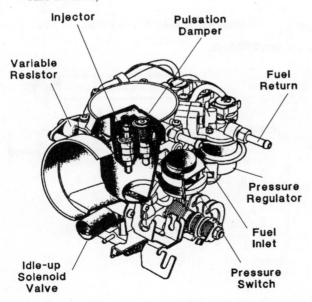

Figure 11C-6 Injection mixer.

- pressure switch;
- fuel pressure regulator.

The injectors are of conventional electromagnetic solenoid design with operating current duration and frequency being determined by the computer. The computer calculates the base fuel quantity from the air flow sensor signal. An increase in intake air volume through the air flow sensor will cause an increase in the frequency (Hz) of the rectangular wave form sent to the computer. The computer uses the wave form as information to alter both **duration** and **frequency** of the injector. Therefore the time the injector is open (duration) and the number of times per second (frequency) that the injector delivers fuel is related to the load condition of the engine.

A brief examination of this system appears to indicate less fuel is injected for greater engine load. An increase in engine load **increases** the number of air sensor signal pulses that occur between injector openings. When the engine is under light load, injection occurs every two sensor pulses yet, at heavy engine load, the computer requires six pulses from the air flow sensor before triggering the injection.

Remember, as the air quantity passing through the air flow sensor increased, the frequency of the wave form pulses also increased. Therefore, for any given time period, more wave form pulses occur at large intake air flows than will occur at low intake air flows (see Figure 11C-7).

The frequency of the air flow sensor pulses and the injection points in the example were for illustrative purposes. A general guide to the frequency bands which can be expressed as light, medium or heavy loads are:

- light—less than 500 to 600 sensor pulses per second;
- medium—greater than 500 to 600 but less than 1000 to 1100 sensor pulses per second;
- heavy—greater than 1000 to 1100 sensor pulses per second.

The broad range and the overlapping of the load categories means the example in Figure 11C-7 is possible but is unlikely to occur. The perfect frequency of injection between the light, medium and heavy load conditions will occur

EXAMPLE ONLY

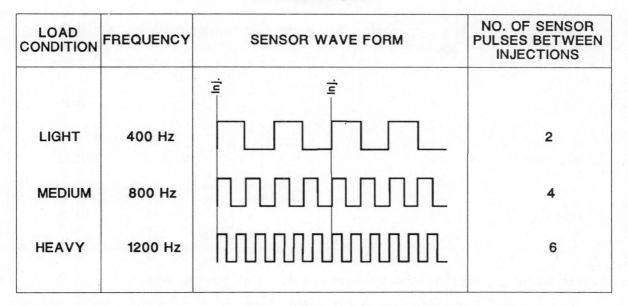

LOAD CONDITION	FREQUENCY	SENSOR WAVE FORM	NO. OF SENSOR PULSES BETWEEN INJECTIONS
LIGHT	400 Hz		2
MEDIUM	800 Hz		4
HEAVY	1200 Hz		6

Figure 11C-7 The frequency of sensor pulses varies with the intake air flow.

only at the sensor Hz values shown. An air sensor frequency of 600 Hz could be identified as a medium load condition by the computer and injection would occur every fourth sensor pulse. If this frequency is compared to the graph it can be seen that only six pulses would occur in the example time frame (one-hundredth of a second). Because injection occurs each four sensor pulses, the injection points will be further apart than occurs during light load conditions.

Increased engine load requires the delivery of additional fuel to the engine to maintain power. The previous text described how the computer modifies the time between injections but this action will not supply the additional fuel required. To provide the fuel needed the computer **modifies the injector duration** in addition to the time between injector openings.

Figure 11C-8 illustrates how the injection duration is modified as the engine load increases. The time the injector is open at light load is doubled for medium load or tripled when heavy load conditions are encountered. Therefore, even though the time between injections may increase slightly as engine load increases, the required air-fuel mixture is maintained by increasing the injection duration.

The chart (see Figure 11C-9) indicates how the engine load condition, air flow sensor pulse frequency, injector timing and fuel quantity injected are linked to produce a suitable air-fuel ratio. This system has two injectors in the injection mixer which are triggered alternately at the required intervals.

Note: Dropping resistors are not fitted in series with the injector windings and current flow is limited by circuitry within the computer. As with other injectors of this type, the connection of a battery directly across the injector terminals will damage the injector.

The full throttle and fast idle mechanism has a thermo-wax plunger interconnected to the mixer mechanical linkage (see Figure 11C-10). When the engine is cold the fast idle cam is positioned to provide increased engine idle r.p.m. Additionally, a pin on the fast idle cam interacts with the full throttle inhibitor lever to prevent full throttle operation while the engine is cold.

The variable resistor (see Figure 11C-11) allows manual adjustment of the idle fuel mixture. Turning the adjusting screw alters the voltage on the sensing connection. The computer senses this varied voltage and alters the quantity of fuel injected.

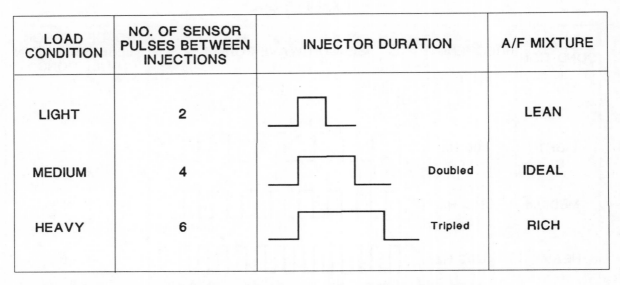

LOAD CONDITION	NO. OF SENSOR PULSES BETWEEN INJECTIONS	INJECTOR DURATION		A/F MIXTURE
LIGHT	2			LEAN
MEDIUM	4		Doubled	IDEAL
HEAVY	6		Tripled	RICH

Figure 11C-8 Injector duration increases as intake air flow increases.

	Light Load Condition	Medium Load Condition	Heavy Load Condition
Air flow sensor signal (Pulse frequency)	Less than 500 - 600 Hz	Between 500 - 600 Hz and 1000 - 1100 Hz	More than 1000 - 1100 Hz
Injection timing	One injection per two pulses	One injection per four pulses	One injection per six pulses
Pulse frequency			
Injector No 1			
Injector No 2			
Amount of fuel injected per one injection	Half of fuel necessary for ideal combustion	Fuel necessary for ideal combustion	One and half times fuel necessary for ideal combustion

Figure 11C-9 Under heavy load, the pulse frequency and the injector duration both increase to provide the required fuel.

The throttle sensor (see Figure 11C-12) is operated from a cam connected to the throttle shaft. The value of the sensor resistance changes as the throttle is opened. The change in voltage at the sensing connection is detected by the computer. The injector pulse frequency and duration are altered to suit the load condition.

The idle switch (see Figure 11C-13) provides a signal to the computer to indicate the throttle valve is at the idle position.

FAST IDLE MECHANISM

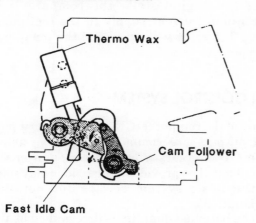

FULL THROTTLE INHIBITOR MECHANISM

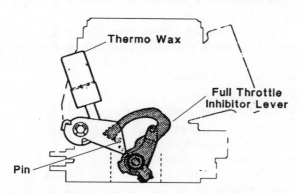

Figure 11C-10 Cold engine fast idle and full throttle 'block-out' are achieved by the use of a thermo wax plunger and mechanical linkage.

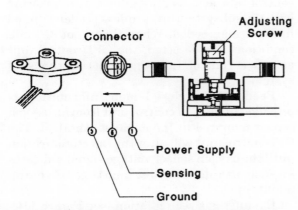

Figure 11C-11 Idle fuel mixture variable resistor.

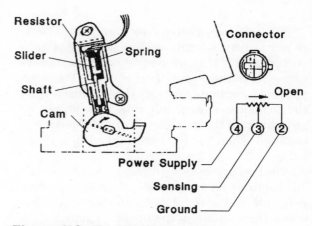

Figure 11C-12 Throttle sensor.

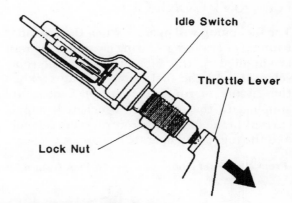

Figure 11C-13 Idle switch.

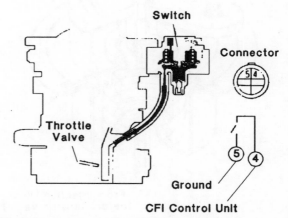

Figure 11C-14 Pressure switch.

The pressure switch (see Figure 11C-14) is required because this vehicle is fitted with a turbocharger. If an over-boost condition occurs (pressure in the mixer rises above 59 kPa. or 8.5 p.s.i.) the computer detects the condition from the pressure switch and cuts fuel supply by switching off the injectors. Normal injection is restored when the over-boost condition drops away.

The fuel pressure regulator (see Figure 11C-15) operates on the excess pressure over flow principle and is regulated to a constant value above the vacuum in the injection mixer.

FUEL PUMP CONTROL

The fuel pump will operate when the engine is running or cranking. During cranking, current is supplied to the fuel pump control relay solenoid, and the fuel pump operates. When the engine is running, ignition pulses are supplied to the computer which will supply current to the fuel pump control relay solenoid and the fuel pump operates.

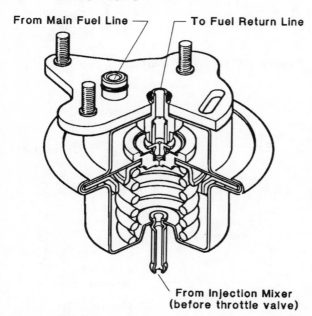

From Main Fuel Line — To Fuel Return Line

From Injection Mixer
(before throttle valve)

Figure 11C-15 The fuel pressure regulator maintains fuel pressure at a constant value above the vacuum in the injection mixer.

If the engine stalls the computer will discontinue current supply to the fuel pump relay. The relay will 'open' and the pump will not operate.

CFI CONTROL SYSTEM

Many of the computer (CFI control unit) inputs and outputs are common to other systems. Where this occurs a brief description of the function will be provided. The reader may refer to the basic system section for detailed explanations.

The base fuel quantity is determined by the air flow sensor input, however correction of this base quantity is derived from several inputs (see Figure 11C-16).

The CFI control unit (see Figure 11C-17) provides, not only the fuel injection system but also some engine management features. The control unit is linked to the:

- feedback system;
- EGR control system;
- idle up control system;
- knock control system;
- turbocharger system;
- self-test code outputs. ('on board diagnostics').

Unlike a number of totally integrated engine management systems, this system allows the EGR, idle-up, knock control and turbocharger to operate as separate subsystems. The CFI control unit acts as a monitor, not interfering with the subsystem units unless predetermined limits are exceeded. When these 'out of limit' conditions are detected, the CFI control unit assumes control (management) of the subsystem.

Feedback or closed loop information is provided to the CFI control unit from an oxygen sensor mounted in the exhaust manifold. The CFI control unit will alter the injector duration until the oxygen sensor voltage signal indicates an acceptable air-fuel mixture is being maintained.

Exhaust gas recirculation (see Figure 11C-18), while effective in reducing NO_x emission levels, can affect the drivability of the vehicle

Correction	Input Sensor	Reason
Engine coolant temperature correction	Water Thermo Sensor	To improve starting efficiency
Acceleration volume increase on warm-up	Water Thermo Sensor	To prevent stalling immediately after starting To improve driveability while cold
Air density correction	Intake Air Temperature Sensor (mounted in air flow sensor)	To correct any deviation of air/fuel mixture caused by changes of air density
Acceleration volume increase	Throttle Sensor	To improve response during acceleration
Deceleration volume decrease	Idle Switch Ignition Pulse Water Thermo Sensor	To improve fuel economy To improve deceleration driveability
Battery voltage correction	Battery voltage	To prevent fluctuations of needle valve open time
Engine over speed volume decrease	Ignition Pulse	To prevent engine over speed
Overboost volume decrease	Pressure Switch	To prevent turbocharger over boost
Idle mixture	Variable Resistor	To adjust idle mixture manually

Figure 11C-16 A suitable air/fuel ratio is obtained by correction to the basic injection quantity. This correction is obtained by changing both the injection duration and the injection frequency in response to information from the sensors as shown in the chart.

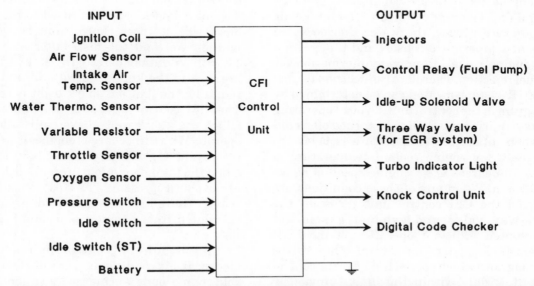

INPUT

Ignition Coil
Air Flow Sensor
Intake Air Temp. Sensor
Water Thermo. Sensor
Variable Resistor
Throttle Sensor
Oxygen Sensor
Pressure Switch
Idle Switch
Idle Switch (ST)
Battery

CFI Control Unit

OUTPUT

Injectors
Control Relay (Fuel Pump)
Idle-up Solenoid Valve
Three Way Valve (for EGR system)
Turbo Indicator Light
Knock Control Unit
Digital Code Checker

Figure 11C-17 The ECU will not assume control of the EGR, Idle-up, knock control and turbocharger systems unless pre-determined operating ranges are exceeded.

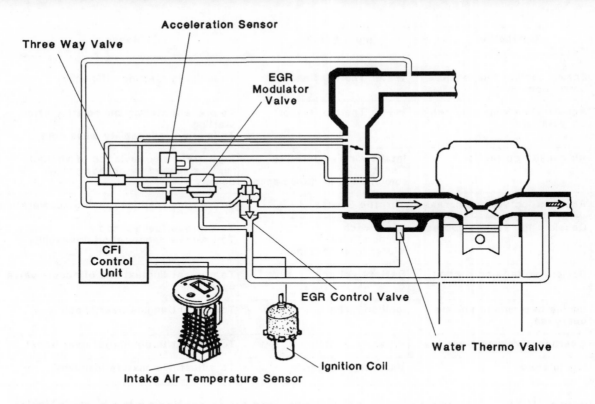

Figure 11C-18   Exhaust gas recirculation system.

under certain conditions. This effect would be particularly noticeable during low ambient or low engine temperature conditions. The CFI control unit is already linked to the intake air temperature sensor, the water thermo valve (coolant temperature sensor) and the ignition coil (engine r.p.m.) to provide information for the fuel injection. Therefore the decision to turn the EGR system on or off can be determined by a program in the control unit. Typical operating limits requiring the EGR to be turned off would be when intake air temperature is below 16 degrees C. or engine coolant temperature is below 70 degrees C. or engine speed is above 3000 r.p.m. When any of these conditions are detected the CFI control unit turns off the three-way valve and exhaust gas is not recirculated. Efficient operation of the EGR system is dependent on several other valves (see diagram) which are not linked to the control unit. It is only within the limits previously described that the control unit overrides normal EGR system operation.

Detonation is a problem which can affect all engines, but is of particular concern when the engine is turbocharged. The knock (detonation) control system fitted is usually of conventional design. A knock sensor is fitted to the engine block with a knock control unit between the ignition coil and the distributor (see Figure 11C-19). Vibrations detected by the knock sensor are sent to the knock control unit and, where required, the ignition spark point is retarded. The CFI control unit sets the limits within which the knock control system is allowed to operate. With turbocharged engines it is common to prevent spark retard below 1650 r.p.m. or when the intake manifold pressure is less than atmospheric pressure. A MAP sensor is not fitted, therefore the manifold pressure is calculated from engine r.p.m. and intake air flow.

Engine idle-up control in response to electrical, air conditioning, power steering and cold engine loads is achieved by a solenoid valve fitted into a throttle bypass passage. Figure 11C-20 shows how power from the ignition switch is directed through the solenoid valve to

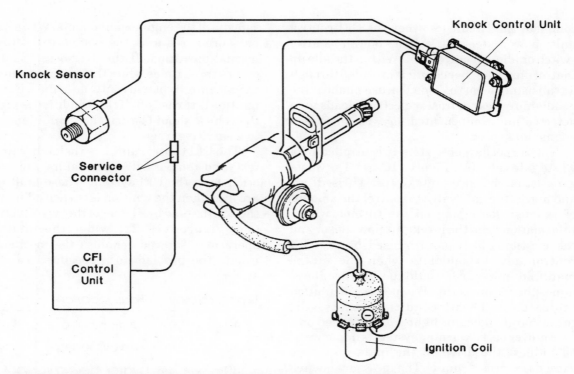

Figure 11C-19 Knock control unit fitted to a turbo-charged engine.

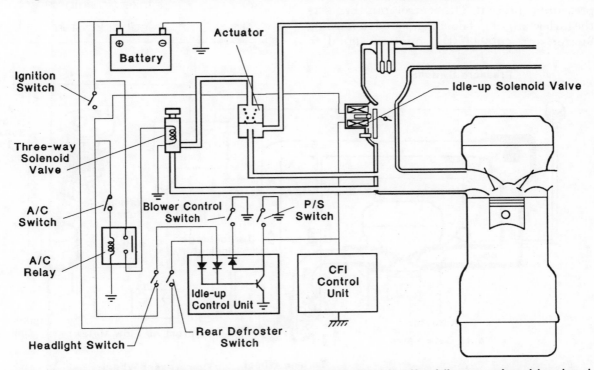

Figure 11C-20 Increased engine idle speed is supplied by the idle-up solenoid valve to compensate for electrical, air conditioning, power steering and cold engine loads.

complete the circuit in either the CFI control unit, power steering switch, blower control switch or idle up control unit. Within the idle up control unit, a further circuit is provided through a transistor to earth. This feature enables the headlamp or rear defroster electrical loads to be detected and compensated for with increased engine idle speed.

The turbocharger system is controlled by a waste gate valve (see Figure 11C-21). The waste gate actuator senses intake (pressurised) air and moves the gate valve to control the amount of exhaust gas acting on the turbine wheel. Information from the intake air flow sensor and the ignition coil (r.p.m.) is used by the CFI control unit to calculate when the intake manifold pressure would be at, or above atmospheric pressure. When the calculation indicates a 'boost' condition exists, the turbocharge indicator light will be turned on.

An 'over boost' warning system is implemented to alert the driver if the intake pressure exceeds a preset limit. The pressure switch signals the CFI control unit when excess pressure is present. The control unit then sets the turbocharge light flashing, sounds a warning buzzer and cuts off the injector signal to

safeguard the engine components. When intake pressure is reduced, the system will return to normal operation. If the over-boost condition occurs three times while the engine is running, the warning light will continue to flash until the ignition is turned off. If this condition is present the vehicle should be driven at minimum speed to a service garage.

The CFI control unit has the ability to 'self-test' and temporarily store faults which may occur with the CFI system. These fault codes are lost when the ignition is turned off however they can be read by a tester of the type illustrated (see Figure 11C-22) when the engine is operating. The code enables the mechanic to isolate the problem to a specific area of the system, e.g.

Digital display	Fault location
10	Ignition pulse
20	Air flow sensor
30	Water thermo sensor
40	Intake air temp. sensor
50	Feedback system

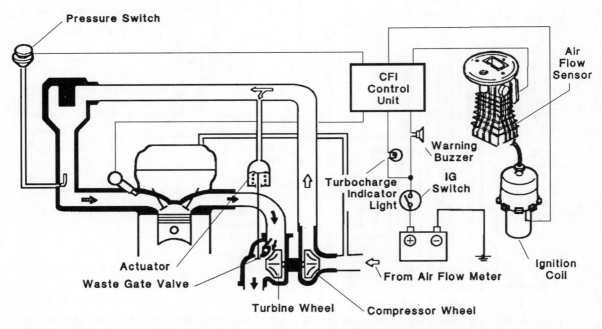

Figure 11C-21 Turbocharger, waste gate and over-boost warning systems.

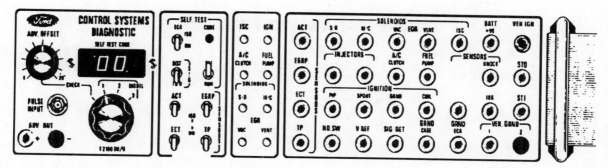

Figure 11C-22 Diagnostic code reader.

It is possible for the vehicle to have one or more sensor faults yet continue to operate. If the control unit logs a code for 20, 30 or 40, a substitute value will be supplied in place of the malfunctioning sensor. The engine will continue operating but with reduced efficiency.

SYSTEM 2 MULTI-POINT INJECTION WITH A MAP SENSOR

Computer control of ignition timing and fuel injection as separate functions was only an interim step towards a single electronic control unit. The microprocessor fitted to this system controls the:

- ignition timing;
- fuel injection system;
- feedback system (closed loop);
- carbon canister purge ;
- idle variation for load from:
    - —air conditioning;
    - —transmission;
    - —power steering;
- trip computer via an output link;
- self-testing and storing of system malfunctions;
- effects of battery voltage variation.

The air intake and fuel subsystems perform the same functions in an engine management system as when fitted to the individual systems already described. This unit, therefore, will focus on the electronic control of the system.

Figure 11C-23 may appear complex but, before considering all the various sensors supplying correction information to the ECU, it will be helpful to locate and understand the base inputs.

Figure 11C-24 has been reduced to the primary information required by the ECU to:

- position the ignition timing;
- provide a base fuel input quantity.

Ignition timing input is provided to the ECU from the profile ignition pulse 'PIP' signal. The PIP rotor (see Figure 11C-25) has six blades—one for each cylinder. When the rotor turns, the blades pass through a Hall effect sensor. As the leading edge of a rotor blade enters the Hall effect sensor, the signal line voltage rises to the same voltage as the power line. This change in voltage is detected by the ECU and the crankshaft position is then known. It is important to understand that the signal for crankshaft position was derived from the distributor PIP signal. If the distributor is not positioned correctly relative to the engine crankshaft, the ECU spark timing calculation will be added from an incorrect base point.

Example: Set the distributor position, to manufacturer's specifications. In this example the base timing will be set at 10 degrees BTDC. The PIP rotor blade leading edge will signal the ECU on entry into the Hall effect sensor. The ECU will calculate the ignition timing advance required (based on other inputs) and then, when the correct time arrives, send a signal to the TFI module to fire the next ignition pulse. Assume the ECU calculation results in an 'advance' of

ENGINE MANAGEMENT SYSTEM

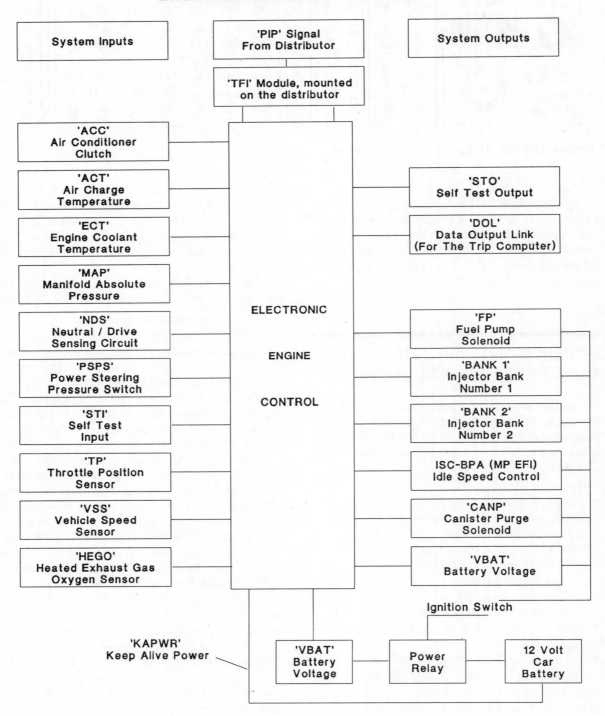

Figure 11C-23 Typical inputs/outputs to an engine management system.

ENGINE MANAGEMENT SYSTEM

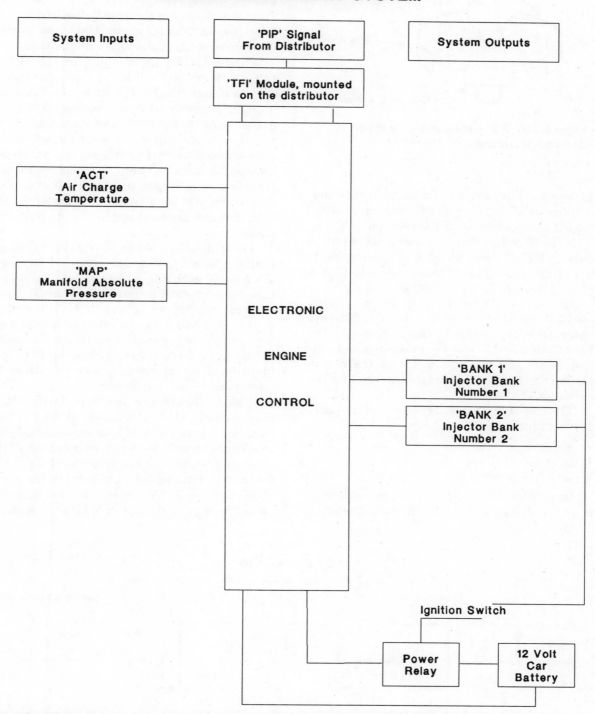

Figure 11C-24 The base fuel quantity and ignition timing can be supplied from these primary sensors.

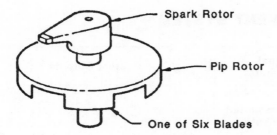

Figure 11C-25 Crankshaft position is derived from the PIP signal.

15 degrees. The next spark plug, in firing order, will fire at 25 degrees BTDC (10 degrees base timing plus 15 degrees ECU calculation).

If the base timing was incorrectly set at 16 degrees BTDC, the next spark plug firing would be at 31 degrees BTDC (16 degrees base timing plus 15 degrees ECU calculation).

Incorrect ignition timing problems will not occur if the manufacturer's instructions are observed when installing the distributor. The base timing point must be reset every time the distributor is repositioned. Check the manufacturer's specific model repair manual for the correct procedure.

The base fuel quantity calculation requires several inputs:

- engine speed (r.p.m.);
- inlet manifold pressure;
- inlet air density.

Instantaneous engine speed can be derived from the PIP sensor signal. When considering ignition timing, emphasis was placed on the leading (rising voltage) edge of the rotor blade. The trailing edge of the blade also produces an effect; as the blade leaves the Hall effect sensor, the signal voltage falls to ground voltage.

Figure 11C-26 represents the signal line voltage rising and falling as the rotor blades enter and leave the Hall effect sensor. The rotor is manufactured with equal spacing of the blade and the gap, therefore the voltage signal will be high for 50 per cent and low for 50 per cent of time when the rotor is moving at constant speed. Voltage 'on time' is equal to voltage 'off time', resulting in a 50 per cent duty cycle and an output signal in the shape of a square wave. The ECU can calculate the engine r.p.m. from this signal.

The rotor will be designed with slight changes if the injector banks are required to inject alternately each engine revolution, instead of both banks simultaneously. To identify No. 1 cylinder, one rotor vane will be reduced in width. This will have the effect of producing a 'signature' (a single shorter 'on' time pulse width) in the wave form output which the ECU can identify. The injection points can then be referenced to No. 1 cylinder.

A manifold absolute pressure 'MAP' sensor is connected to the inlet manifold via a rubber tube (see Figure 11C-27). This sensor is designed to operate on a 50 per cent duty cycle and the resulting signal is sent to the ECU. The frequency of the signal varies proportionally to absolute pressure. Variations in inlet manifold pressure cause a change in MAP sensor signal

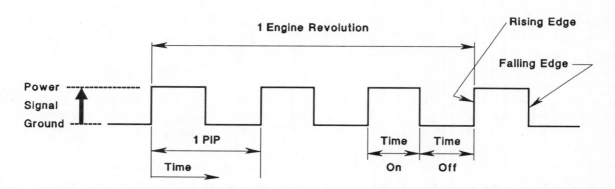

Figure 11C-26 Profile ignition pick-up wave form (sensor line voltage).

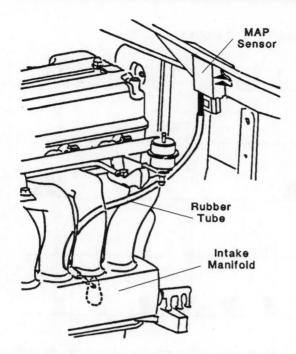

Figure 11C-27 The MAP sensor is usually located away from the engine.

frequency so the ECU is aware of conditions in the manifold at all times the engine is operating. The sensor has a useful function even before the engine is cranked. As the ignition key is turned 'on', the MAP sensor measures the manifold pressure. Before the engine cranks, the manifold pressure will be equal to atmospheric pressure, therefore the sensor is reading atmospheric pressure. This information enables the ECU to compensate for the effect weather conditions and altitude have on engine operation.

If volumetric efficiency of an engine cylinder was 100 per cent, the task of the ECU would be relatively simple. Engine speed, intake air pressure and cylinder size are known and the air quantity entering the engine could be calculated. However volumetric efficiency is not 100 per cent and does not remain as a constant. The quantity of air entering the engine cylinder will vary through a range of manifold pressures and engine speeds. To solve the problem, the exact volumetric efficiency for each engine operating point was established and the resulting data was stored in the ECU. For any

combination of engine speed and manifold pressure, supplied from the sensors, the ECU references the stored data and derives a value equivalent to the quantity of air entering the engine.

The density of air will vary depending on temperature. Therefore before calculating the mass of a quantity of air, the temperature of the air must be known. This information is sent to the ECU from an air charge temperature sensor located in the inlet manifold.

The ECU can now calculate the exact mass of air entering the engine cylinders from the PIP signal, air charge temperature sensor, MAP sensor and ECU stored data.

Optimum burning of the fuel is achieved at an air-fuel mass ratio of approximately 14.7:1 (stoichiometric ratio). The ECU operates the engine at stoichiometric by proportioning the mass of fuel injected against the calculated mass of the air entering the engine. Internal programming of the ECU allows richer or leaner mixtures to be supplied on demand, e.g. lean cruise mode or wide open throttle.

INJECTOR FIRING MODES

The fuel injectors are of conventional electromagnetic design. Each inlet manifold branch contains an injector (one for each cylinder) positioned so the fuel spray is towards the inlet valve. The injectors operate at a nominal pressure of 250 kPa.

The injectors are arranged, electrically, in two banks. Bank 1 controls the injectors for cylinders 1, 3 and 5. Bank 2 controls the injectors for cylinders 2, 4 and 6. During normal operation banks 1 and 2 are fired alternately on each third rising edge PIP signal (see Figure 11C-28). When the engine is cranking on the starter motor, both injector banks are fired simultaneously on each PIP rising edge signal.

The next group of units to consider are those that have direct power supply from the battery or power relay (see Figure 11C-29).

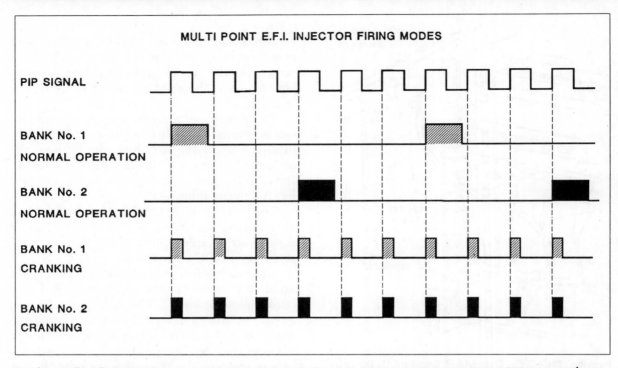

Figure 11C-28 Injection occurs in banks of three injectors at each third PIP signal (one engine revolution). During cranking the normal operating pattern is replaced and all injectors operate simultaneously on each PIP signal.

BATTERY VOLTAGE

An important concept to understand, in relation to voltage supply, is that all the outputs shown (lower right of ECU) are current sinking. This means 12 volts is supplied to each output and the second or grounding wire to complete the circuit is linked to the ECU. The ECU controls the operation of each output by switching the ground (earth) connection on or off.

Battery voltage is supplied to the ECU via the ignition switch, power relay and a direct line (KAPWR). From previous study, you will be aware that a vehicle battery is a 'nominal' 12 volt. However, in actual operation, the system voltage may vary considerably from the nominal value. This variation, if not compensated, will affect the system operation. As an example, consider the operation of a fuel injector during a period when supply voltage is low. The injector solenoid requires a very short time to react when it is switched on or off. Because the solenoid is voltage sensitive, the reaction times (known as opening offset and closing offset) will increase

as battery voltage decreases. When the ECU calculates the injector duration, the opening and closing offsets are added to the required fuel delivery time. Therefore by monitoring 'VBAT' battery voltage, the ECU can, if required, alter the offset times to compensate for variation in battery voltage. This monitoring operation, on some systems, is also used to compensate the ignition system for low battery voltage.

The 'KAPWR' (keep alive power) wire enables the ECU to retain 'memory' after the ignition key is switched off. This feature is necessary for the ECU to retain fault codes and develop learning strategies.

CANISTER PURGE (CANP)

Evaporative emission systems include a canister to store fuel vapors accumulated from the fuel tank. When the canister is purged, the fuel vapors are drawn into the engine. During some periods of engine operation it is undesirable to have fuel vapor entering the

ENGINE MANAGEMENT SYSTEM

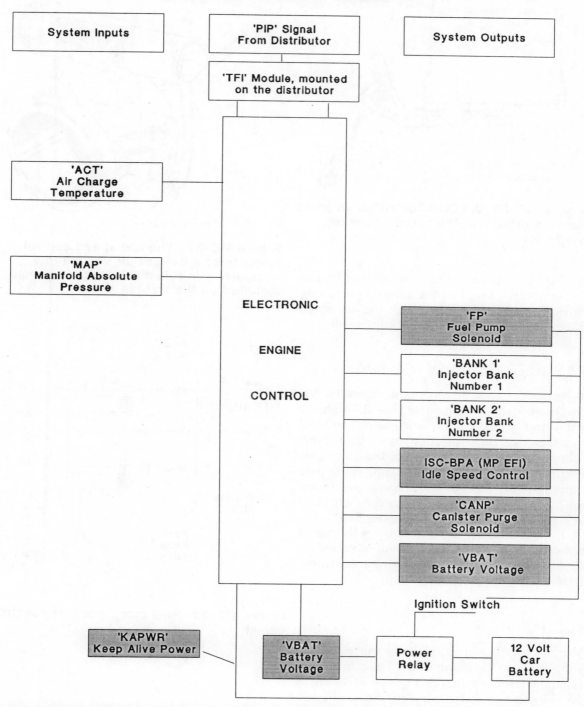

Figure 11C-29 Battery voltage is supplied to many components other than the injectors.

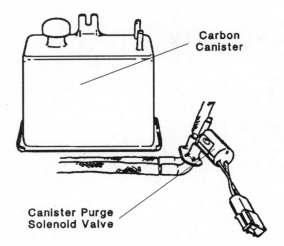

Carbon
Canister

Canister Purge
Solenoid Valve

Figure 11C-30 An ECU controlled solenoid valve is fitted into the carbon canister purge line.

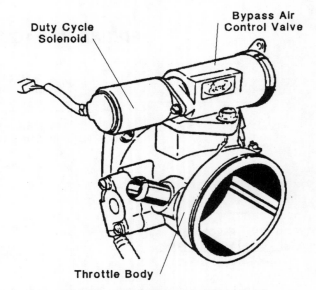

Duty Cycle
Solenoid

Bypass Air
Control Valve

Throttle Body

Figure 11C-31 The idle speed control consists of a bypass air valve and a computer controlled duty cycle solenoid mounted on the throttle body.

engine. A canister purge solenoid (see Figure 11C-30) valve fitted in the purge line can be turned on and off by the ECU, e.g. no purge during wide open throttle operating mode.

IDLE SPEED CONTROL (ISC-BPA)

A bypass air control valve is mounted on the throttle body, allowing air to pass through the valve and bypass the throttle plate. The ECU controls the valve position by a duty cycle solenoid attached to the end of the valve (see Figure 11C-31). Current control, from zero (minimum air flow) to one ampere (maximum air flow), is provided by varying the duty cycle of a 160 Hz square wave inside the ECU.

In addition to the standard idle speed correction, the base idle speed can be varied for loads applied by the air conditioner, transmission and the power steering.

FUEL PUMP (FP)

Voltage to the pump relay closing circuit is supplied from the power relay. The pump relay closing circuit is controlled by the ECU which provides the ground to complete the circuit (see Figure 11C-32). The ECU controls the pump relay during:

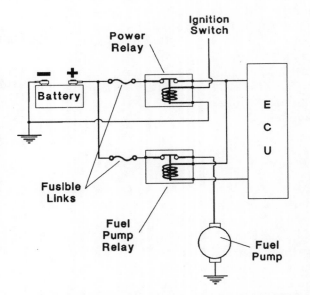

Figure 11C-32 Fuel pump electrical control circuit.

- fuel pressurisation;
- engine cranking;
- engine running;
- engine stalled.

Turning the ignition key 'on' causes the power relay points to close. To complete the pump relay circuit, a ground is provided by a timer circuit in the ECU. The pump relay points close and the pump starts to operate. If the ignition key is not turned to the engine cranking (start) position, within one second, the pump relay ground is opened within the ECU and the pump stops operating. This preliminary pump operation has the effect of pre-pressuring the fuel system.

The ECU also completes the relay circuit when the ignition key is turned to 'start'. After the engine starts, and the key is returned to the 'on' position, the fuel pump obviously must be able to continue operating. The ECU senses engine speed and, if in excess of 120 r.p.m., the fuel pump relay ground circuit is maintained and the fuel pump continues to operate.

For safety reasons, the fuel pump will cease operation if the engine stops or falls below 120 r.p.m. The ignition key may still be 'on' (engine stall or accident) but the ECU will not complete the relay circuit. The system will resume normal operation when an engine cranking signal is supplied.

ENGINE IDLE LOAD SENSORS

Compensation for idle speed is required for specific engine loads such as the air conditioning, power steering and transmission engagement (see Figure 11C-35). This section describes the methods used to advise the ECU that these loads are present and acting on the engine idle speed. The ECU will send a signal to the idle speed control valve as described in the section 'ISC-BPA'.

AIR CONDITIONER CLUTCH (ACC)

Air conditioning systems, when operating, require significant engine torque to rotate the compressor. This load on the engine could cause

the engine to stall at idle. By linking to the air conditioner compressor clutch power line, the ECU can monitor when and for how long the compressor clutch is engaged (see Figure 11C-33). Prolonged periods of compressor 'on' time indicates a high air conditioning load is present and the idle speed is slowly increased. Loss of the compressor clutch power signal during engine idle indicates, either the air conditioner has cycled or has been turned off. Under these conditions the ECU will react quickly to retain a smooth idle.

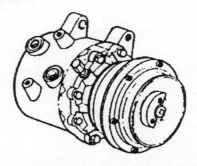

Figure 11C-33 A/C compressor clutch engagement is monitored by the ECU.

POWER STEERING PRESSURE SWITCH (PSPS)

A pressure switch is fitted to the power steering box to signal the ECU when high steering loads are present (see Figure 11C-34). This information is required to prevent the possibility of the engine stalling when parking with the engine idling. The ECU alters the air bypass valve position and the engine idle speed is increased to compensate for the steering load.

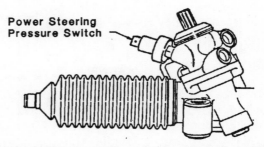

Power Steering Pressure Switch

Figure 11C-34 The ECU is advised of high steering loads by the pressure switch.

ENGINE MANAGEMENT SYSTEM

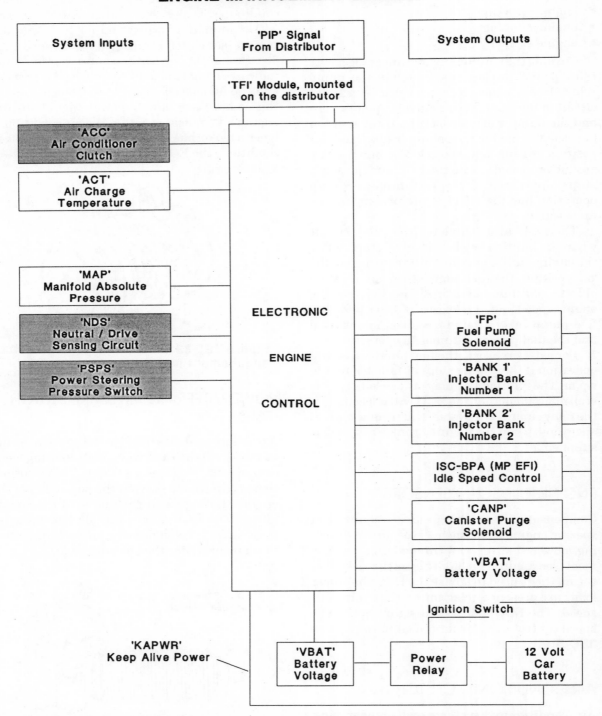

Figure 11C-35 The system is advised of external loads by the ACC, ACT, NDS and PSPS inputs.

NEUTRAL/DRIVE SENSING CIRCUIT (NDS)

The NDS signal is obtained from the ignition inhibitor switch on the automatic transmission (see Figure 11C-36). When Drive or Reverse is selected, the ECU monitors the change in switch position. The ECU is aware that transmission load has been applied to the engine and the air bypass valve position is altered to maintain smooth engine idle. The NDS signal is also used to cancel deceleration fuel shut-off and lean cruise mode if the selector lever is moved to neutral.

Manual transmission vehicles have two switches wired in parallel. The clutch switch can detect clutch engaged or disengaged. The second switch is mounted on the transmission to detect an 'in gear' or neutral position. The signal is not required for idle compensation but is used to cancel deceleration fuel shut-off and lean cruise mode if the engine drive is disconnected from the rear wheels.

TEMPERATURE, SPEED, THROTTLE POSITION AND OXYGEN SENSORS

AIR CHARGE TEMP. SENSOR (ACT) AND ENGINE COOLANT TEMP. SENSOR (ECT)

These sensors (see Figure 11C-38) provide information, in the form of a voltage signal,

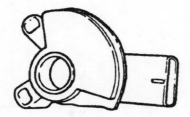

Neutral Switch (Auto Trans)

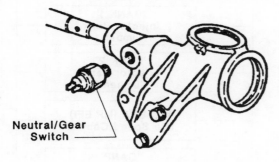

Neutral/Gear Switch

Neutral Switch (Manual Trans)

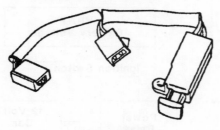

Clutch Switch

Figure 11C-36 Switches located at the transmission (manual/auto) and the clutch are used to advise the ECU of a neutral, drive or clutch disengaged condition.

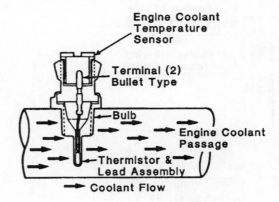

Engine Coolant Temperature Sensor

Terminal (2) Bullet Type

Bulb

Engine Coolant Passage

Thermistor & Lead Assembly

Coolant Flow

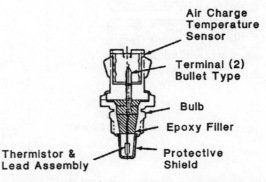

Air Charge Temperature Sensor

Terminal (2) Bullet Type

Bulb

Epoxy Filler

Thermistor & Lead Assembly

Protective Shield

Figure 11C-38 The ACT and ECT sensors provide reference signals to the ECU for air and coolant temperatures.

ENGINE MANAGEMENT SYSTEM

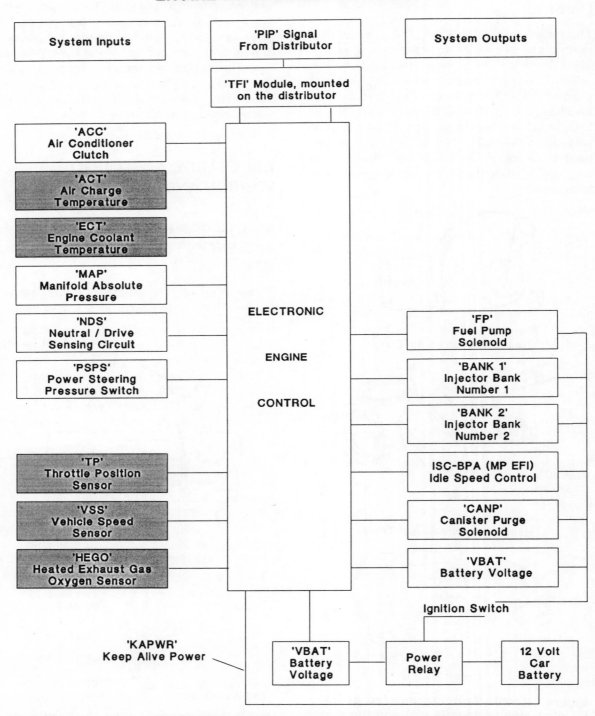

Figure 11C-37 Temperature, speed, throttle position and feedback information is provided by the sensors indicated.

directly relative to the temperature of the air in the inlet manifold and the engine coolant temperature. The ECU monitors the sensors closely and uses the information to:

- modify air-fuel ratio during warm-up;
- adjust spark timing for optimum cold engine performance;
- deny, under cold engine conditions, the operation of:
    —canister purge control;
    —deceleration fuel shut off;
    —closed loop fuel control.

VEHICLE SPEED SENSOR (VSS)

The ECU requires information related to vehicle speed when determining the relevant vehicle operating mode, e.g. lean cruise mode or deceleration fuel shut off. Vehicles using this system tend to be fitted with electronic speedometers, therefore the ECU can be linked to the transmission speedo transducer.

THROTTLE POSITION SENSOR (TP)

The throttle position sensor is a variable resistor and mounts to the end of the throttle butterfly shaft. A reference voltage is applied to the resistor and is grounded through the signal return line. The movable contact 'Vtp' supplies

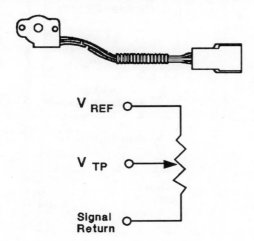

Figure 11C-39 Throttle position sensor.

a voltage to the ECU (see Figure 11C-39). The voltage value is dependent on the movable contact position on the resistor. In this way the ECU detects the throttle position: closed, part or fully open. The rate of change of voltage is also sensed and the ECU is aware of how quickly the accelerator is being depressed. This information, in conjunction with other sensor inputs, is used to:

- control deceleration fuel shut off;
- provide acceleration enrichment;
- assist in preventing/clearing a 'flooded' engine—wide open throttle position during engine cranking causes the fuel to be shut off;
- determine the engine operating mode, idle, cruise etc.

HEATED EXHAUST GAS OXYGEN SENSOR (HEGO)

The role of oxygen sensors has been described in previous systems within this chapter. The reader will be aware of the sensor operating characteristics and the terms 'open and closed loop'. This system contains a number of refinements to improve the operation of the sensor and to provide engine 'aging' information to the ECU.

The sensor (see Figure 11C-40) contains an electrical heating element which reduces the time between engine startup and the sensor becoming operational. The heating element prevents the oxygen sensor cooling down during extended engine idle. A 'cool' sensor cannot supply the ECU with a closed loop (feedback) signal. Heated type sensors are readily identified as they require three wires—one wire for the ECU signal and two wires to power the heating element.

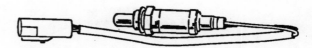

Figure 11C-40 HEGO - Heated exhaust gas oxygen sensor.

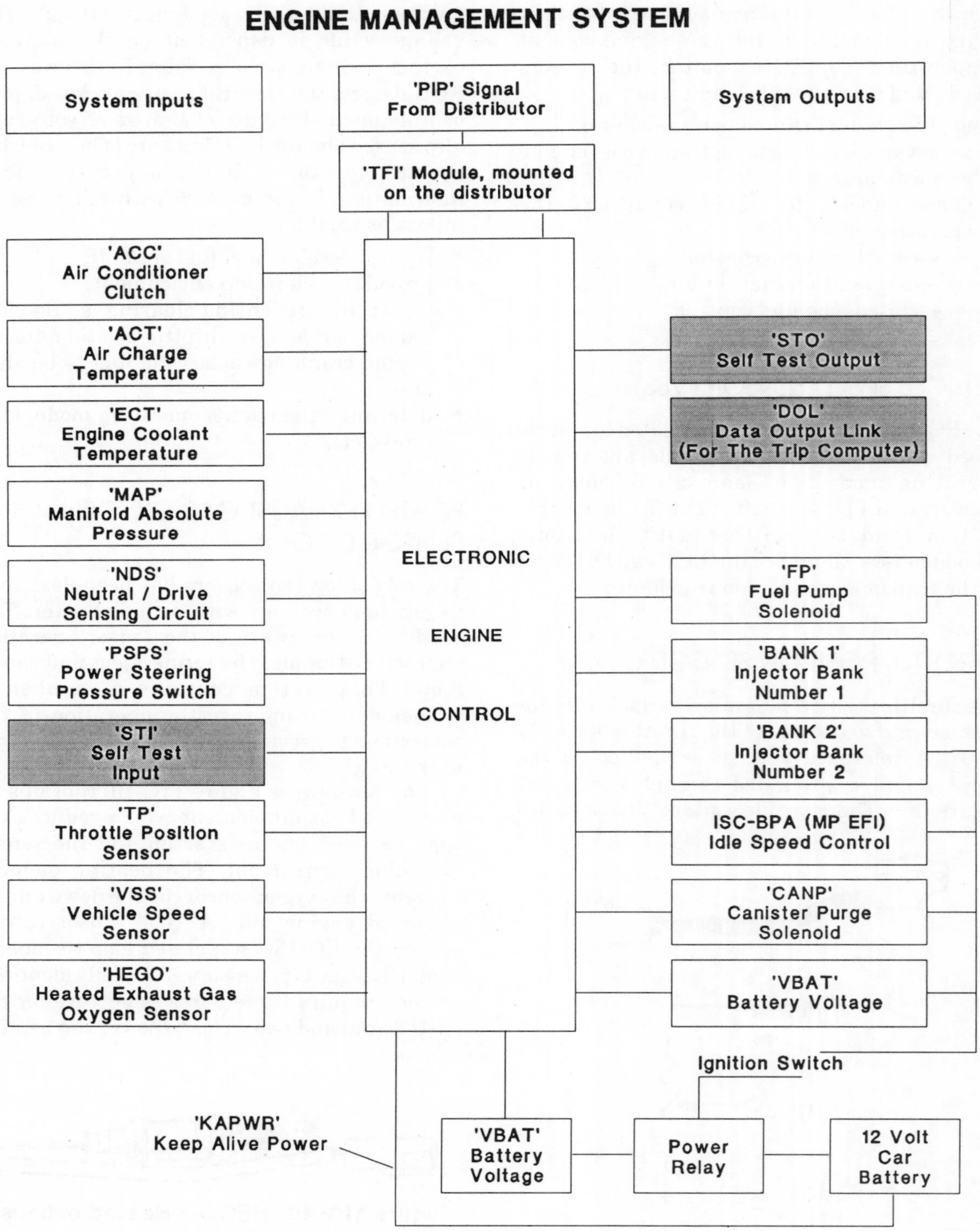

ENGINE MANAGEMENT SYSTEM

Figure 11C-41 The system is completed by the addition of the diagnostic and data output links.

During wide open throttle, lean cruise mode or cold engine operation the fuel requirement of the engine will not be stoichiometric—air-fuel ratio of approx. 14.7:1. During these operating conditions, the ECU changes to open loop (no feedback) and adjusts the fuel delivery to either rich or lean, as programmed, from values stored in the ECU.

The system also contains a learning strategy to compensate for aging of engine components, deterioration of the fuel system and individual differences between engines. When the engine is operating in the closed loop mode, the actual fuel requirement to maintain optimum engine operation is compared to a predicted fuel requirement. The predicted fuel requirement is based on the theoretical amount of fuel required if the system was open loop—no oxygen sensor feedback. If a difference exists between the actual and predicted fuel requirements, the ECU adjusts the predicted values for open loop operation. The corrected values are stored in a section of the ECU known as KAM (keep alive memory). This learning process continues through the life of the system, therefore the open loop modes; cold engine, wide open throttle and lean cruise, are effectively adjusted by information 'learnt' during the closed loop operating mode.

DIAGNOSTIC AND DATA OUTPUT LINKS

SELF-TEST INPUT (STI)

This is a service technician-controlled input which is grounded during fault finding diagnostic procedures. The relevant repair manual, for the vehicle under test, should be consulted before grounding this terminal (see Figure 11C-41).

SELF-TEST OUTPUT (STO)

Connections to this output terminal should only be made by a product trained service technician using test units that cannot cause accidental damage to the vehicle systems. The fault codes to diagnose system faults are emitted from this terminal.

DATA OUTPUT LINK (DOL)

This is a communication port to the trip computer. The trip computer requires fuel usage details from the ECU to provide a base for information it will display to the driver. The information needed is related to fuel use only and is not concerned with the problems of opening or closing the injector. The ECU fuel injection duration signal contains opening and closing offsets and also includes further compensation for battery voltage variation. Therefore the pulse width of the DOL signal will closely resemble the injector signal but will have less 'on' time because of the absence of the offsets and battery compensation times.

SYSTEM 3   MULTI-POINT INJECTION WITH A HOT WIRE AIR SENSOR

The microcomputer based control unit in this engine management system has six main functions providing control of the:

- fuel injection;
- ignition;
- idle speed;
- fuel pressure;
- fuel pump;
- self-diagnosis.

To avoid unnecessary repetition, system components which have similar construction and operation to units previously described will receive minimal detail. This approach will enable the reader to concentrate on components with specific operating differences.

The components selected for detailed explanation in this system are the:

- 'hot wire' type air flow meter;
- air bypass valves for idle control;
- fuel pump control;
- fuel pressure control;

- ignition timing control;
- crank angle sensor fitted with LEDS and photo-diodes;
- fuel injection control;
- ECU onboard diagnostics.

AIR INTAKE SYSTEM

The air intake system consists of an air cleaner, air flow meter, throttle body and air bypass valves (see Figure 11C-44).

AIR FLOW METER (HOT WIRE TYPE)

The air flow meter (see Figure 11C-42) does not contain any moving parts. A thin platinum wire is looped in a triangular pattern and fitted inside a plastic tube. An electrical current is passed through the wire and the wire increases in temperature. Because of this feature, the air flow meter is commonly described as a 'hot wire' air mass meter. Air entering the engine passes over the heated wire, removing heat from the wire. Cooling of the wire lowers its resistance, causing a change in the voltage relationship between the heating wire and the internal circuitry of the meter. The control circuit detects this change and increases the

current flowing through the heating wire, restoring the wire to its original temperature. This correction occurs rapidly (a few milliseconds), therefore the temperature of the heated wire is virtually constant regardless of change in the air flow rate. A direct relationship between the current required to heat the wire and the quantity of air entering the engine is established. Based on this relationship, an electrical signal is sent to the electronic control unit advising the quantity of air entering the engine.

The air flow meter (see Figure 11C-43) also contains:

- an air temperature compensation resistor which compensates for changes in air intake temperature. The low mass of the platinum film resistor, chosen for this application, ensures rapid response to air temperature variation. The output signal is matched to air temperature within seconds of a change occurring.;
- a self-cleaning control circuit. If the 'hot wire' becomes dirty, air passing over the surface of the wire would not cause the same change in electrical value of the wire.

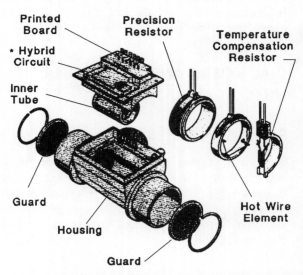

* In addition to the resistors of the bridge circuit, the hybrid circuit also contains the control circuit for maintaing a constant temperature and the self-cleaning circuit.

Figure 11C-43 Hot wire air mass meter.

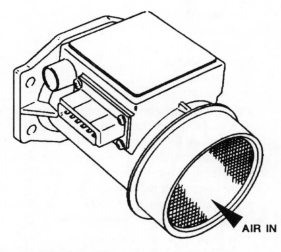

Figure 11C-42 The air flow meter is a 'hot wire' type. Intake air passes through the tube and across a heated wire.

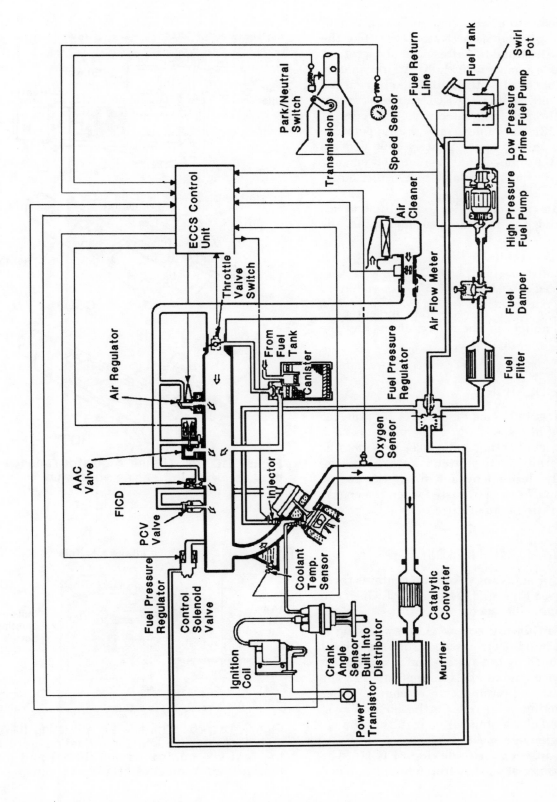

Figure 11C-44 Electronic fuel injection fitted with 'hot-wire' type air flow meter.

To ensure the wire remains clean a self-cleaning circuit is activated each time the engine is shut down after normal operation. (> 1500 r.p.m. or > 20 km/h). Approximately six seconds after the ignition key is turned to the 'off' position the ECU causes a current flow through the heated wire. This current is of sufficient value to raise the temperature of the wire to approximately 1000 degrees Celsius for one second. This 'burn off' ensures the wire remains free of dust or dirt build-up.;

- an idle potentiometer (not illustrated), enabling the mechanic to adjust the idle mixture.

AIR BYPASS VALVES

Air can bypass the throttle plate and enter the intake manifold at the air regulator and the idle air adjusting unit.

AIR REGULATOR

The air regulator (see Figure 11C-45) provides increased idle speed when the engine is cold. An electrical heating coil causes a bi-metallic strip, acting on a slotted shutter, to progressively close off bypass air as the engine warms up. No air is bypassed when the engine is at operating temperature.

IDLE AIR ADJUSTING UNIT

The idle air adjusting (IAA) unit contains three devices (see Figure 11C-46), each of which can affect engine idle speed:

- **Idle adjusting screw**. A screwed plug is fitted into the air passage. The screw is not controlled by the system and is used to set the base engine idle to specifications. The plug can be rotated by a mechanic, either restricting or increasing the air flow past the end of the plug with a resulting change in engine idle speed.
- **Fast idle control device (FICD)**. The FICD is a solenoid operated plunger. When

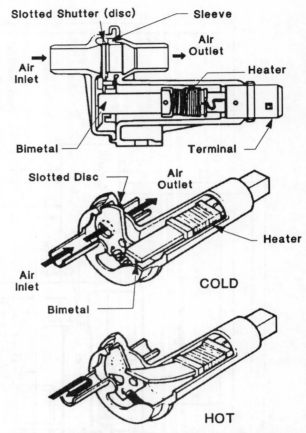

Figure 11C-45 The air regulator reduces bypass air as the engine warms up.

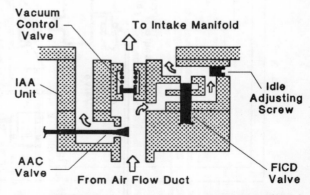

Figure 11C-46 The idle air adjusting (IAA) unit contains the idle air adjusting screw, the fast idle control device (FICD) and the auxiliary air control (AAC).

the air conditioning system is turned 'on', the solenoid is activated and a plunger is withdrawn from the air passage. The increased bypass air flow will be detected by the air flow meter. The ECU will increase the injector opening duration and the idle speed will rise to compensate for the load placed on the engine by the air conditioner.

- **Auxiliary air control (AAC) valve**. The AAC valve is positioned by a duty cycle solenoid operating at a constant 160 Hz frequency. The valve admits air through a bypass passage into the intake manifold. The increased air flow will be detected by the air flow meter and the idle speed will alter in a similar manner as caused by the FICD. The operation of the valve is relatively simple; but more complex are the sensor inputs and electronic control to accurately position the valve.

Engine idle speed is not a constant value for all conditions. By monitoring the range of sensors (as illustrated in Figure 11C-47), an optimum idle speed can be selected. As the engine warms up, the ECU calculates the most suitable idle speed based on engine coolant temperature and the gear position selected. The actual idle speed and the calculated idle speed are compared. Any difference between the two speeds causes a change in the duty cycle signal sent to the AAC valve. Greater 'on' time and less 'off' time results in increased engine idle speed.

The duty cycle signal continues to be sent to the valve after the engine reaches operating temperature. By applying correction factors to the base duty cycle (varying the on/off time) the valve can fulfil several roles.

- Hot engine starting is assisted; the valve moves to maximum value for 10 seconds after a start signal is received and the key is returned to the 'on' position.
- During normal starting the valve is at its base duty cycle (nominally 80 per cent 'on' time).
- Correction to the times the valve will be 'on' also occur during deceleration to allow the engine r.p.m. to reduce smoothly and also prevent the engine stalling.
- A constant engine idle speed can be maintained with compensation for loads such as power steering, electrical and automatic transmission. Any actual idle speed variation greater than 25 r.p.m. from the calculated idle speed (after base duty cycle and gear position correction) causes a feedback correction circuit to be activated. The feedback correction will bring the idle r.p.m. to the required speed.

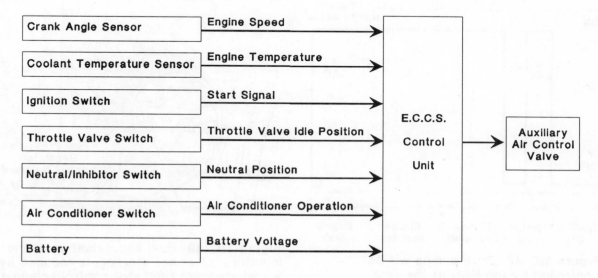

Figure 11C-47 The duty cycle of the auxiliary air valve is varied in response to a range of input sensors.

- Idle speed will be increased by 100 r.p.m. if the battery voltage sinks below 12 volts for more than 1.3 seconds. The increased idle speed will raise alternator output.
- Idle r.p.m. lower limit can be preset from the mode selection switch of the ECU. The mechanic should be aware rotation of this switch clockwise in stages has the effect of raising the idle lower limit in steps of 25 r.p.m. to a maximum of 150 r.p.m.

FUEL DELIVERY SYSTEM

The fuel system contains a fuel tank, two pumps, pulsation damper, filter, pressure regulator and multi-point fuel injectors inserted into the intake manifold (see Figure 11C-42).

Fuel delivery is provided by two pumps:

- a low pressure (20—40 kPa) electric centrifugal unit mounted inside the fuel tank and used to prime the externally mounted high pressure pump;
- a high pressure (in excess of 250 kPa) electric roller type pump, mounted externally, supplying the requirements of the injection system.

Fuel Pump Signal

Figure 11C-48 Both pumps will be controlled by the ECU at the time intervals and operating conditions shown in the graph.

Safety control of the pumps is derived from the ECU (see Figure 11C-48). Both pumps are wired in parallel and supplied from the fuel pump control relay. The ECU completes the circuit to energise the pump relay after the ignition key is turned to the 'on' position. If the key is not turned to 'start' within 5 seconds the relay circuit will be opened and the pumps will cease operation. To maintain the relay circuit the ECU requires a continuing signal from the crank angle sensor. One second after the crank angle signal ceases (engine stall) the ECU will open the relay circuit and the pumps cease operating.

FUEL PRESSURE CONTROL

In addition to a conventional spring-loaded diaphragm pressure regulator, a fuel pressure regulator control solenoid valve is fitted to the system. The solenoid valve is turned on and off by the ECU and is designed to minimise fuel vaporising at the injectors during hot engine starting.

The solenoid valve is mounted to the intake manifold (see Figure 11C-49) and the pressure regulator vacuum supply tube is connected to a fitting on the valve. Vacuum to the pressure regulator must pass through the solenoid valve.

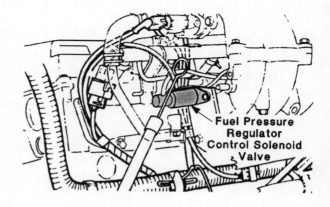

Figure 11C-49 Fuel vapourisation at the injectors, after hot starting, is reduced by a fuel pressure regulator control solenoid valve.

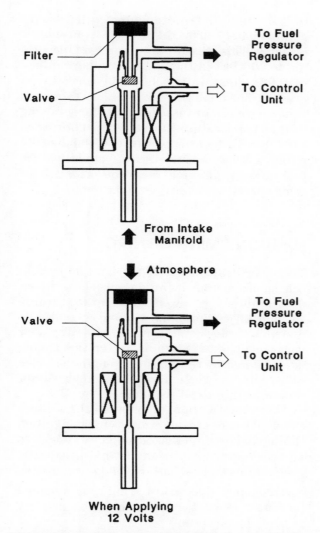

Figure 11C-50 The fuel pressure regulator control solenoid valve can supply either intake manifold vacuum or atmospheric pressure to the fuel pressure regulator.

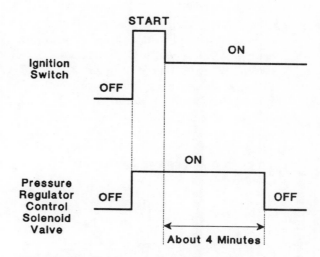

Figure 11C-51 At coolant temperatures in excess of 95 degrees C, the solenoid valve is turned on for approx. 4 minutes after starting.

The solenoid valve (see Figure 11C-50), when activated by the ECU, pulls down a valve which closes the intake manifold passage. Outside air can now enter through a filter, pass through the valve and out to the fuel pressure regulator. Therefore atmospheric pressure is applied to the diaphragm of the regulator.

The ECU will only activate the valve if, during starting, the engine coolant temperature is greater than 95 degrees Celsius (see Figure 11C-51). The 'on' signal will continue for approximately 4 minutes before the valve is turned 'off' and normal system operation will be resumed.

As stated previously, the valve is designed to minimise the possibility of vapour forming in the fuel at the injectors. A simple method of reducing vapor formation is to increase the pressure acting on the fuel in the fuel distributor pipe. Before describing how this is achieved a quick review of pressure regulator operation may be helpful.

The intake manifold vacuum is linked to the pressure regulator diaphragm. Variations in manifold pressure (vacuum) will cause changes in regulated pressure (see Figure 11C-52). The fuel injector pressure will rise and fall but a constant 250.1 kPa pressure difference will be maintained between the pressure in the fuel distributor pipe and the pressure in the intake manifold.

During the period (left-hand side of graph) (see Figure 11C-53) the valve is turned 'on' by the ECU, the connecting vacuum pipe to the pressure regulator is vented to the atmosphere. Because there is no correction to the regulator from the intake manifold vacuum, the pressure

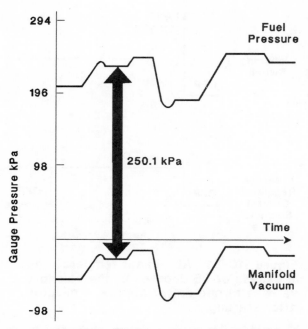

Figure 11C-52 Fuel pressure is maintained at 250.1 kPa above manifold vacuum during normal engine operation.

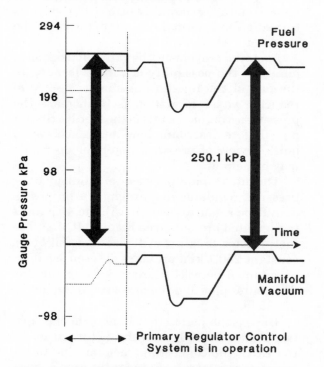

Figure 11C-53 Fuel pressure is increased by venting the regulator vacuum intake to the atmosphere.

in the fuel distributor pipe will rise to approximately 250.1 kPa above atmospheric pressure. This pressure is a significant increase (up to 50 kPa) on the 250.1 kPa above intake manifold pressure which would normally be present. The increased fuel pressure decreases vapour formation caused by engine heat, raising the fuel temperature. The valve must be turned 'off' after the fuel temperature is stabilised (4 minutes allowed after starting) otherwise the fuel quantity injected would be incorrect for some operating conditions of the engine.

IGNITION TIMING CONTROL

The basic components of the ignition system consist of a moulded ignition coil, a gear driven distributor and a power transistor for turning 'on' and 'off' coil primary winding current. The distributor does not contain centrifugal or vacuum advance mechanisms. Ignition timing is calculated by the ECU and engine position referencing is supplied by a crank angle sensor located within the distributor.

From information supplied by the sensors the ECU can provide variations in ignition timing and dwell period (see Figure 11C-54) to suit ordinary operation, cranking, idling, deceleration and low battery voltage conditions.

- **Ordinary operation**. Timing is supplied from stored data based on injection quantity and engine speed.
- **Cranking**. Timing is supplied from stored data but two correction possibilities are available. If engine cranking speed is normal (greater than 100 r.p.m.) but coolant temperature is less than 0 degrees Celsius, the ignition timing is advanced. The delay in combustion pressure rise caused by the cold condition is compensated by the increased advance. Alternatively, if cranking speed is low (less than 100 r.p.m.) the combustion pressure may rise before the piston has time to pass TDC. This pressure rise will slow the engine further by attempting to reverse the direction of crankshaft rotation. Under these conditions the ignition timing is delayed.

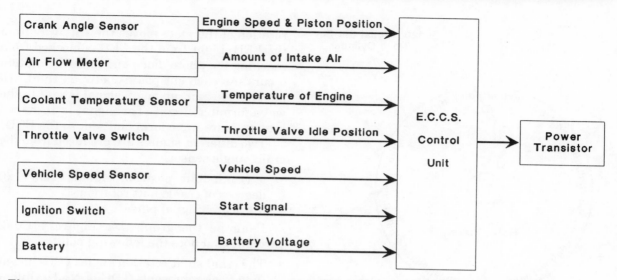

Figure 11C-54 Ignition timing and dwell period are supplied by the control unit after evaluation of each of the input sensors.

- **Idle and deceleration** timing data is supplied when the ECU detects a closed throttle signal from the throttle switch.
- **Dwell control**. Ignition coils have inductance, and a time is required for the coil to reach maximum magnetic saturation.(for maximum secondary output). If the voltage supplied to the coil is reduced, the time required by the coil to 'saturate' will increase. The ECU can detect low battery voltage conditions and compensates by turning the ignition power transistor 'on' earlier. This action increases the dwell period of the coil.

CRANK ANGLE SENSOR

Engine speed and engine piston position (crankshaft rotational position) signals are supplied to the ECU by a crank angle sensor (see Figure 11C-55). The sensor assembly is contained within the lower half of the ignition distributor and consists of three sections:

- rotor plate;
- sensor unit;
- signal wave forming circuitry.

The rotor plate, which is mounted on and turns with the centre shaft, has two sets of slits cut through it. The outer set consists of 360 slits set at 1 degree intervals. The inner set, starting

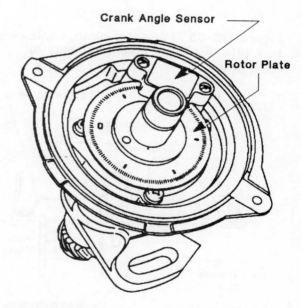

Figure 11C-55 The crank angle sensor supplies engine speed and engine piston position signals to the ECU.

with a slightly larger No. 1 cylinder reference slit, has six slits set 60 degrees apart. (corresponding to 120 engine degrees).

The sensor unit (see Figure 11C-57) contains two LED and photo-diode pairs mounted across a small gap. The slits in the rotor plate are

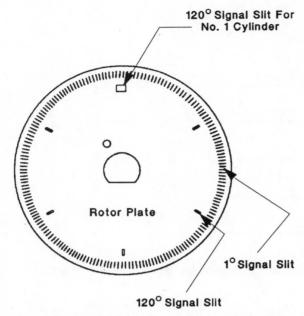

Figure 11C-56 The rotor plate, containing two sets of slits, is mounted inside the distributor.

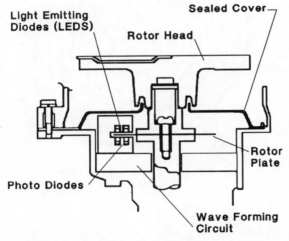

Figure 11C-57 The slits in the rotor plate pass through the gap between a pair of light emitting diodes and photo diodes.

positioned so they pass through the gap between the LED and the photo-diode as the plate rotates. When a slit aligns with one of the LED/photo-diode pairs, light from the LED falls on the photo-diode, causing an 'on' signal. Continued

rotation of the plate moves the slit away from the LED. Light from the LED is blocked from reaching the photo-diode and an 'off' condition occurs. As each slit passes between the LED/photo-diode pair, an on/off signal is sent to the wave forming circuit. The signal is reshaped into a square wave form and sent to the ECU.

Two different signals are produced from the crank angle sensor:

- a crankshaft position signal every 120 degrees of crankshaft rotation;
- a 1 degree signal pulse.

Examine the graph (see Figure 11C-58) carefully and note the following points.

- The crankshaft rotation angle (TDC) line at 0 degrees represents TDC of No.1 cylinder on compression. Each following 120 degrees is the TDC point of the next cylinder in firing order. (120 = No.5, 240 = No.3, etc).
- The crank angle sensor 120 degree signal is positioned so that the on/off signal will act as a reference point of 70 degrees BTDC for each respective cylinder. The signals are 120 degrees apart, but *do not* occur at cylinder TDC.
- The 1 degree signal pulse count is provided so the ECU can count down from the 70 degree reference point to place the ignition spark at the correct firing point.

EXAMPLE

Information from various engine sensors is fed to the ECU. The ECU calculates the ignition timing advance required is 40 degrees BTDC. Obviously it is not possible for the ECU to detect TDC then count backwards to position the timing because the event would have already occurred. The crank angle signal makes the ECU aware that a cylinder will be at TDC compression in 70 degrees of crankshaft rotation. The ECU accepts the crank angle signal, then counts 1 degree signal pulses until the difference between 70 degrees and 40 degrees is reached (70—40 = 30). In this example, the count would be 30 (including the 4 degree on/off signal width). The ECU turns off the coil primary current power transistor

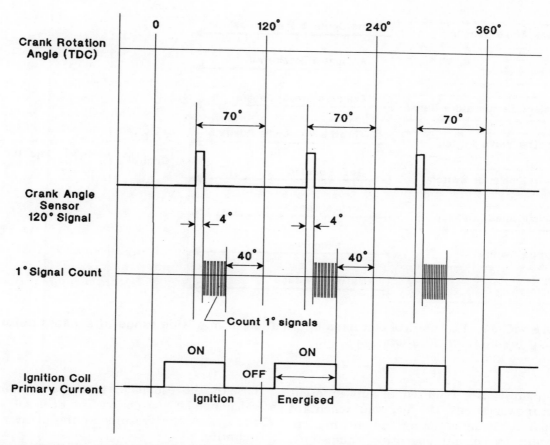

Figure 11C-58 On detection of the 120 degree indicator (set at 70 degrees BTDC) a countdown from 70 degrees is implemented to accurately position the ignition spark.

and the high voltage spark will be delivered at precisely 40 degrees BTDC.

FUEL INJECTION CONTROL

The system is of the multi-point type with the injectors fitted into the intake ports. Injection occurs once each crankshaft revolution with all injectors 'firing' simultaneously. The single crankshaft revolution interval between injections means the injectors will spray fuel into the intake ports twice during the four stroke cycle of the engine. The total fuel required for a cylinder working stroke is therefore supplied in two equal parts and the injector opening time per injection is kept to a minimum.

Injection timing is referenced from the crank angle sensor, but is independent of the working strokes of the engine cylinders. Fuel may be injected into the ports while the intake valve is closed. The air-fuel mixture in the port is then drawn into the engine cylinder the next time the intake valve opens.

FUEL INJECTION QUANTITY

Figure 11C-59 depicts the range of sensors supplying information to the ECU. As with many other systems, the basic fuel injection quantity is developed from information related to engine speed (crank angle sensor) and engine load (air flow meter). Corrections to the basic

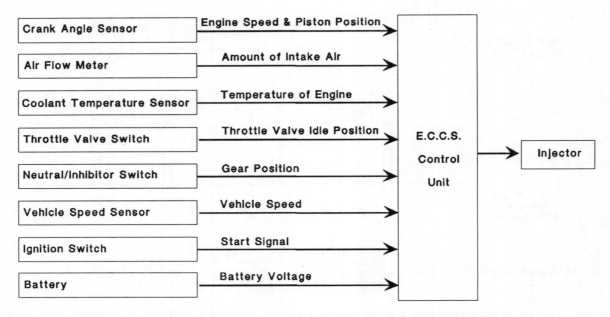

Figure 11C-59 The control unit uses information from a wide range of sensors before setting the injector duration.

fuel quantity are required to suit different operating conditions. The fuel requirements of an engine, during cranking and starting, are different to normal operating conditions, therefore the description of injection quantity control will be explained in two parts:

- Engine starting.
- Other conditions (engine running).

ENGINE STARTING

When the ignition switch is turned to the 'start' position there is a brief moment before the engine starts to turn. During this period there is no crank angle sensor or air flow meter signal available to the ECU. A 'first time injection' value based on engine coolant temperature and taken from data stored in the ECU determines the fuel quantity to be injected (fuel quantity is greater if coolant temperature is lower).

After the engine starts to rotate the 'first time injection' quantity must be corrected. Two possible options are programmed into the ECU. The decision to use either option 1 or option 2 is based on the result of the fuel injection quantity calculated by each option. The ECU will select the option that results in the greater fuel quantity.

Option 1. The injection quantity is calculated by the ECU using the normal fuel injection quantity as a reference. A further addition to the injector opening period will be required if battery voltage correction is required.

Option 2. The fuel injected is based on the 'first time injection' quantity. The 'first time injection' quantity is modified by two reducing coefficients that reduce the fuel quantity injected as cranking speed and time increase.

As ignition occurs, the engine rotational speed causes the air intake quantity to increase. The increased air flow generates a signal from the air flow meter. This signal feeds into the normal injection quantity component of option 1. The fuel injection quantity calculated by option 1 will increase. When the option 1 quantity exceeds the option 2 quantity, the ECU will adopt option 1 to determine engine fuel requirements.

ENGINE RUNNING

The 'normal' injection quantity is the result of correction factors being applied to the basic fuel injection quantity (basic quantity is derived from engine speed and load). These correction factors are necessary because engines are required to accelerate, decelerate, run smoothly both hot and cold, and not over-speed. The operating conditions, which are considered by the ECU to require correction to the basic fuel quantity, are listed below.

- **High speed/heavy load operation**. A mixture ratio correction coefficient referenced to engine speed and basic injection quantity is stored in the ECU. Whenever high speed or heavy load operation is detected the mixture ratio coefficient increases the quantity of fuel injected.
- **Low engine coolant temperature**. The fuel quantity is increased whenever the engine coolant temperature is below normal operating temperature.
- **High engine coolant temperature**. Detonation could occur under some operating conditions if the mixture strength is not enriched. Ten per cent additional fuel will be injected at temperatures greater than 95 degrees Celsius during off-idle operation.
- **After starting**. The mixture is enriched after the engine starts. The enrichment will cease after a short period and is required to allow the engine to stabilise after starting.
- **Vehicle starting from rest**. Mixture enrichment is provided to enable a smooth transition from idle to off-idle operation. The enrichment value determined will be modified further by input from the engine coolant sensor. The enrichment will be progressively reduced as the vehicle moves away, however if the throttle is returned to the idle position the enrichment will cease immediately.
- **Acceleration**. Two compensation factors must be considered. Sudden acceleration from engine idle position can cause an 'over response' from the air flow meter. The false voltage signal generated is decreased in value (see Figure 11C-60) to prevent the

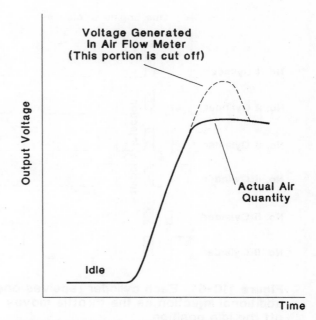

Figure 11C-60 Under acceleration, from idle, the 'over response' from the air flow meter is decremented allowing the fuel injected to be reduced to the correct quantity.

fuel quantity calculations producing an incorrect result. This correction feature ceases immediately if the throttle is returned to the idle position or the engine speed is in excess of 3200 r.p.m.

The second factor is the need to overcome the lean period immediately following the throttle opening from the idle position. A 'once only' additional injection (see Figure 11C-61) occurs for all cylinders simultaneously. The standard injection timing is resumed immediately following the additional injection.

- **Deceleration**. If normal injection quantities continued during deceleration, fuel would be wasted and emission levels would rise. To achieve the fuel reduction required, the ECU performs a calculation with inputs from throttle position, engine speed, vehicle speed and engine coolant temperature. A slight delay occurs before the fuel quantity reduction is applied. When the throttle is

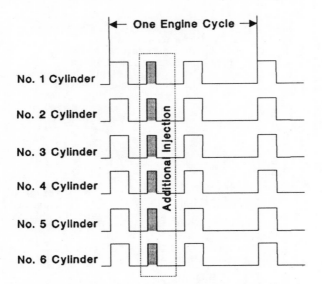

Figure 11C-61 Each cylinder receives one additional injection as the throttle moves off the idle position.

released, the vehicle starts to 'push' the engine. If the fuel reduction occurred at the same time as the change in engine load, simultaneous torque shocks would be applied to the engine. To smooth the transition from acceleration to deceleration the fuel reduction signal is delayed by 0.6 second for automatic transmission equipped vehicles. Manual transmission equipped also utilises a delay in fuel reduction, but spaces the reduction into two parts. Cylinders 1, 2 and 3 reduce fuel after 0.3 seconds, then a further 0.3 seconds later, fuel is reduced on cylinders 4, 5 and 6.

The fuel reduction will continue until the fuel recovery zone is reached or the throttle is reopened. The fuel recovery zone (r.p.m. range at which normal injection will be restored) is a variable. The engine coolant temperature, transmission type or air conditioner are factors which are considered in determining the fuel recovery zone, therefore the engine r.p.m. at which normal injection is restored will vary.

Deceleration fuel reduction will operate in any gear, therefore a vehicle speed lower limit below which the fuel reduction will not occur is preset in the ECU. Vehicle road speed is advised to the ECU from a signal generated by the vehicle speed sensor (see Figure 11C-62). This unit is mounted within the speedometer assembly and rotation of the speedo cable causes a pulsed signal to be sent to the ECU.

- **Overspeed**. Injection pulse width is controlled to prevent the engine exceeding 6200 r.p.m.

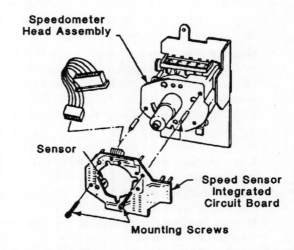

Figure 11C-62 For each revolution of the speedometer cable a two pulse signal is sent to the ECU.

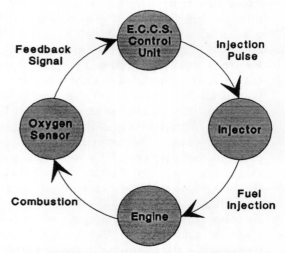

Figure 11C-63 The oxygen sensor is a vital part of a closed loop feedback system.

- **Low battery voltage**. Any reduction in voltage, below the design specification, results in slow injector plunger movement. The ECU compensates for low voltage by increasing the injector 'on' time.
- **Closed loop feedback control**. A heated oxygen sensor is fitted into the exhaust manifold to supply a 'feedback' signal to the ECU. The system reverts to open loop when the engine is starting, idling, operating at high speed or heavy load. The feedback signal will also be disregarded when the vehicle is decelerating with the fuel reduction feature operating and if a 'too lean' condition continues for more than 10 seconds (fault condition is present).

ELECTRONIC CONTROL UNIT

The ECU is microprocessor based and has 'on-board' diagnostic capabilities. The ECU casing has apertures which allow access to a diagnostic mode selector and a pair of inspection lamps (LEDs) (see Figure 11C-64). The LEDs, when flashed in a coded sequence, provide diagnostic information to the mechanic.

The range of inputs and outputs serviced by the control unit is depicted in Figure 11C-65. The sensors and switches form the inputs to the ECU. The ECU has stored a set of possible

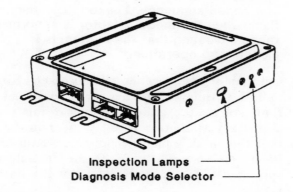

Inspection Lamps
Diagnosis Mode Selector

Figure 11C-64 On-board diagnostics are fitted to the ECU.

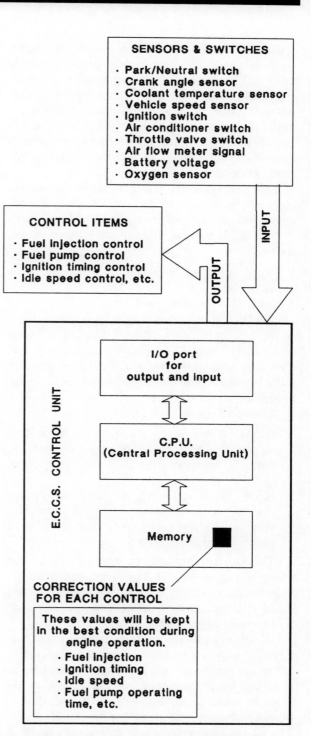

SENSORS & SWITCHES

- Park/Neutral switch
- Crank angle sensor
- Coolant temperature sensor
- Vehicle speed sensor
- Ignition switch
- Air conditioner switch
- Throttle valve switch
- Air flow meter signal
- Battery voltage
- Oxygen sensor

CONTROL ITEMS

- Fuel injection control
- Fuel pump control
- Ignition timing control
- Idle speed control, etc.

INPUT

OUTPUT

E.C.C.S. CONTROL UNIT

I/O port for output and input

C.P.U. (Central Processing Unit)

Memory

CORRECTION VALUES FOR EACH CONTROL

These values will be kept in the best condition during engine operation.
- Fuel injection
- Ignition timing
- Idle speed
- Fuel pump operating time, etc.

Figure 11C-65 The control unit has a stored 'back-up' value for each of the major controls. This value can be supplied to the output if a sensor input signal falls outside a pre-determined range.

operating values for each input. If an input signal is outside the permissible operating range, the ECU assumes a fault condition is present and substitutes a *back-up* value for the faulty sensor. This feature allows the vehicle to continue operating until repairs can be effected. The only sensor input, for which the control unit cannot supply a substitute value, is the crank angle sensor.

ECU outputs are the actuators (injectors, etc) which are used to control the engine operation, as determined by the ECU program.

The ECU can be divided into three major parts:

- input/output—I/O port;
- memory;
- central processing unit.

The I/O ports are a processing area where conversion of the signal form takes place. Input signals are changed to a form acceptable to the CPU. In a similar way, information from the CPU is converted to a signal form suitable for operating the actuators.

The memory has two areas:

- **ROM**. Read only memory: where the operating program and the preset calibration data for injection, ignition timing etc. are stored.
- **RAM**. Read and write memory: where the results of calculations derived from the sensor inputs, operating program and correction data can be temporarily stored.

The central processing unit (CPU) is the controlling area of the microcomputer. Information from the input port, ROM and RAM will be processed, then the computed result will be sent to the output port.

A self-diagnostic system is mounted 'on-board' the ECU and provides information to the servicing mechanic, via two LEDs operating in a coded flash sequence. A diagnostic mode selector (see Figure 11C-66) can be rotated to place the system into normal or diagnostic mode. The LEDs, one red and one green, may flash in both modes. In normal mode the LEDs flash as a result of the air-fuel mixture. When the system is operating in the closed loop mode (oxygen sensor feedback), both or one LED may flash.

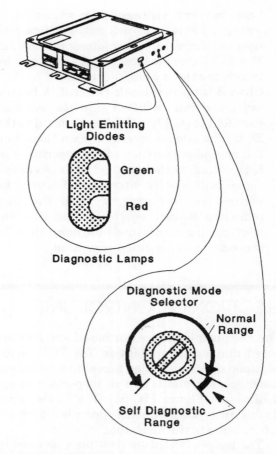

Figure 11C-66 The on-board diagnostics are accessed by rotating the mode selector fully clockwise (do not use excessive force). The LED's will flash in a coded sequence. These LED's also flash when the mode selector is in normal range indicating that the system is in closed loop operation.

Optimum engine operation will cause both LEDs to flash in time with each other. Variations in the flashing rate of either LED indicate mixture strength variation or possibly an indication that the air flow meter requires adjustment. Refer to the manufacturer's repair manual for specific details of how to interpret the flash rate.

Rotating the diagnostic mode selector fully clockwise automatically starts the ECU self-diagnosis system. The red LED assumes the value of 'tens' and the green LED becomes the 'ones'. Any faults detected during system operation will be captured and stored in the ECU. The self-diagnostic system checks for the presence of fault codes and displays the information as a series of coded flashes, e.g. one red flash, then both LEDs out for a 2.4 second period followed by three green flashes indicates a fault code of 13. In this system, the fault code indicates a malfunctioning coolant temperature sensor, connection or associated wiring. As previously indicated, it is essential to follow the manufacturer's repair manual instructions for specific vehicles. On start-up this particular system could display up to four coded signals, yet no fault is present in the system. The manufacturer check sequence requires the mechanic to drive the vehicle above a determined speed, move the throttle, move the transmission selector and operate the air conditioner. The fault codes will be erased if faults are not present in the vehicle systems.

Remember, a fault code does not necessarily mean the component is at fault. Prior to component replacement the location of the fault should be confirmed by unit testing and wiring checks.

SYSTEM 4 MULTI-POINT INJECTION WITH DIRECT FIRE IGNITION

The engine management system described in this section has been applied to a V6 engine. The engine is fitted with multi-point fuel injection with the injectors spraying into the intake ports. The system does not require an ignition distributor and uses an ignition type known as direct fire ignition (DFI). The management function has extended beyond the fuel, ignition and idle load areas into the automatic transmission and the engine cooling system. Figure 11C-67 illustrates the range of the operating conditions sensed and the control functions provided by the system computer.

In a similar way to the systems described previously, this system will be separated into air, fuel, ignition and electronic control subsystems. Emphasis will be placed on those components which have not already been described, particularly the ignition system.

AIR INTAKE SYSTEM

The air intake system contains an air cleaner, ducting, manifold absolute pressure (MAP) sensor, manifold air temperature (MAT) sensor, throttle body and an idle air control (IAC) valve.

MANIFOLD ABSOLUTE PRESSURE (MAP) SENSOR

Engine load and r.p.m. changes will cause a variation in the intake manifold vacuum. By linking the MAP sensor to the intake manifold vacuum, by a length of vacuum tubing (see Figure 11C-69), it is possible for the ECM to be aware of the engine load condition. The electronic control module (ECM) provides a 5 volt electrical source to the MAP sensor. A second wire provides a return path to the ECM where the output voltage from the MAP sensor is monitored. The third wire is for sensor earthing within the ECM.

When the throttle is opened, manifold vacuum will decrease (the pressure in the manifold will rise towards atmospheric pressure). The rise in pressure causes the MAP sensor output voltage to also rise towards the supply voltage value. The ECM monitors the voltage rise, knows the throttle has been opened and the engine will need more fuel, and so extends the injector 'on' time.

When the vehicle is decelerating, the intake manifold vacuum becomes very strong. The manifold absolute pressure will decrease and the MAP sensor output voltage will also decrease. The ECM knows the engine fuel requirement is low and reduces the injector 'on' time.

The relationship between the MAP sensor, the manifold absolute pressure and the outside atmosphere enables the ECM to make corrections for altitude and climatic changes.

The MAP sensor acts as a major input to the

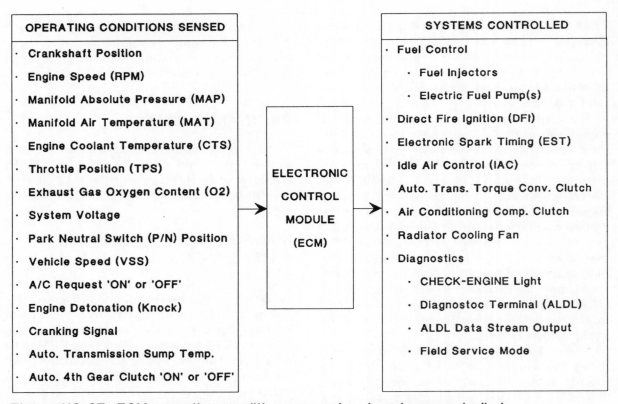

OPERATING CONDITIONS SENSED	SYSTEMS CONTROLLED

OPERATING CONDITIONS SENSED

- Crankshaft Position
- Engine Speed (RPM)
- Manifold Absolute Pressure (MAP)
- Manifold Air Temperature (MAT)
- Engine Coolant Temperature (CTS)
- Throttle Position (TPS)
- Exhaust Gas Oxygen Content (O2)
- System Voltage
- Park Neutral Switch (P/N) Position
- Vehicle Speed (VSS)
- A/C Request 'ON' or 'OFF'
- Engine Detonation (Knock)
- Cranking Signal
- Auto. Transmission Sump Temp.
- Auto. 4th Gear Clutch 'ON' or 'OFF'

ELECTRONIC
CONTROL
MODULE
(ECM)

SYSTEMS CONTROLLED

- Fuel Control
  - Fuel Injectors
  - Electric Fuel Pump(s)
- Direct Fire Ignition (DFI)
- Electronic Spark Timing (EST)
- Idle Air Control (IAC)
- Auto. Trans. Torque Conv. Clutch
- Air Conditioning Comp. Clutch
- Radiator Cooling Fan
- Diagnostics
  - CHECK-ENGINE Light
  - Diagnostoc Terminal (ALDL)
  - ALDL Data Stream Output
  - Field Service Mode

Figure 11C-67 ECM operating conditions sensed and systems controlled.

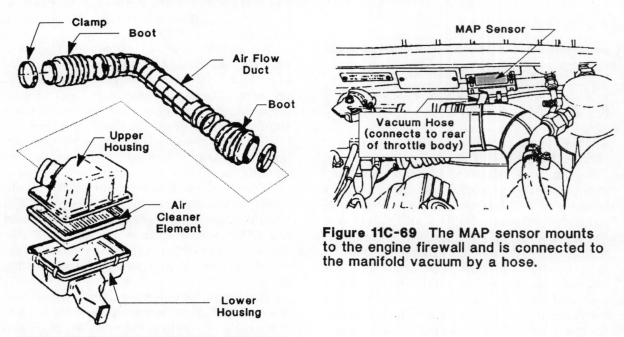

Clamp
Boot
Air Flow Duct
Boot
Upper Housing
Air Cleaner Element
Lower Housing

Figure 11C-68 Air cleaner assembly and ducting.

MAP Sensor
Vacuum Hose (connects to rear of throttle body)

Figure 11C-69 The MAP sensor mounts to the engine firewall and is connected to the manifold vacuum by a hose.

ECM's ability to provide correct injection quantity and ignition timing.

MANIFOLD AIR TEMPERATURE (MAT) SENSOR

The temperature of the air entering the engine must be considered when determining the amount of fuel to be supplied to the engine. A thermistor type MAT sensor is fitted into the intake manifold air stream to sense the air temperature. A supply wire, with a 5 volt potential, connects the MAT sensor to the ECM. The circuit is completed by a second wire between the sensor and the ECM. The return wire provides a sensor earth within the ECM (see Figure 11C-70).

Cold air will cause a high resistance within the MAT sensor; the ECM will detect a voltage, close to supply voltage, between the supply and return lines. As intake air temperature increases, the MAT sensor resistance value will decrease and the voltage monitored by the ECM will also decrease.

The voltage signal developed by the MAT sensor is one of the inputs used by the ECM in determining injector 'on' time.

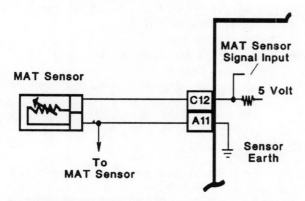

Figure 11C-70 Manifold air temperature sensor (MAT).

THROTTLE BODY

The throttle body (see Figure 11C-71) is fitted with a potentiometer known as a throttle position sensor (TPS). The potentiometer supply line has a reference voltage of 5 volts. The output (signal) voltage at the sliding contact of the potentiometer will vary as the throttle is opened. The voltage ranges from 4.5 volts at wide open throttle to less than 1.25 volts at closed throttle. From this voltage signal, the ECM can detect if the throttle is closed, partly open, fully open, moving and the rate of movement. From this information the ECM can determine fuel requirements based on driver demand.

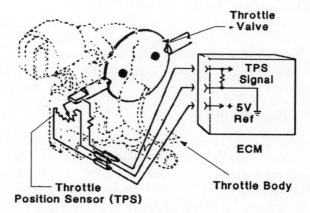

Figure 11C-71 The throttle position sensor (TPS) is mounted on the end of the throttle valve shaft.

IDLE AIR CONTROL (IAC) VALVE

The idle air control valve (see Figure 11C-72), when open, allows air to bypass the closed throttle plate. The bypass air will cause the engine idle speed to increase. A large opening provides high idle speed and a small opening will reduce idle speed. Accurate positioning of the valve is achieved by using a 'stepper' motor with 256 possible positions. The ECM sends a pulse, or pulses, to the motor; the motor moves the valve one position for one pulse. The ECM will continue to send pulses until the desired valve position is obtained. The valve position is

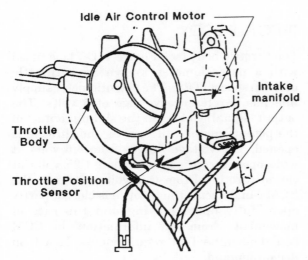

Figure 11C-72 The idle air control valve is mounted to the throttle body.

the result of an ECM calculation based on battery voltage, engine coolant temperature, engine load and actual engine r.p.m.

FUEL SYSTEM

The fuel system contains a fuel tank, lines, filters, fuel pumps, fuel pump relay, pressure regulator, fuel rail and the injectors. Many components in this system operate in the same manner as other units which have already been described in detail. A low pressure pump, mounted inside the fuel tank, feeds an externally mounted pump. The external pump supplies the system working pressure. The system pressure is controlled by a pressure regulator which bypasses excess fuel through a pressure relief return line to the fuel tank. The pressure regulator is linked to the intake manifold vacuum so a constant pressure relationship between intake manifold absolute pressure and fuel rail pressure can be maintained. The injectors are 12 volt electro-magnetic solenoids and are 'earth switched' within the ECM.

Electrical control of the parallel wired fuel pumps is shown in Figure 11C-73. If the ignition switch is turned to the 'on' position without the engine running, the ECM will turn the pump relay on for a period of 2 seconds. Fuel pressure

will build up rapidly but if the engine is not cranked the ECM will turn the relay 'off'. When the engine is cranking the ECM is sent a signal from the crankshaft reference sensor and the pump relay will close allowing the pump to operate.

An interesting 'back up' circuit exists in this system in case the pump relay failed. The engine oil pressure switch will close when oil pressure reaches approximately 28 kPa. The pressure switch is wired parallel to the fuel pump contacts of the pump relay. If the pump relay failed to close, the fuel pumps would still operate with current supplied through the oil pressure switch.

IGNITION SYSTEM

This system is known as direct fire ignition (DFI). The name is appropriate, as the system does not have a distributor. The high tension leads are linked directly from multiple ignition coils to the spark plugs. The DFI module and the ECM provide ignition control, ignition timing and detonation control. The system description, which follows, is for a V6 engine and contains:

- three dual ended ignition coils;
- crankshaft sensor and interrupter rings;
- direct fire module (DFI);
- detonation sensor;
- electronic control module (EST portion of the module only);
- sensor inputs—engine, transmission, vehicle speed.

IGNITION COILS

The ignition coil secondary winding is not earthed, as normally occurs in a conventional ignition coil. Each end of the secondary winding is connected to an external HT terminal. Each coil, therefore, has two high tension output terminals. A high tension wire is connected from each coil terminal to a single spark plug. This arrangement allows one ignition coil to supply high tension voltage to two spark plugs. Three direct fire ignition coils are required to supply the firing voltage to six spark plugs.

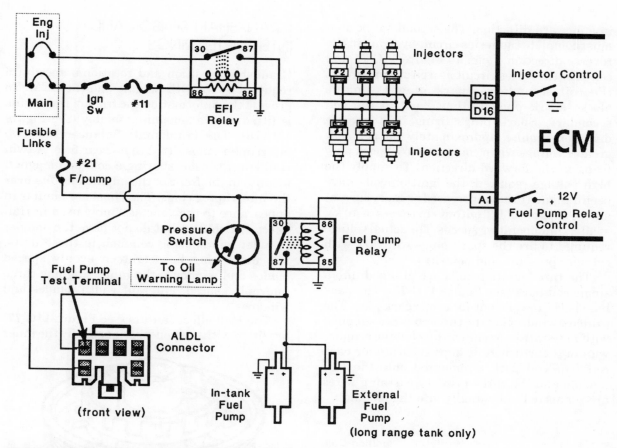

Figure 11C-73 Fuel pump test and control circuit.

A V6 engine cylinder and piston arrangement has the pistons operating in pairs (see Figure 11C-74). When one piston is at TDC, a second piston will also be at TDC. The cylinder pairs (companion cylinders) in which the pistons arrive at TDC together are cylinders 1 and 4, 2 and 5, and 3 and 6. As the piston of one cylinder is on compression TDC, the piston in its companion cylinder will be at TDC at the end of the exhaust stroke. When the ignition coil discharges, high voltage current fires both spark plugs. Only one of the spark plugs can fire into a combustible mixture (cylinder on compression) and the second plug firing is wasted into exhaust gas. The second spark is known as a waste spark.

The polarity of the ignition coil primary and secondary windings is fixed. When the coil discharges, one spark plug will fire with current

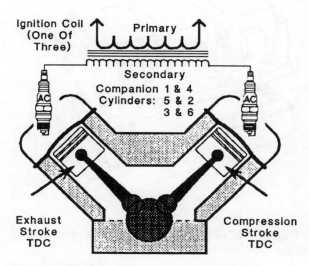

Figure 11C-74 Waste spark ignition. Companion cylinders have pistons at TDC at the same time.

in a forward direction. The second spark plug must complete the series circuit by firing in the reverse direction. Refer to Figure 11C-74 and consider the series circuit as a circle starting at the coil secondary—spark plug 1—engine block—spark plug 2—then back to the coil secondary. Spark plugs firing in a reverse direction require approximately 30 per cent greater voltage when compared to a spark plug firing in the forward direction. To supply the high voltage required, the ignition coils have been designed to have increased primary current and be capable of HT outputs in excess of 40 kV at all engine operating speeds. The actual voltage required to fire the spark plugs will vary with cylinder pressure and polarity.

The three ignition coils are grouped into a single coil pack (see Figure 11C-75). The pack has six HT towers, one for each spark plug. The primary windings of the three coils are supplied with 12 volts from a common link but the primary windings earth wire is kept separate for each coil. The coil pack is mounted onto the DFI module and the three primary earthing wires are connected individually into the module.

CRANKSHAFT SENSOR AND INTERRUPTER RINGS

Crankshaft position and rotational speed are required by the DFI module and the ECM. To provide this information the front of the engine is fitted with a sensing assembly (see Figure 11C-76). The crankshaft balancer has two interrupter rings fitted to its rear face. A twin Hall effect sensor, sharing a common magnet, mounts to the front of the engine and fits over the interrupter rings. The rings are similar in appearance to the vane assembly used in Hall effect distributors but do not have the one per cylinder gap spacing common to these units. The outer ring has eighteen evenly spaced blades and windows; the inner ring has three unevenly spaced, different width, blades and windows.

The Hall effect sensors (see Figure 11C-77) are fitted with one sensor triggered by the inner

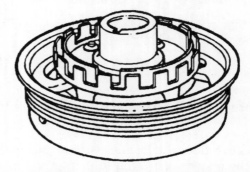

Figure 11C-76 Crankshaft balancer with interrupter rings.

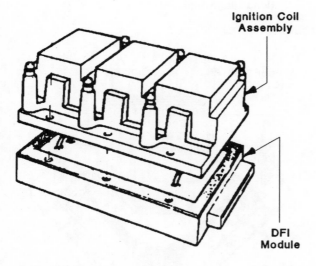

Figure 11C-75 Ignition coil pack and direct fire ignition module.

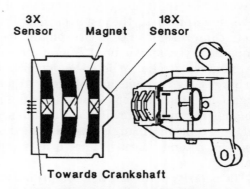

Figure 11C-77 Crankshaft sensor.

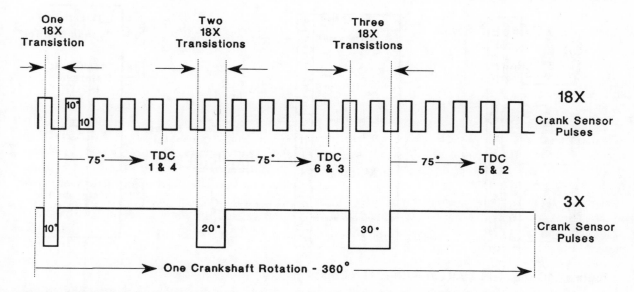

Figure 11C-78 18X and 3X crank sensor pulses during one crankshaft revolution.

ring and the other sensor triggered by the outer ring. As the crankshaft rotates, the eighteen-blade ring will produce eighteen pulses from the 18X sensor in one crankshaft revolution. Simultaneously the three-blade ring will produce three pulses from the 3X sensor in one crankshaft revolution.

The ignition coils must be sequenced so the correct coil is fired to suit the crankshaft position. Examine the illustration (see Figure 11C-78) and note the following points:

1 The 3X ring vane enters the 3X sensor 75 degrees BTDC for each pair of companion cylinders.

2 Identification of cylinder pairs is possible from the size of the 3X pulse width (window). The 10 degree pulse relates to cylinders 1 and 4, the 20 degree pulse is cylinders 6 and 3, and the 30 degree pulse is for cylinders 5 and 2.

3 By measuring the number of 18X transitions (either 'on' or 'off' signals) that occur within one 3X pulse width, the DFI module has an indication of the crankshaft position. The DFI module can use this information to sequence the coils.

3X No. of pulse width	18X transitions	Next firing pair
10 degrees	1	1 and 4
20 degrees	2	6 and 3
30 degrees	3	5 and 2

4 The DFI module must have both the 3X and 18X pulse to fire the ignition coils.

DFI MODULE

The DFI module (see Figure 11C-79) performs several functions: it provides power for both Hall sensor internal circuits and the 3X and 18X voltage signal. The Hall switches turn these voltages 'on' and 'off' from earth, which generates the 3X and 18X sensor pulses within the DFI module. As described in the text related to the crank sensor, the 18X and 3X signal pulses are used to sequence the ignition coil firing. The sequencing action is detected at start-up and, after the engine is running, the correct sequence is 'remembered' by the module. Sequencing of the coils has been described but

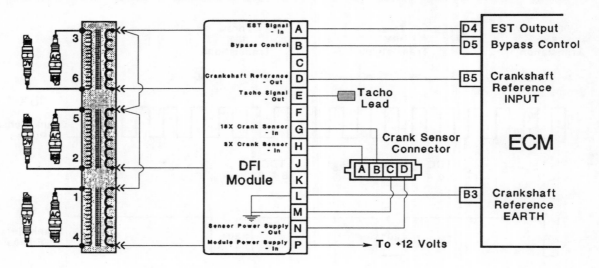

Figure 11C-79 DFI system with EST circuits.

the correct ignition point and engine r.p.m. are required before the system is functional.

The crankshaft position or reference signal does not require a separate sensor, but is developed inside the DFI module from the 18X and 3X sensor inputs. Figure 11C-80 shows the crankshaft reference signal positioned below the 18X and 3X sensor pulses. The reference pulse is a 60 degree 'on' and 60 degree 'off' signal with the falling edge of each signal occurring 70 degrees BTDC for each cylinder pair. This signal is obtained by dividing the 18X pulse by 6, producing 3 reference pulses per crankshaft revolution. The division process is started by the 3X signal, therefore if either the 18X or 3X signal is missing, the crankshaft reference signal can not be developed and sent to the ECM. The ECM interprets the crankshaft reference signal into two parts; the engine r.p.m. and the engine crankshaft position from which the EST ignition advance can be based. If the ECM does not receive the crankshaft reference pulses, the fuel injectors will not operate and the engine will not run.

SPARK TIMING CONTROL

Ignition control is a shared responsibility between the DFI module and the EST section of the ECM. The choice of which component is

in control at any given time is determined by the engine operating condition.

During engine cranking, spark timing is controlled by the DFI module. The module sets the ignition timing at 10 degrees BTDC and has a predetermined dwell period for the coil. This operating condition is known as 'module mode' and will maintain module control of the ignition until the engine starts and exceeds 450 r.p.m. The system will remain in, or revert to, module mode if a specific system fault code is detected.

As the crankshaft rotates, the DFI module sends a crankshaft reference signal to the ECM. As the ECM receives crankshaft reference signal pulses, it issues electronic spark timing pulses to the DFI module (refer to EST output terminal on Figure 11C-79). Below 450 r.p.m., or when a fault is present, the EST output pulses are earthed within the DFI module and the system remains in module mode. When 450 r.p.m. is reached by the engine, the ECM applies 5 volts to the bypass control line (see Figure 11C-79). The DFI module detects the bypass voltage and removes the earth link which has been grounding the EST pulses. Control of the ignition firing point and the dwell period has now been assumed by the EST portion of the ECM. This operating condition is known as 'EST mode'.

The ECM calculates the most suitable spark timing and coil dwell time from information supplied by a range of sensors. The major spark

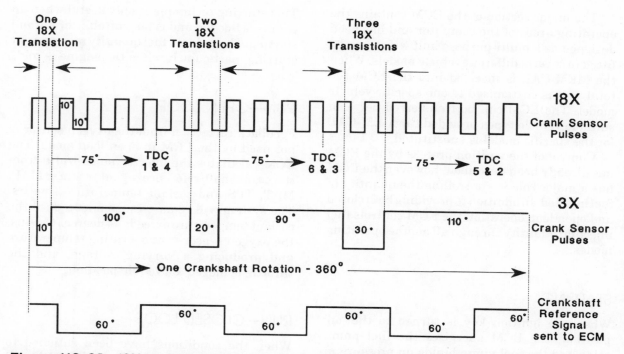

Figure 11C-80 18X and 3X crank sensor pulses generate a crankshaft reference signal.

advance inputs are from engine r.p.m. (crankshaft reference signal), engine load (MAP sensor) and engine coolant temperature sensor. A detonation sensor is fitted to the engine block and if detonation is detected, the spark advance may be reduced up to 8 degrees by the ECM. When detonation ceases, the ECM slowly adds back the 'lost' advance. Other inputs which may have an effect on the ignition advance point or dwell period are the crankshaft position (crankshaft reference signal), throttle position switch, Park/neutral switch, vehicle speed sensor, system voltage and the diagnostic request input terminal.

To summarise the previous text: Ignition spark timing may be positioned by the ECM:

- as an EST output provided 5 volts is present on the bypass control; or
- will be fixed at 10 degrees BTDC operating in module mode if 5 volts are not present on the bypass control.

It is important to note that ignition coil sequencing remains under the control of the DFI module in both Module and EST modes.

ELECTRONIC CONTROL MODULE (ECM)

The ECM fitted to the system contains a detachable section known as a MEM-CAL (see Figure 11C-81). The MEM-CAL is a memory unit and contains calibration data for one specific model vehicle. The type of data is related to vehicle weight, engine type, transmission, rear axle ratio etc.

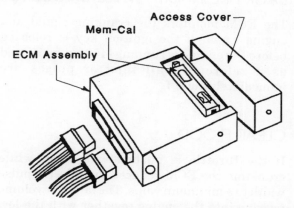

Figure 11C-81 Electronic control module fitted with a 'Mem-Cal' memory unit.

The major section of the ECM contains the operating areas of the computer and has been designed as a multi-purpose unit which can be fitted to several different vehicle models. When the MEM-CAL is inserted into the ECM the total unit is customised to one specific vehicle model. The ECM can be transferred between suitable vehicle types provided the MEM-CAL for the specific model is substituted.

Control of the ignition process by the ECM has already been described, however the ECM has a major role in controlling the quantity of fuel injected. In addition to providing both closed and open loop operation the ECM controls fuel injection quantity through all engine operating modes.

STARTING

When the ignition key is turned to the 'on' position, the ECM energises the fuel pump relay and the fuel pump builds up pressure in the fuel rail. The quantity of fuel required by the engine during cold start cranking will be greater than when the engine is at operating temperature. The ECM checks the coolant temperature via the sensor signal and adjusts the injector pulse width accordingly. The starting mode is maintained until the engine reaches 300 r.p.m. or the mode is cancelled by a 'clear flood' mode signal.

BACK FIRE INHIBIT—WHILE CRANKING

The ECM monitors the cranking signal and engine r.p.m. If the ignition key is released before the engine is running (450 r.p.m. is used as a reference), all injector pulses are immediately cut off.

CLEAR FLOOD

If the throttle is pushed wide open while cranking, the ECM reduces the injector pulse width to a minimum value. The large air volume passing into the engine together with the low fuel quantity should clear a flooded engine.

This starting technique is only useful when an engine is flooded and is not suitable for normal starting. The reduced fuel quantity may prevent starting, particularly when the engine is cold.

RUN—OPEN LOOP

Feedback correction from the oxygen sensor is not used by the ECM in open loop mode. The ECM calculates the injector pulse width from the crankshaft reference signal (r.p.m.), MAP, MAT, TPS and coolant temperature sensors. The system will remain in open loop until the coolant temperature exceeds 44 degrees Celsius, the oxygen sensor is at operating temperature and producing a varying voltage, and the throttle is moved above idle position.

RUN—CLOSED LOOP

When the conditions have been satisfied to leave open loop (refer open loop), the ECM will still use the same sensors to calculate injector pulse width. The final injector signal will be modified in response to feedback correction from the oxygen sensor. The air-fuel ratio will be maintained at exactly 14.7 to 1 to ensure maximum efficiency from the catalytic convertor.

IDLE

An engine requires a slightly richer mixture for suitable idle quality. The ECM detects a closed throttle switch (TPS) signal but must also check that the vehicle speed sensor (VSS) indicates less than 5 km/h. If both conditions are present, the ECM ignores the oxygen sensor signal and adjusts the injector pulse width to suit the idle r.p.m.

If the engine has been operating in closed loop mode as the vehicle rolls to a stop, the changeover to idle mode will be delayed approximately 25 seconds.

ACCELERATION

Opening the throttle for acceleration causes rapid changes in the throttle position switch and the manifold vacuum (MAP sensor). The TPS and the MAP signals are detected by the ECM and the injector pulse width is increased to supply additional fuel. If the acceleration fuel requirements are severe, the ECM causes extra injector pulses to occur between the normal pulses (once per crankshaft revolution).

DECELERATION

During normal deceleration, excess fuel in the manifold will cause a rise in emissions and increase the possibility of back-firing. The ECM detects the deceleration condition, by monitoring the TPS and MAP sensors, and reduces the injector pulse width.

DECELL—FUEL CUT OFF

Several conditions must be met before the ECM will 'cut off' the injector signal under deceleration.

- Engine coolant temperature greater than 56 degrees Celsius;
- Engine r.p.m. greater than 1500;
- Vehicle speed greater than 35 km/h;
- TPS indicates closed throttle;
- MAP signal indicates no engine load (less than 20 kPa.);
- Park/Neutral (auto) switch—indicates 'in gear'.

All the conditions listed above must be met to enter fuel 'cut off' mode. Therefore any change in engine r.p.m., vehicle speed, throttle position, engine load or transmission switch which is not within the limits will cause the ECM to resume the injector signal.

BATTERY VOLTAGE

Low battery voltage causes weaker ignition sparks and increases the mechanical opening time of the injectors. The ECM detects the reduced voltage and compensates the system by increasing idle r.p.m., injector pulse width and dwell time as required.

FUEL CUT OFF

Four conditions can prevent the supply of fuel from the injectors.

1 The ignition is off. No fuel passes the injectors which prevents the engine 'dieseling'.

2 The reference pulse from the DFI module is not detected by the ECM. The absence of this signal indicates the engine is not running and fuel should not be supplied.

3 The engine r.p.m. is greater than 5400. Injection will be restored when the r.p.m. drops below 5000.

4 The vehicle speed is greater than 220 km/h. Injection will be restored when the vehicle speed drops below 210 km/h.

TRANSMISSION AND COOLING FAN CONTROL

The ECM, in this system, extends its control features to the operation of the transmission convertor clutch and the engine cooling fan operation.

TORQUE CONVERTOR CLUTCH (TCC)

The vehicle automatic transmission is fitted with a 'lock-up' torque convertor clutch. Electrical control of the lock-up is provided by routing the earthing circuit of the TCC solenoid into the ECM (see Figure 11C-82). The ECM monitors the engine coolant temperature sensor (greater than 44 degrees C), vehicle speed sensor (greater than 72 km/h) and the throttle position sensor. High throttle settings delay the TCC engagement until higher road speeds are attained, however if the throttle is closed the ECM will open the earth circuit and release the TCC.

Two switches are provided within the transmission as a form of over-temperature

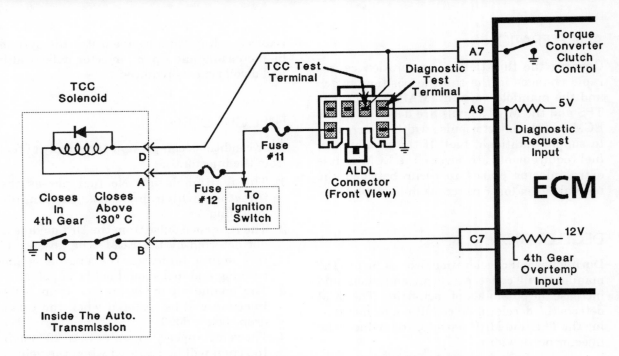

Figure 11C-82 Torque converter electrical supply, test and over-temp circuits.

control. The switches are wired in series, requiring two conditions to be met before the circuit is earthed. In normal operation, when the vehicle transmission moves into fourth gear, one switch will be closed. The second switch will close only if the temperature of the transmission fluid exceeds a pre-set value. With both switches closed the over-temperature input becomes active. Under these conditions the ECM ignores the engine coolant (ECT) and vehicle speed (VSS) inputs and energises the TCC solenoid at any throttle setting in excess of 4 per cent opening.

COOLING FAN

The radiator cooling fan is driven by an electric motor which is controlled by the ECM (see Figure 11C-83) or the air conditioning system.

The fan is switched on by the ECM when the engine coolant temperature exceeds 100 degrees Celsius or if a coolant temperature sensor failure is detected. The fan will also be turned on if the system is in diagnostic mode with the engine running.

If the vehicle is stationary, or at a road speed

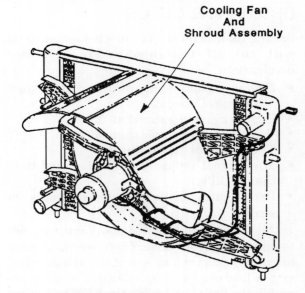

Figure 11C-83 The cooling fan electric motor is controlled by the ECM.

below 55 km/h. when the air conditioner is turned on, the cooling fan will also turn on. The fan will remain 'on' for speeds below 55 km/h but will turn 'off' when the road speed is greater than 55 km/h.

ON-BOARD DIAGNOSTICS

The ECM has significant capabilities for self-diagnosis and storage of any system malfunction detected. A check engine light (CEL) on the instrument panel advises the driver if a system malfunction has occurred. The CEL will illuminate when the ignition switch is turned to the 'on' position. The CEL should go out when the engine starts. The vehicle should be taken to a service station for checking, if the CEL remains illuminated, or is delayed in going out.

An assembly line diagnostic link (ALDL) (see Figure 11C-84) connector is located next to the ECM and will assist the mechanic when diagnosing the system. The ALDL was used by the assembly plant to test the system prior to

the vehicle leaving the plant. The link can be used to connect factory approved diagnostic units into the system when difficult problems are encountered. Provision has also been made at the link for the servicing mechanic to access the vehicle system.

The mechanic can place the system into diagnostic mode by bridging the ALDL diagnostic terminal to the earth with the simple tool shown in Figure 11C-85.

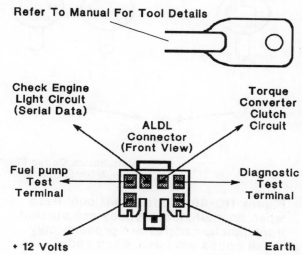

Figure 11C-85 6-pin ALDL connector.

When ALDL test terminal is earthed, the ignition is 'on' and the engine is *not* running, malfunction codes will be flashed from the dash mounted engine check light. The flashing codes can be decoded by referring to the manufacturer's service manual. The circuit indicated can be tested by following the procedure described in the manual.

With the system in the diagnostic mode; the cooling fan relay, A/C control relay, TCC and the IAC valve will all be energised. A code 12 (see Figure 11C-86) indicates that no faults have been stored in the ECM memory. The check light will flash a code 12 signal until the ALDL testing link is disconnected. If a single

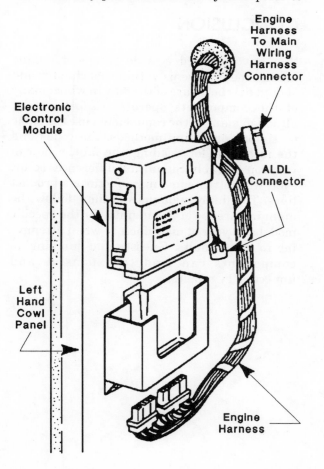

Figure 11C-84 ALDL connector location.

DIAGNOSTIC MODE
. Ignition On
. Engine Not Running
. ALDL 'Diagnostic Test
   Terminal' Earthed

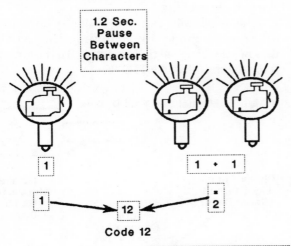

Figure 11C-86 Code 12 will only flash when no malfunction codes are present. If malfunction codes are present, only those codes will flash. Each code flashes three times.

fault is 'logged' the check light will continue to flash the fault code, however if multiple faults are 'logged' the flashing code will cycle to the next fault after flashing the first fault three times.

The mechanic must be aware that starting the engine with the ALDL test terminal earthed causes the system to enter the 'field service mode'. In this condition the CEL flashes will *not* be malfunction codes. The flashing sequence is useful for diagnosing fuel mixture faults as it indicates whether the system is operating in open loop, closed loop, rich or lean mixture. While in this mode, the ignition timing will be fixed at 10 degrees BTDC when the engine is operating at less than 2000 r.p.m. When driving the vehicle the engine performance will be less than expected and the CEL will be mostly:

• on while accelerating;
• off when decelerating.

CONCLUSION

You have studied a wide range of engine management systems within this chapter and should now be aware of the way in which many of the components operate. Unfortunately, although many of the components use the same or similar operating methods, the codes and the technique for 'reading' the fault code can vary considerably. Significant differences occur, even within the same manufacturer's product lines. To minimise diagnostic time, follow the manufacturer's service manual for the specific model and year of manufacture when attempting to locate faults. Guesses and 'jumping' to components can lead to expensive and unnecessary repairs.

12

CHARGING SYSTEMS

INTRODUCTION

The charging system of the motor vehicle was possibly the first system to be influenced by electronics. The DC generator, which had mechanical rectification (commutation), was replaced by the AC generator (alternator). The early alternator system included a dry-plate bridge type rectifier and a double contact regulator. The rectifier, constructed from plates of selenium or magnesium sulphide, was a bulky unit. The regulator, consisting of three relays which controlled the voltage, current and load, was about the same size as the DC generator regulator.

Advances in electronics provided a compact charging system through the introduction of solid state devices (diodes and transistors) into the rectification and control circuits. This chapter reviews the changes that the solid state devices have made to the automotive charging system.

CHARGING SYSTEM

The charging system consists of four sections. These sections are the:

- alternator;
- rectifier and exciter;
- voltage regulator;
- battery, ignition switch and warning light.

ALTERNATOR

LOCATION

The alternator is a three-phase AC current device that is driven by the engine through a belt. It is rigidly attached to the engine block (see Figure 12-1) near the front of the engine and connected by wires into the vehicle's charging system.

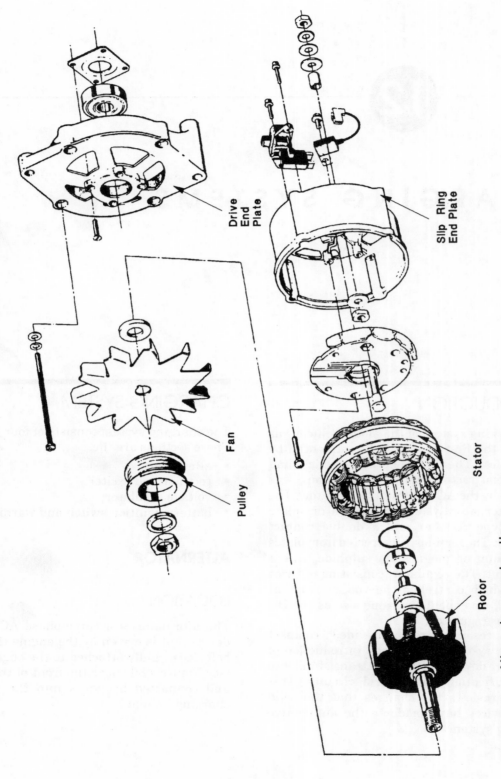

Figure 12-1   Alternator construction.

Drive
End
Plate

Slip Ring
End Plate

Fan

Pulley

Stator

Rotor

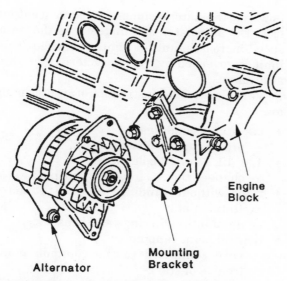

Figure 12-2 Location of an alternator.

CONSTRUCTION

The alternator (see Figure 12-2) consists of a:

- stator;
- rotor;
- drive end plate;
- slip ring end plate;
- fan and pulley assembly.

The stator has three windings (phases) which are wound onto a laminated ring and connected in a star configuration. The windings are spaced at 120 degrees to one another around the laminated ring.

These windings may be connected end to end to form a triangle (delta) shape or one end of each winding may be connected to form a star shape. Star connected windings are a common configuration (see Figure 12-3).

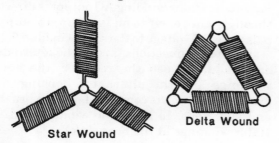

Figure 12-3 Common stator winding configurations.

The rotor has a field winding enclosed by twelve pole pieces which are attached to a drive shaft (see Figure 12-4). Each end of the field winding is connected to a slip ring. The two slip rings are insulated and attached to the drive shaft. The drive shaft, supported on two ball bearings, has a fan and a pulley, keyed to its end opposite to the slip rings.

The rotor drive pulley and fan are formed from pressed steel or cast from steel. A cooling air stream, produced from the fan, is directed over the internal components of the alternator.

Both end plates are cast from aluminium alloy. The slip ring end plate provides a mount for a diode heat sink, an end bearing and a regulator. The drive end plate houses an end bearing and provides mounting points for the alternator.

Both end plates, when bolted together over the stator, form the complete alternator housing. The rotor is free to spin inside the stator.

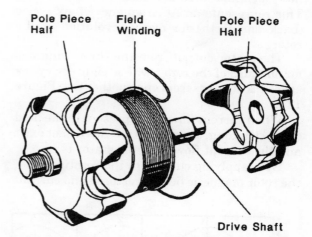

Figure 12-4 Rotor construction.

OPERATION

When the field (rotor) winding is connected to the battery, the current flow produces six strong magnetic fields between the pole pieces (see Figure 12-5). These magnetic fields spin at the same speed as the rotor. As each field cuts across a stator winding, voltage is produced across the winding and current flows in the stator circuit. Due to a north pole piece being followed closely by a south pole piece, the current

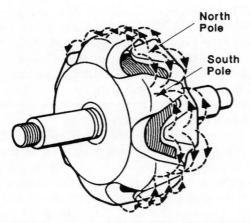

Figure 12-5 A magnetic field established round the field winding.

will flow in one direction and then flow in the opposite direction through the stator winding. One turn of the rotor will cause the current to flow back and forth six times in one winding. This represents an AC frequency of six cycles (back and forth) for every revolution of the rotor.

The stator output from the three windings for one turn of the rotor, will be eighteen cycles spaced 20 degrees from one another (see Figure 12-6).

When current flow is not present in the field (rotor) winding, the magnetic fields do not exist and the current flow from the stator is zero.

By reapplying current to the rotating field, the rotor magnetic field is restored and current

flow from the stator resumes. The current output from the alternator will increase as the rotor speed increases, until a limit is reached. This limit is governed by the characteristics of the alternator.

Variations to the construction of an alternator are:

- two stator windings;
  — produce a two phase output;
- 4, 6, 12, 14 or 16 pole pieces;
  — changes the AC frequency output;
- rotor without windings and a stationary excitation field;
  — eliminates slip rings and brushes;
- neutral point tapping;
  — wire connected to the common point (centre of star);
  — used to run components with an AC voltage;
  — provides additional power output;
- a 'W' terminal;
  — wire connected to one of the stator windings;
  — provides a pulsating DC voltage (half-wave rectification) to drive a tachometer fitted to a diesel engine.

RECTIFIER AND EXCITER

LOCATION

The rectifier and excitation section, formed on a heat sink, is attached to the slip ring end plate inside the alternator (see Figure 12-7). The rectifier changes the AC output of the stator winding into a pulsating DC output. Part of the stator winding AC output is directed through the excitation section to supply pulsating DC voltage to the rotor winding (field).

Internal connections (inside the alternator) are made between the rectifier, the stator windings and the alternator housing. The excitation section is connected, via the brushes and slip rings, to the rotor winding and the stator windings at the heat sink.

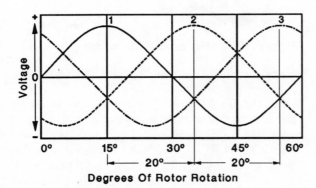

Figure 12-6 The three field winding outputs for a third of a turn of the rotor.

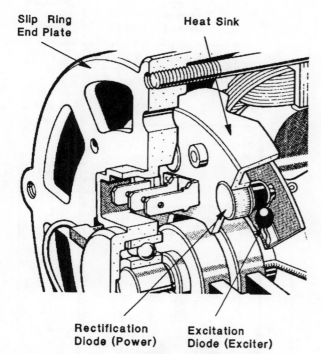

Figure 12-7 The location of the rectifier and the excitation section.

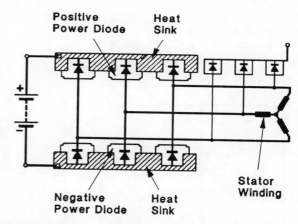

Figure 12-8 The construction of the rectifier.

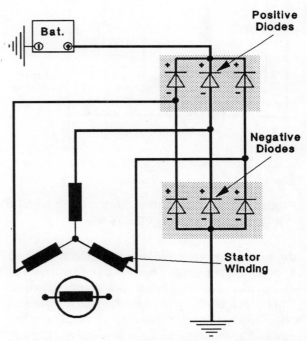

Figure 12-9 The connection of the power diodes in a rectifier.

CONSTRUCTION

The rectifier is of a bridge type, consisting of six diodes (power diodes) (see Figure 12-8). Three diodes, with their negative terminals connected internally to their bases, are pressed or glued to one heat sink. The other three diodes, with their positive terminals connected internally to their bases, are pressed or glued to another heat sink. These diodes are called 'negative' and 'positive' power diodes respectively.

The negative heat sink is bolted directly to the slip ring end plate and provides an earth (ground) for the rectifier.

The positive heat sink is bolted to, but insulated from, the slip ring end plate. One of the bolts passes through the housing to provide a connection point for the battery.

The pin terminal of a negative diode is connected by a wire to the pin terminal of a positive diode. This wire is also connected to the end of a stator winding (see Figure 12-9).

The other diodes and stator windings are connected in a similar manner to complete the rectifier circuit.

Excitation is achieved through three small diodes (excitor diodes) which are attached to one heat sink (see Figure 12-10). Each excitor

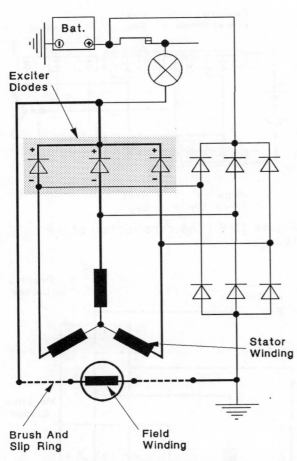

Figure 12-10 The connection of the exciter diodes in the excitation circuit.

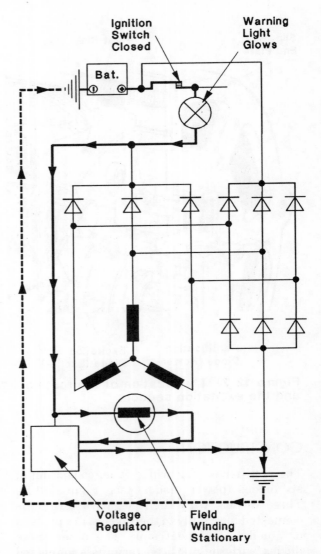

Figure 12-11 Pre-excitation of the field winding.

diode negative terminal is connected to the end of a stator winding. The positive terminals of the diodes are connected to a common point. This point is then connected, via a brush and slip ring, to the rotor winding (field). The other end of the rotor winding is connected through another slip ring and brush to a voltage regulator which provides an earth (ground) to complete the excitation circuit.

OPERATION

When the ignition switch is 'on', current flows through the:

- warning light;
- brush and slip ring;

- rotor winding (field);
- voltage regulator to earth.

The current flow through this circuit creates the strong magnetic fields at the pole pieces. This action is called **pre-excitation** (see Figure 12-11).

Note: Pre-excitation will ensure that the alternator begins to produce power the instant the rotor spins. Without pre-excitation, the rotor would have to be spun at a speed much higher than the idling speed of an engine.

When the engine starts, the rotor spins inside

the stator. The magnetic fields cut the stator windings to produce AC voltage at the rectifier and excitation input terminals (see Figure 12-12).

A positive pulse from a stator winding (A) arrives at the junction of its positive and negative diodes. The pulse passes through the positive diode to the battery. The voltage in stator winding (B) will be less than zero. Current will then flow from the vehicle earth (zero voltage) through one negative diode and its respective stator winding (B) to the star (centre) point (see Figure 12-13).

Further rotor movement will cause a negative voltage pulse from the stator winding (A) to be present at the junction of its positive and negative diodes (see Figure 12-14). The pulse is negative, therefore cannot pass through the positive diode to the battery. One of the other stator windings (B) now has a positive pulse moving through it and its respective positive diode passes the positive voltage to the battery. Because stator winding (A) was at a voltage below zero, current will flow from the vehicle earth (zero voltage) through A's negative diode into the stator winding.

This action is repeated for the other stator windings and their corresponding power diodes. Each of the rectified pulses overlap and an output at the alternator appears as shown in Figure 12-15. The result is a DC voltage with slight pulsation being sent to the battery. The average value of the DC voltage is controlled by the voltage regulator.

Current to the rotor field coil is obtained by tapping the output of the stator windings near

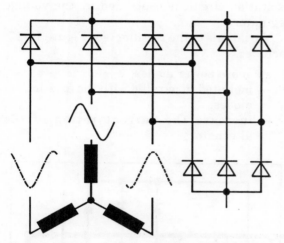

Figure 12-12 An AC voltage produced at the diodes.

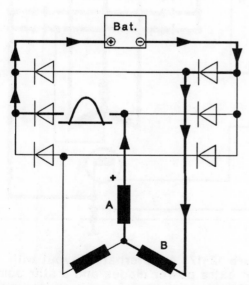

Figure 12-13 The positive section of the AC pulse passes through the power diode.

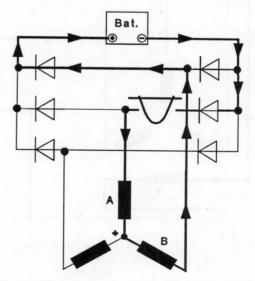

Figure 12-14 The negative section of the AC pulse allows current to flow through the negative power diode to the stator winding (A).

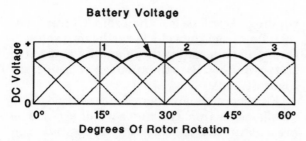

Figure 12-15 The rectified AC voltage from the three field windings, for a third of a turn of the rotor.

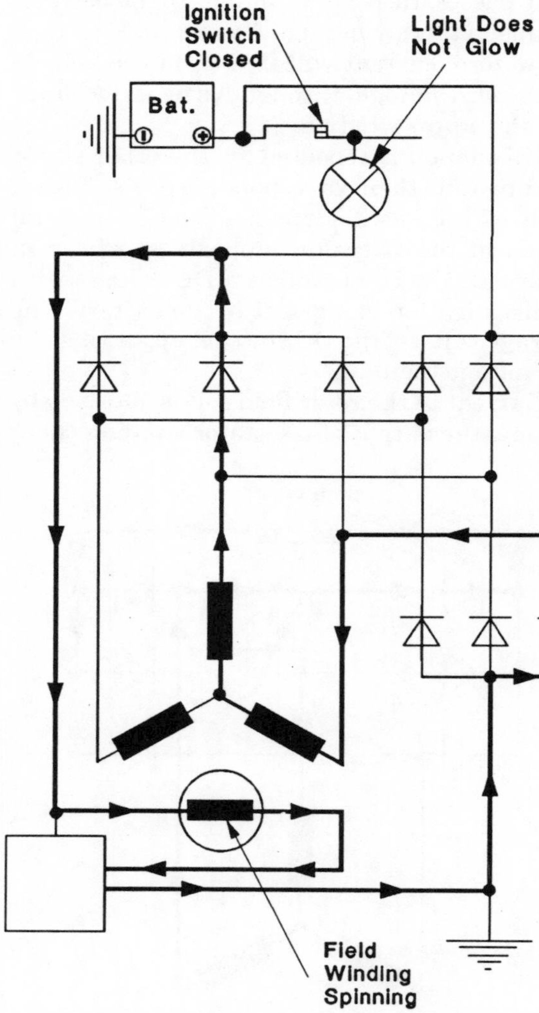

the junction of the positive and negative power diodes. The process of rectifying the AC voltage obtained at this point, as it passes through the exciter diodes into the excitation circuit, is the same as the rectification section.

The exciter voltage opposes battery voltage, turns off the warning light and maintains the current flow through the rotor winding (field). This action is called 'self-excitation' (see Figure 12-16). The amount of current flow in the excitation circuit is controlled by the voltage regulator.

Variations to the rectifier (see Figure 12-17) include:

- six more power diodes;
  - mounted in parallel with the existing six diodes;
  - increase the current rating of the alternator;

Figure 12-16 Self-excitation of the field winding.

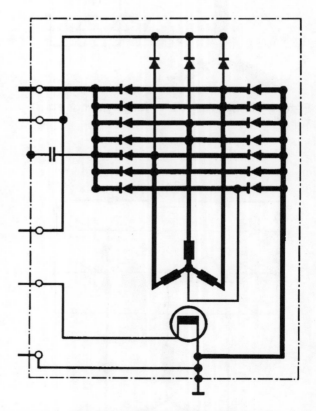

Figure 12-17 An alternator layout with eight extra power diodes and a star point tapping.

- two additional diodes at the star point;
  - connected in a similar manner to the other power diodes;
  - provides extra power output (may be up to 15 per cent).

Variations to the excitation circuit are related to the position of the voltage regulator in the circuit.

A voltage regulator, located between the excitor diodes and the rotor winding (field), controls the supply voltage to the field. For example, a low field voltage will produce a low current flow, which results in the magnetic fields being weak.

A voltage regulator, located between the rotor winding and the earth, controls the voltage drop across the field. For example, a high voltage drop (14 volts) will produce a high current flow, which results in strong magnetic fields.

VOLTAGE REGULATOR

LOCATION

The regulator is an integral part of the brush holder and the brushes. This assembly housing is bolted to, but insulated from, the slip ring end plate housing (see Figure 12-18). Connections into the excitation circuit are achieved by pressure contacts, lugs, eyes or short wires.

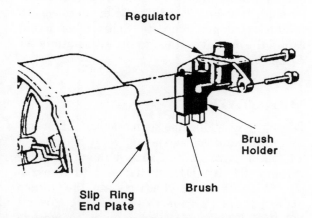

Figure 12-18 The location of the voltage regulator (internal type).

CONSTRUCTION

The regulator consists of several stages (see Figure 12-19). These are the:

- voltage divider stage;
- control stage;
- power stage;
- temperature-compensation stage;
- DC ripple smoothing stage;
- free-wheeling stage.

The voltage divider stage consists of resistors R1, R2, R3, R6 and R7, connected in the manner shown in Figure 12-19. The resistor R7 ensures that transistors' T2 and T3 switching action is fast and precise.

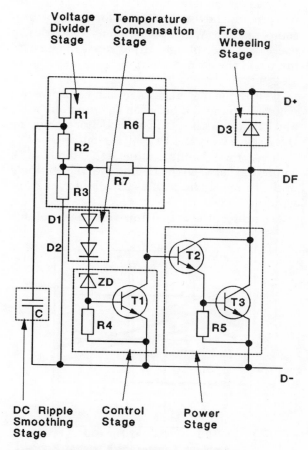

Figure 12-19 The five sections of a voltage regulator.

The control stage comprises of a zener diode (ZD), a transistor (T1) and a resistor (R4). ZD and R4 are connected in series between D2 and D- (earth). The base of T1 is attached to the junction of ZD and T4. The emitter of T1 is joined to D-.

Transistors T2 and T3, and resistor R5 form the power stage. Both the transistor collectors are connected to the DF terminal. The emitter of T2 is coupled to the base of T3 and R4 which is earthed at D-. The emitter of T3 is earthed at D-. The base of T2 is attached to R6 and the collector of T1.

The temperature-compensation stage comprises two diodes (D1 and D2), a zener diode (ZD) and a resistor (R4).

The correct selection of D1, D2, ZD and R4 will provide the required range for temperature-compensation. When the ambient temperature decreases, the voltage output must increase to maintain the correct charging rate (see Figure 12-20).

The DC smoothing stage is a capacitor, C, which is connected between the junction of R1 and R2, and earth at D-. The capacitor C smooths the ripples from the DC voltage output.

The free-wheeling stage is a diode, D3, connected between the D+ and the DF terminals of the rotor winding. D3 is parallel to the rotor winding.

To protect the transistors from high voltage

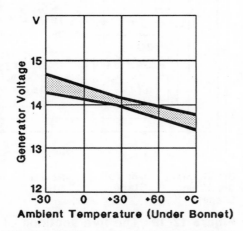

Figure 12-20 The effect of temperature on the output voltage of the alternator.

resulting from self-induction, diode D3 conducts allowing the voltage to rapidly decay.

OPERATION

The voltage regulator is an electronic switch which is voltage sensitive. It works in a pre-set voltage range that will ensure that the voltage at the battery does not exceed a safe limit. In the 'on' state, the regulator allows full excitor voltage to be applied to the rotor winding (field). In the 'off' state, the regulator interrupts the current through the rotor winding (field).

'ON' STATE

When the excitor voltage (alternator output) is less than the set value for the alternator, ZD is non-conductive and no voltage is applied to the base of T1 (see Figure 12-21). With T1 being in a non-conductive state, T2 is conductive because of the voltage applied to its base through R6. Immediately T2 conducts, voltage is applied to the base of T3 to switch it 'on'. With T3 in a conductive state, all the excitor current can flow from the excitor diodes through the rotor winding, terminal DF, transistor T3, terminal D- to earth. The circuit is completed as the current flows from earth through a negative power diode and a stator winding back to the excitor diode. As the rotor speed increases, the output voltage increases and the current flow through the rotor winding (field) increases to produce stronger magnetic fields. This action raises the output voltage to the alternator's set value.

'OFF' STATE

When the set value has been reached, ZD starts to conduct and voltage is applied to the base of T1 (see Figure 12-2). Current flows from D+ through R1, R2, D1, D2, ZD and T1 to earth (D-). With T1 in a conductive state, the voltage at the base of T2 drops to a value that 'turns it off'. With T2 in a non-conductive state, voltage is not applied to the base of T3. Transistor T3 is switched 'off'. The excitor current does not have

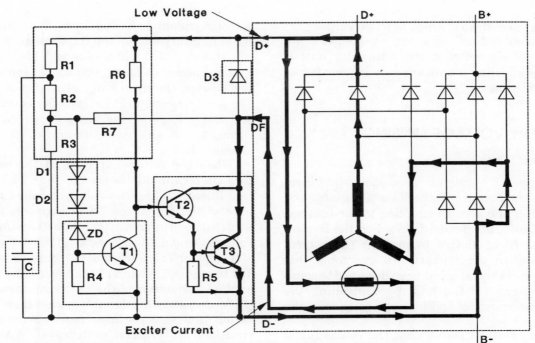

Figure 12-21 The regulator switches the exciter current 'On'.

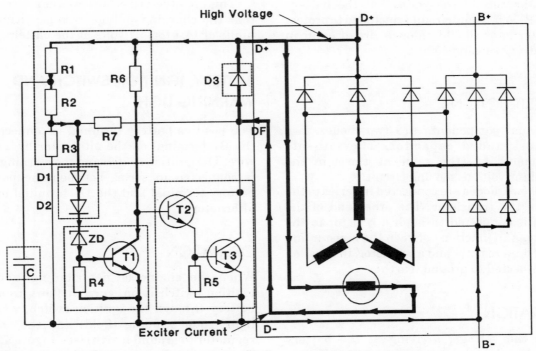

Figure 12-22 The regulator switches the exciter current 'Off'.

a path to earth so it drops to a low value.

The 'on' and 'off' states are repeated at a rate that will ensure that the alternator output voltage remains within the safe voltage limits for the battery.

BATTERY VOLTAGE SENSING REGULATOR

The type of regulator, previously described, senses the output voltage of the alternator and regulates the voltage supplied to the battery. Sometimes, the demand placed on the battery will be high enough to prevent the correct charging of the battery. To overcome this problem, a regulator has been designed to sense the battery voltage and, by controlling the alternator output, will correctly charge the battery.

Figure 12-23 shows that the conventional regulator senses the voltage at the common terminal (A) of the excitor diodes. It is sensing alternator voltage. The battery voltage sensing regulator detects the voltage at the battery (point B). A fail-safe circuit is included to protect the alternator should an open circuit occur in the battery sensing circuit.

CONSTRUCTION

Additional components are two diodes, two resistors and a capacitor. The fail-safe components are the same as those in the alternator voltage sensing circuit.

The two diodes are connected in series to the base of the transistor. The other end of the diodes is connected through a resistor to the battery. At a junction between the resistor and the diode, a resistor and a capacitor, in parallel, are connected to ground (earth).

OPERATION

When the voltage, arriving at the battery through terminal B, is less than 14.1 volts (see Figure 12-24), the voltage applied to the base of the transistor (TS), through the diodes, is not high enough to 'switch it on'. This is due to the voltage divider R1 and R2. Transistor TS is 'off', therefore Tr1 will also be 'off'. With Tr1 in a non-conductive state, voltage is applied to the base of the switching transistor of P.Tr. The power transistor P.Tr continues to conduct, and the voltage to the battery can increase through the alternator power diodes and terminal B.

When the voltage, arriving at the battery through terminal B, is greater than 14.7 volts (see Figure 12-25), the voltage applied to the base of the transistor TS is high enough to 'switch it on'. The transistor conducts and applies voltage through the zener diode (DZ) to the base of transistor Tr1. Sufficient current flow causes Tr1 to conduct. When Tr1 conducts, the voltage applied to the base of the switching transistor (P.Tr) drops to a level that 'switches it off'. This action interrupts the base current of the power transistor (P.Tr) and it 'switches off'. Without a path for the current to flow from the field coil to ground (earth), the voltage at terminal B drops to battery voltage.

This cycle (described above) occurs so rapidly that the charging voltage is kept within the limits of 14.1 to 14.7 volts at the battery.

BATTERY, IGNITION SWITCH AND WARNING LIGHT

The positive battery terminal is connected to the B+ terminal on the alternator by a thick wire. The ignition switch and the warning light is connected in series between the positive battery terminal and the D+ terminal on the alternator.

OPERATION

When the engine is 'not running' and the ignition switch is on, current flows from the battery through the ignition switch, warning light, rotor winding (field) and the voltage regulator to ground (earth) (see Figure 12-26). The current flow through the rotor winding causes a magnetic field to build up around the rotor poles.

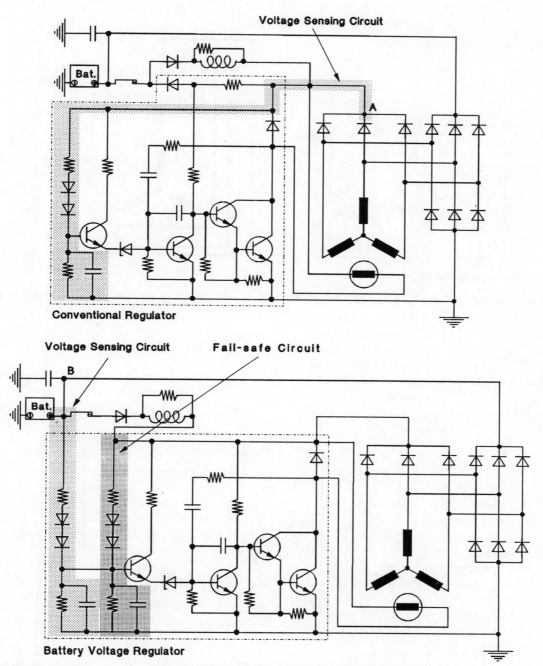

Figure 12-23 The different voltage sensing points for a regulator.

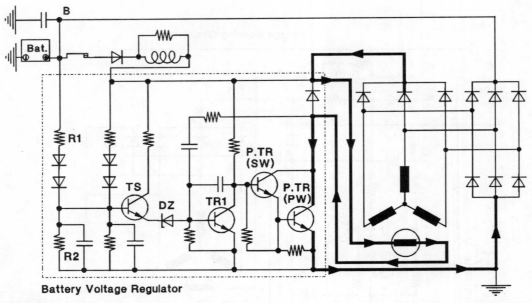

Figure 12-24   The operation when a low voltage is 'sensed' at the battery.

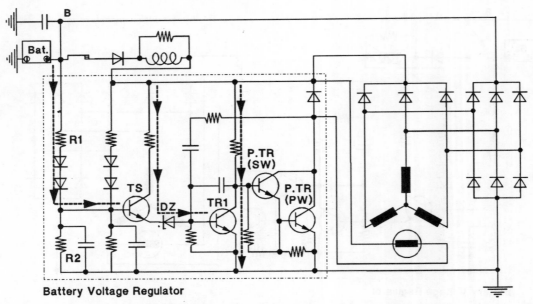

Figure 12-25   The operation when a high voltage is 'sensed' at the battery.

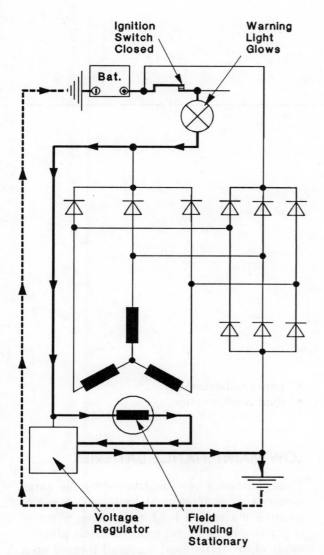

Figure 12-26 The rotor field is excited and the warning light is 'ON'.

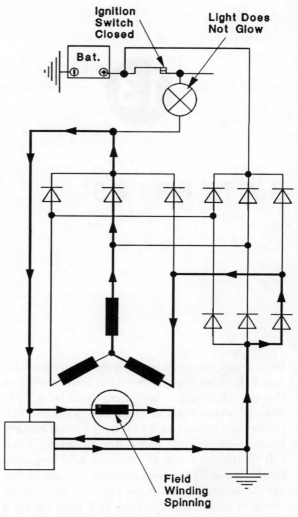

Figure 12-27 The stator voltage across the excitor diodes turns the warning light 'OFF'.

When the engine starts and the alternator begins to produce voltage, the voltage at the excitor diodes is high enough to oppose battery voltage and the warning light is 'turned off' (see Figure 12-27). This indicates to the driver the battery is being charged.

13

BATTERIES

Batteries have been used in motor vehicles for many decades. They have been constructed on the principles of a lead-acid cell. In a fully charged state with no load, the lead-acid cell produces a voltage of 2.1 volts. A 12 volt battery consists of six cells connected in series. Each cell contains a set of positive and negative plates made from a lead material. These plates are submerged within an electrolyte mixture of water and sulphuric acid. In a fully charged state, a 12 volt battery will have an output of approximately 12.6 volts.

This chapter deals with the changes that have occurred in battery construction over the years and battery maintenance.

CHANGES IN BATTERY CONSTRUCTION

The changes that have occurred in battery construction are related to:

- low maintenance batteries;
- maintenance-free batteries;
- panel construction;
- dual configuration.

LOW MAINTENANCE BATTERIES

These batteries are constructed by the same methods as conventional batteries. A set of positive plates are held away from a set of negative plates by separators. Each plate is made of an active lead material formed on a grid. A series of these plates are connected along the top edge.

In this type of battery, the active lead material has been improved, and the grids have a low content of antimony, which reduces gassing during charging. Another advantage of these changes is a reduction in plate contamination.

The space above the plates has been increased to provide more electrolyte volume and more condensation volume. The sump (volume below the plates) has been reduced so that the height of the battery remains the same.

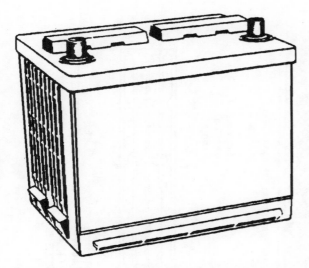

Figure 13-1 A typical 'Low Maintenance' battery.

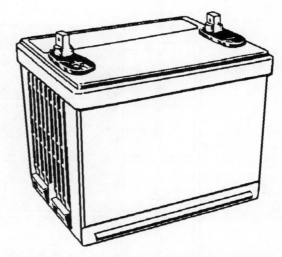

Figure 13-2 A typical 'Maintenance Free' battery.

The cells of this battery, sealed with a conventional type of cap, will vent to the atmosphere when the pressure exceeds a safe limit. Normal 'topping up' of the electrolyte in each cell is reduced, because the loss of water due to gassing has been reduced.

MAINTENANCE-FREE BATTERIES

Structurally, the battery is the same as the low maintenance battery, however the separators between the positive and negative plates have been redesigned to allow the oxygen to move freely between the plates. This type of battery does not contain 'free' electrolyte. The electrolyte is held captive in a specially designed microglass wool. Vent plugs are located in the completely sealed top to release the pressure if it exceeds a safe limit.

Water loss in a conventional battery occurs when the battery is over-charged. An over-charge condition causes hydrogen to form at the negative plates and oxygen to form at the positive plates. Because the hydrogen and oxygen are released from the water, the water in the electrolyte is reduced and the electrolyte level drops. Water must be added to the electrolyte to restore its level.

This effect does not occur in a maintenance-free battery. The active materials in the plates are adjusted so that the positive plates will reach a fully charged state before the negative plates. The oxygen formed at the positive plates will move through the electrolyte and separator system to the negative plates. Reaction between the oxygen and the plate material forms lead sulphate, which is reduced by further charging. This action is called 'recombination electrolyte'.

Because of recombination, the advantages of a maintenance-free battery are:

- safer handling and storage (no acid spillages);
- improved starting performance;
- no water needs to be added to the cells.

PANEL CONSTRUCTION

Panels of active material are formed on a frame constructed of plastic and covered with antimony-free lead mesh (see Figure 13-3). There are three negative panels and three positive panels of active material on each frame (plate). Each positive panel is connected to an adjacent negative panel, across the frame. Two plates must be positioned, one on either side of a separator plate, so that a positive panel aligns with a negative panel. This forms a simple 12

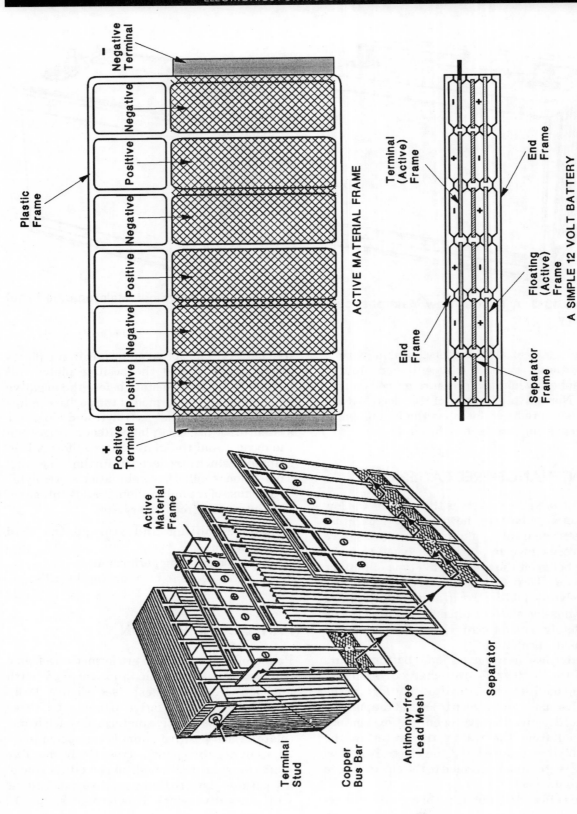

Figure 13-3 A panel constructed battery.

volt battery. An electrolyte mixture of acid and water surrounds the panels. Several of these 'simple' batteries are placed side-by-side and connected by copper bus bars to form a battery suitable for automotive applications.

This battery operates on the same principles as a conventional battery, but water loss does not occur so it is a maintenance-free battery. The advantages of this battery are:

- reduced size;
- light weight;
- rugged construction;
- reduced plate corrosion (due to the use of antimony free lead mesh);
- maintenance free operation (due to the use of antimony free lead mesh);
- a large number of construction extensions (easy to form a 24, 36 or 48 volt battery in one battery of minimum size).

DUAL CONFIGURATION

The dual configuration is an extension to the panel constructed battery. Two 12 volt batteries can be made as one unit (see Figure 13-4). Each

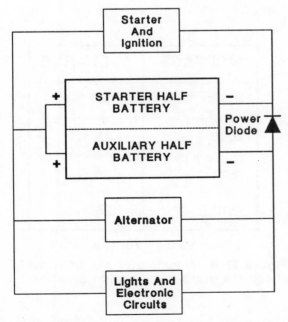

Figure 13-4 A dual battery configuration.

battery can work independently of one another but, due to the incorporation of a power diode between the batteries, they can be charged by the vehicle's alternator. One of the batteries is the 'starting battery' and the other is the 'auxiliary battery'.

When the engine is being started, the starting battery supplies the power. Under extreme starting conditions, the auxiliary battery can assist through the power diode.

When the engine is switched off, the auxiliary battery supplies the power to the other electrical circuits in the vehicle. The starting battery is isolated by the power diode.

One of the main advantages of this battery is that the auxiliary battery could be 'flattened' and the vehicle's engine would still be started. An example of this advantage is when a driver leaves the headlight on for a long period of time.

Further development to this concept is the 'switch battery'. The switch battery has two parts connected by a power diode and a heavy duty switch. One part is the 'main battery' and the other part is the 'back-up battery' (see Figure 13-5).

Normally, the switch is in the 'off' position and the main battery acts the same way as any battery: it supplies power for all vehicle electrical functions. If the main battery goes flat due to the lights being left on or some other mishap, the switch can be turned on and the engine can be started from the back-up battery. One disadvantage with this type of battery is that the switch must be turned off after the engine has been started. Failure to turn off the switch may cause both batteries to go flat due to a mishap.

PRECAUTIONS WITH LOW OR MAINTENANCE-FREE BATTERIES

1 They must never be fast charged.
2 The charging voltage must not exceed 14.6 volts.
3 At least 48 hours must be allowed to recharge a flat battery.
4 The alternator voltage must be accurately set.

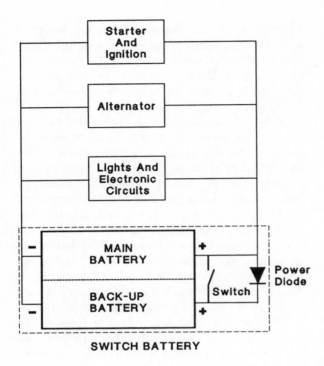

SWITCH BATTERY

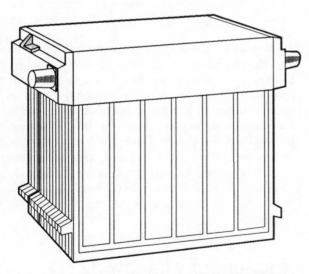

Figure 13-5 A switch battery configuration.

TESTING LOW OR MAINTENANCE FREE BATTERIES

There are two tests that can be conducted on low maintenance or maintenance-free batteries. These tests are:

- state of charge;
- high rate discharge.

STATE OF CHARGE

For those batteries which have cell caps, the state of charge can be determined by using a hydrometer and following the normal method.

For those batteries that are completely sealed, a digital voltmeter placed across the terminals of an unloaded battery will give the voltage output. The voltage output can be converted to a percentage charge condition. This is shown in Figure 13-6. It is important to note that an incorrect voltage reading will result if the battery is tested immediately after disconnecting it from a battery charger. The battery should be subjected to a light load or allowed to 'stand' for about an hour before carrying out the test.

VOLTAGE	% CHARGE
12.72 - 12.60	100 %
12.48	75 %
12.27	50 %
12.12	25 %

OPEN CIRCUIT VOLTAGE TEST

Figure 13-6 A voltmeter can be used to check the state of charge of a battery.

HIGH RATE DISCHARGE

This test is carried out with a high rate discharge meter connected to the battery terminals. The battery is subjected to a high current draw until a reading can be taken from the meter. The load *must* only be applied for a short period of time, to prevent permanent battery damage. The amount of current draw can be determined from one of two codes which are placed on the battery. One code is the amp/hour capacity and the other is the cold cranking amps. The cold cranking amps may be expressed as:

- CCA;
- cranking power;
- cranking current;
- starting amps;
- starting power.

When the cold cranking amps method is used to determine the current draw, the battery should not be subjected to more than 50 per cent of the cold cranking amps.

When using the amp/hour capacity of the battery, the current draw should not exceed three times the amp/hour capacity.

14

INSTRUMENT DISPLAYS AND SYSTEM INFORMATION PANELS

One of the more noticeable differences between the vehicles of today and the motor cars of yesteryear, is in the instrument display panels. The older type vehicles have a minimum amount of gauges, switches and warning lights in their 'dash units'.

The types of instrument display panels incorporated in motor vehicles consist of:

- electronic controlled instrument systems;
- printed circuits;
- a multi-coloured array of analogue gauges and digital meters;
- bar graphs, symbols, icons;
- flashing LEDs, visual and audio warning signals;
- a variety of switches;
- control and monitoring units.

The instrument display panel provides the driver with complete and accurate information on the various operating systems within the vehicle. It can also provide displays on a priority basis, to warn the driver of any potentially dangerous condition by activating a warning mode.

In this chapter, the function, construction, operation and system layouts of some of the more popular types of instrument display units are detailed.

Most electronic instrument display units consist of:

- input sensor signals;
- electronic processing units (EPU);
- output displays and/or audio signals.

Figure 14-1 shows the various input signals (left side), being sent to the processing units (centre) and then to the display unit assembly (right side).

INPUT SENSOR SIGNALS

The input signals to a system of this type are obtained from a:

- pulse generator;
- ignition coil;
- fuel gauge sender;
- coolant temperature sender;
- oil pressure sender.

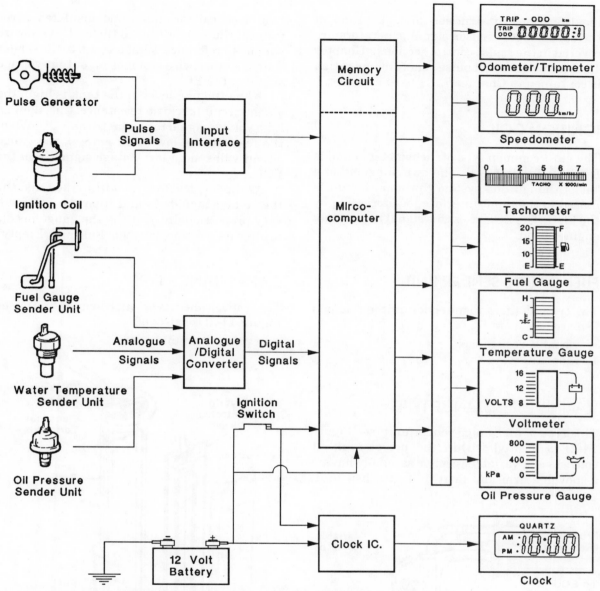

Figure 14-1 A typical layout of an electronic instrument display unit.

PULSE GENERATOR

The pulse generator can be in the form of a/an:

- inductive pulse;
- Hall effect generator;
- infra-red sensor;
- flexible drive cable and micro switch assembly

The output signal from the sensor is routed through an input interface to condition the signal before processing by the ECU.

The ECU uses this signal as an input for vehicle speed information.

The pulse generator can be located within the transmission or housed within the speedometer display unit. Information on identification, construction and operation of the speedometer sensor signal generators are covered in the cruise control section in Chapter 15. Specific details of pulse generator operation is contained in Chapter 7.

IGNITION COIL

The most common type of tachometer signal is from the ignition coil. The switching 'off' and 'on' of the ignition system's primary circuit is pulsed to the input interface for processing by the ECU. The ECU uses this signal to calculate engine speed.

FUEL GAUGE SENDER UNIT

Two types of fuel gauge sender unit in common use are based on:

- variable resistance;
- capacitance.

VARIABLE RESISTANCE TYPE

This type of fuel gauge sender unit (see Figure 14-2) is located within the petrol tank. It consists of a variable resistor assembly having a movable arm and float unit attached to it,

two electrical terminals and insulated wires connect the sender unit to the ECU. One wire supplies a reference voltage signal to the sender unit the other wire supplies an output voltage signal to the ECU.

When the fuel level in the tank is high, the gauge circuit has little resistance; this allows a high voltage signal to be sent to the ECU. When the fuel level is low (tank empty), the gauge circuit will send a low voltage signal from the ECU.

As the fuel level varies within the tank, the the voltage signals to and from the ECU will vary proportionally, causing the gauge needle to alter its position between 'Full' and 'Empty'.

CAPACITANCE TYPE

The capacitance type fuel sender unit (see Figure 14-3) consists of:

- a printed circuit board;

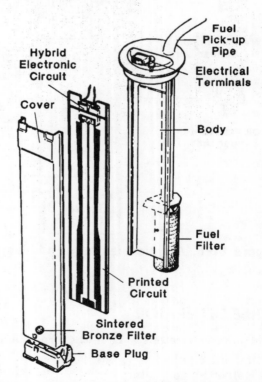

Figure 14-3 A capacitance type, fuel sender unit.

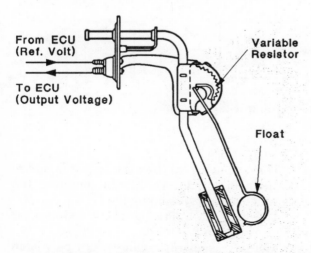

Figure 14-2 A variable resistance type, fuel gauge sender unit.

- protective cover metal plates and base plate assembly;
- fuel filter and pick up pipe unit;
- a hybrid electronic circuit encased in epoxy resin.

The **printed circuit board** (see Figure 14-4) consists of the following:

- connection points to the hybrid IC for the positive plates of measuring capacitor and the reference capacitor;
- contact rails to cover;
  - the cover forms the negative plates of the measuring capacitor and the reference capacitor;
- measuring capacitor's positive plates;
  - note the plates vary in area, this is to compensate for the shape of the fuel tank. The capacitor produces an output voltage in relation to the level of fuel and the shape of the fuel tank, to ensure a linear movement of the fuel gauge;
- reference capacitor's positive plate.

The circuit board is housed in, and is 'earthed' by the sealed metal casing. The function of the reference capacitor is to maintain the calibration of the measuring capacitor.

As the quality of fuels vary, this causes the value of the dielectric to change and the resultant output voltage would alter. The output voltages from the reference capacitor is sent to the hybrid IC located in the upper section of the sender unit.

The **hybrid electronic circuit** unit (see Figure 14-5), consists of the electronic components which are encased in an epoxy resin mixture.

Its function is to monitor and process the varying capacitance and provide an output voltage signal.

A **sintered bronze filter** is fitted to the base of the cover. This filter will:

- prevent the inflow of any dirt and water, which can affect the quality of the dielectric.
- control the rate of fuel flowing in and out of the measuring chamber. This control causes a dampening effect, which ensures the unit is not sensitive to the variance in the fuel level as the vehicle brakes or corners.

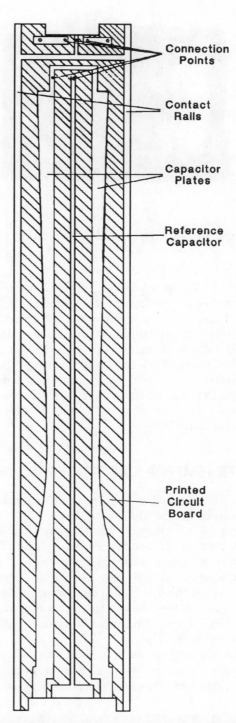

Figure 14-4 The constructional features of a printed circuit board.

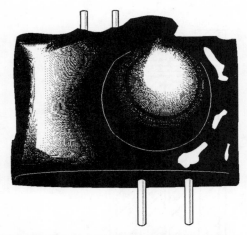

Figure 14-5 A hybrid electronic unit.

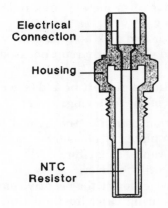

By varying the level in the fuel tank, the capacitance of the capacitor also changes. The capacitance value is monitored and processed by the hybrid electronic circuit; the resultant output voltage signal is sent to the fuel gauge.

When the tank is empty, a zero output voltage is signalled to the fuel gauge unit. The output voltage increases as the fuel level increases, until the maximum voltage (4.3 volts) is achieved when the tank is full.

WATER TEMPERATURE SENDER

One of the more common types of temperature sender units is the thermistor type.

It consists of a threaded metal body assembly, which is screwed into the cylinder head. This assembly houses a sensing element at the 'wet' end and an insulated unit with two electrical terminals at the other end. One terminal is for the sensor's reference voltage input signal and the other one is the sensor's signal voltage output to the ECU.

The sensing element consists of a semiconductor material which has a negative temperature coefficient (NTC) (see Figure 14-6).

As the temperature of sender unit is increased its resistance decreases, and conversely as the temperature decreases, its resistance increases.

When the sensor element is cold, a low voltage

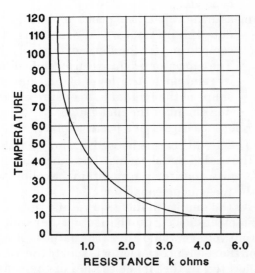

Figure 14-6 An engine coolant temperature sender unit.

signal is sensed by the ECU and a high voltage is sensed when the element is hot. The value of the sensor's output voltage signal to the ECU will vary proportionally to any changes in the coolant temperature.

Oil pressure sender unit, variable resistance type, consists of a threaded metal body and housing assembly, which is screwed into the cylinder block and engine oil gallery (see Figure 14-7).

The housing contains a flexible diaphragm which is subjected to engine oil pressure at one side and a variable resistor at the other side.

The variable resistor (rheostat type, a wire

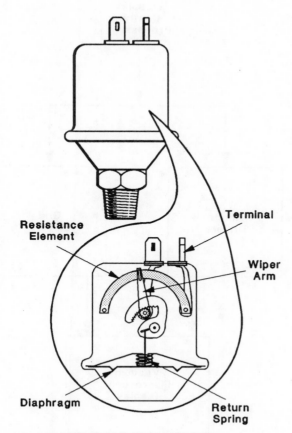

Figure 14-7 This resistance type, oil pressure, sender unit will supply a low voltage (high resistance) to the ECU when the oil pressure is low and a high voltage (low resistance) when the oil pressure is high.

wound coil and wiper arm) has two insulated electrical connector terminals. One terminal is for the sensor's reference voltage input signal and the other one is the sensor's voltage signal output to the ECU.

When the engine oil pressure acts on the flexible diaphragm, moving the diaphragm and wiper arm, the value of the voltage signal from the variable resistor to the ECU is changed. The value of the output voltage signal to the ECU will vary proportionally with any changes in engine oil pressure.

ELECTRONIC CONTROL UNITS (ECUS)

ECUs vary in size, shape and design; some are small and are mono-purpose units, other ECU assemblies are large, complex and multifunctional. In this chapter the multifunctional-type ECU is detailed.

This type of ECU module may consist of two or three printed circuit boards housed in an aluminium casing, which provides heat sinking and shielding from radio interference.

The function of the components mounted on the printed circuit board are to supply switching power, voltage regulation, analogue to digital signal conversion, signal processing and a multiplexing facility for the display unit.

The ECU can be divided into two main sections.

The **power supply, micro computer printed circuit board,** consists of the:

- input signal conditioning stage (which protects the A to D convertors, and other electronic devices from transient voltage spikes. Voltage spikes can be generated by switching on and off any high inductive loads such as the air conditioning compressor clutches, fan motors and some solenoids);
- A to D convertors (which change the analogue signals from the fuel, temperature and oil pressure sender units for processing within the CPU);
- main central processing unit (which has 2 kilobytes of ROM. and handles all gauge functions).

This circuit board also contains components to retain the electronic odometer reading, even when the battery is disconnected. A memory integrated circuit, non-volatile random access memory (NOVRAM) is used, it constantly updates and stores the odometer reading when the vehicle is in motion. When the ignition is switched off, the CPU senses this condition and writes the odometer reading into NOVRAM. When the vehicle is restarted, the odometer reading is relayed back to the CPU. The NOVRAM

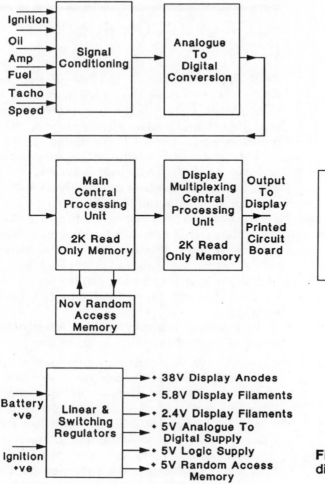

Figure 14-8 The function of the main printed circuit boards of a multifunctional ECU.

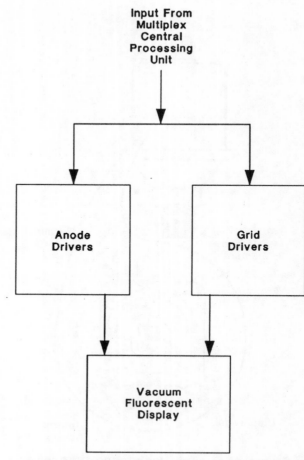

Figure 14-9 The schematic layout of a display printed circuit board.

circuit can store an odometer reading for a minimum period of ten years.

• display multiplexing CPU unit (which contains 2 kilobytes of ROM and its function is to provide the output signals to the display printed circuit board);

• linear and switching regulators section (which contains the switching power supply, transformers and voltage regulators which supply the correct voltage requirements for the various circuits—circuit voltages range from 2.4 volts to 38 volts).

This circuit board has a back up power supply, in the form of a large capacitor, which is employed when the battery voltage drops during cold starting of the engine. The back up system ensures that the micro computer has the correct voltage supply at all times.

The **display printed circuit board** receives its inputs from the display multiplexing CPU. This printed circuit board contains the anode and grid driver integrated circuits for the vacuum fluorescent display (see Figure 14-9).

This is the circuit board to which the display gauges, bar graphs, icons and digital meters are connected. The display assembly informs the driver of the operating condition of the various systems within the vehicle.

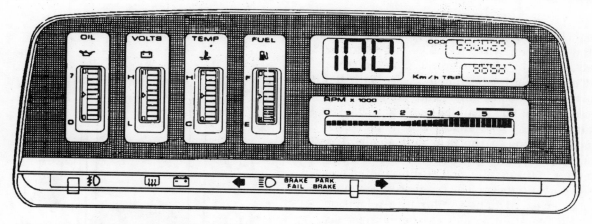

Figure 14-10 A typical vacuum fluorescent display unit.

OUTPUTS

The outputs from the various system processors results in a communication to the driver by audio or visual means. These outputs will take the form of warning devices, instrumentation and diagnostic information. Figures 14.10 and 14.15 are examples of typical electronic display units, showing a combination of analogue, digital and bar graph displays.

TYPES OF DISPLAY UNITS

Visual displays can be represented by:

- vacuum fluorescent units;
- liquid crystal assemblies;
- analogue meters.

FLUORESCENT TYPE

The instrument display unit (see Figure 14-10) consists of several different vacuum fluorescent applications. The fluorescent segments can be formed into numbers, horizontal bar graphs and vertical columns. Displays with these characteristics can be used as speedometers, tachometers, visual warning and level indicators (oil, temperature, volts, fuel).

The vacuum fluorescent display assembly (see Figure 14-11) consists of three sections: a wire filament (cathode), a grid (a thin honeycomb

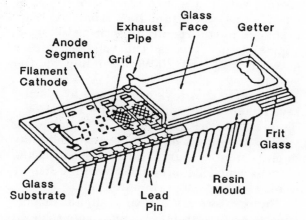

Figure 14-11 A vacuum fluorescent display assembly.

mesh) and an anode. The anode is a variety of conductive segments, also symbol and bar sections used for the output display. The conductive segments are insulated and coated with phosphor. The three sections are enclosed in a glass casing which is subjected to a vacuum.

A metal plate separates the vacuum fluorescent display unit from the printed circuit board; the function of this plate is to provide 'heat sinking' from the display unit.

OPERATION

The vacuum fluorescent display unit has a supply voltage range from 2.4 to 38 volts, and its anode and grid driver circuits receive input signals from the ECU. The cathode is an oxide-coated tungsten wire, which emits electrons when heated by electric power. The electrons emitted from the heated filament wire (cathode) pass through the grid and strike the phosphor-coated anodes. The vacuum fluorescent display is activated by the ECU, which energises specific sections of the grid and anode. The anode and grid circuits determine which segments of the display will be illuminated and the display intensity. The fluorescent segment (anode) gives off light similar to a television tube; the resultant display can be in any of six different colours.

The vacuum fluorescent display unit can operate within a wide temperature range. It has a low power consumption, is quick to react to change and the required operating voltage is approximately 40 volts (This voltage can vary between 12V and 100V depending on the system).

Numeric displays are formed by placing seven fluorescent units into the shape of a figure eight (see Figure 14-12). By illuminating various combinations of these segments, any numeral from zero to nine can be formed. Placing these assemblies side by side allows the display to register units, tens, hundreds etc.

Bar graphs are a type of digital display which consists of a number of rectangular shaped segments (see Figure 14-13). These segments

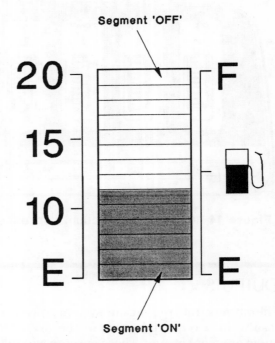

Figure 14-13 A display unit with fourteen rectangular segments.

can be of the same size, or of different sizes; they may be positioned vertically or horizontally within the instrument display panel. The segments are illuminated by fluorescent lighting and one bar graph display can contain segments illuminated in different colours. Bar graphs are used to display units of temperature, oil pressure, fuel quantity and voltage.

LIQUID CRYSTAL TYPE

Liquid crystal display (LCD) units are being used in a number of vehicles. A widely used type of LCD unit consists of two flat sections of glass having a thin liquid crystal layer sandwiched between them (see Figure 14-14). Transparent conductor segments are etched onto the inner surface of the front glass and form the display pattern. Each front glass panel conductor segment is connected to the ECU to form an electrode. The number of electrodes on the front glass is dependent on the display. A transparent electrode is coated onto the inner

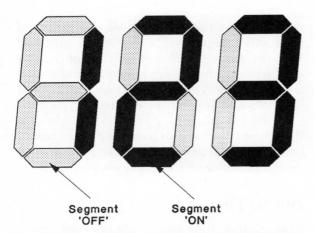

Figure 14-12 Seven segments are required to form an eight in a numerical display.

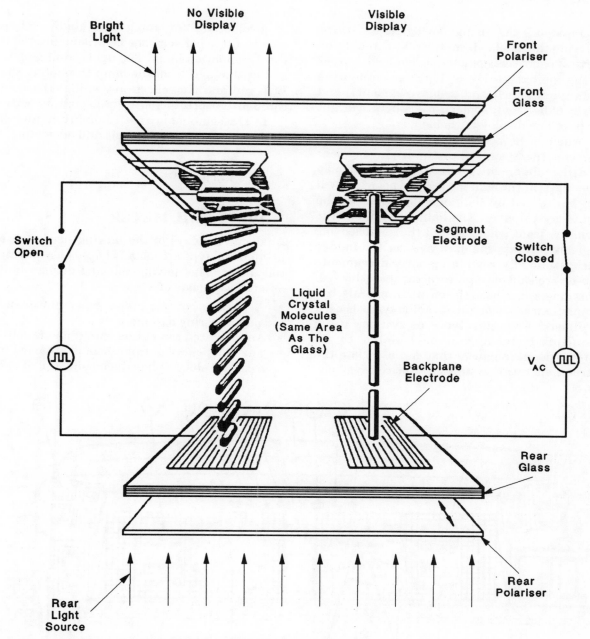

Figure 14-14 A typical Liquid Crystal Display (LCD).

surface of the rear glass. Treatment of the surfaces causes the liquid crystal molecules to point in the same direction and lie parallel to the surface of the glass. This group of plates and crystals form an LCD cell.

The LCD cell is placed between two polarising filters. A polariser can be likened to a series of slits in a plate. These slits could be arranged vertically (allowing only vertical lines of light) or horizontally (allowing only horizontal lines of light). A horizontally oriented polariser is placed to the front of the cell and a vertically oriented polariser at the rear.

The left hand part of the illustration, Figure

14-14, shows the display switched 'off'. Light entering the display from the rear through the vertical polariser is twisted through 90 degrees by the liquid crystal layer. It passes through the front glass panel (and display segments) and the front horizontal polariser. It shows a bright light and no display is visible.

When the ECU applies current to one display segment, the crystals below the display segment lose their alignment. Light entering the display from the rear through the vertical polariser will not be twisted by the energised areas of the liquid crystal layer. The light will not emerge from the front polariser and the switched 'on' segment will appear dark against the lighter background. By combining several segments the desired display pattern can be achieved. This display can be in the form of numerals, bar graphs or icons. Icons are graphic symbols used to illustrate specific types of systems being monitored. Icons can be likened to an international language that can associate the various operating systems by the icon symbols.

A coloured filter, and a 'light guiding' plate may be fitted between one polarising filter and the illumination lamp. The light guiding plate guides the light from the lamp through to the LCD 'cell' and also reflects any sunlight passing through the cell to increase the display intensity.

LCD assemblies are used for trip computers, clocks, climate control panels and odometers.

ANALOGUE TYPE

MOVING COIL METER

Analogue displays of the moving needle on a printed dial type are usually based on a moving coil meter. The moving coil and display unit assembly consists of:

- A permanent magnetic base plate and magnetic ring assembly.
- An insulated coil of fine wire (moving coil).
- The balance weight platform is attached to a central shaft, it has the moving coil at one

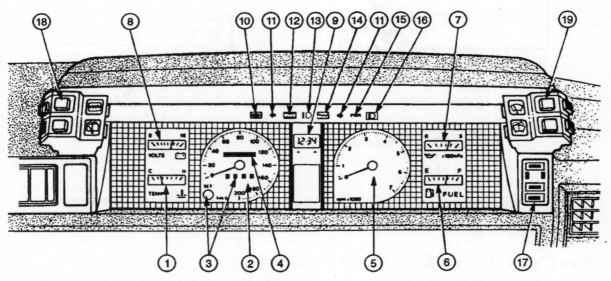

1. Coolant Temperature Gauge
2. Speedometer
3. Tripmeter and Reset Button
4. Odometer
5. Tachometer
6. Fuel Gauge
7. Oil Pressure Gauge
8. Voltmeter
9. Digital Clock
10. Generator Warning Lamp
11. Turn Signal Indicator Lamp
12. Brake Failure Warning Lamp
13. High Beam Indicator Lamp
14. Park Brake Indicator Lamp
15. Power Mode Indicator Lamp
16. Lamps 'ON' Warning Lamp
17. Cruise Control Switch
18. 'Satellite Switches' - Tailgate Wiper, Tailgate Washer, Rear Window Demister and Antenna Height Control
19. 'Satellite Switches' - Windscreen Wiper, Windscreen Washer

Figure 14-15 A typical analogue display unit.

end and a balancing counterweight at the other end.

- An indicating needle is indirectly attached to the balance weight platform.
- Two hair spring coils, insulated from each other, and each having one end attached to the moving coil and the other end attached to a central shaft.
- The central shaft is fitted into an insulated bearing, which allows the shaft assembly to rotate.

Operation of a moving coil meter is similar in most applications. When voltage is not present at the moving coil, the indicating needle will point at the zero position due to the tension on both of the hair springs acting on the central shaft.

Voltage input to the gauge is connected through electrical wiring to the two hair spring coils. The voltage applied to the hair spring coils and the moving coil unit, causes a magnetic field to form around the moving coil. This magnetic field will react with the permanent magnetic field surrounding the magnetic ring. The resultant force will rotate the balance weight platform, and the indicator needle away from zero.

A varying input voltage signal regulates the position of the indicator needle. To obtain a steady needle reading, the rotation of the shaft may be dampened by a silicone fluid.

The indicating needle assembly can be driven by a processor, ECU or directly from an input sensor. Displays of this type are used for speedometers, tachometers, and fuel, temperature, volts, amps and oil gauges.

The moving coil meter (see Figure 14-16) is widely used but analogue gauges of similar appearance may be constructed with different coil configurations. These include cross coil and balance coil types.

CROSS COIL TYPE ANALOGUE GAUGE

The cross coil type analogue gauge (see Figure 14-17) consists of:

- a dial, indicating pointer, shaft, a movable magnet, and a retraction magnetic ring;

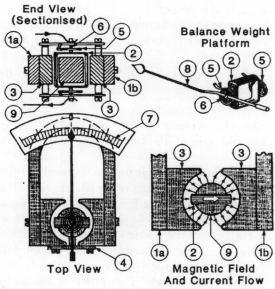

1a. Magnet - North Pole
1b. Magnet - South Pole
2. Coil
3. Pole Piece (Soft Iron)
4. Base Plate
5. Hair Spring
6. Central Shaft
7. Scale
8. Needle
9. Magnetic Ring (Soft Iron)

Figure 14-16 The construction and operation of a moving coil meter.

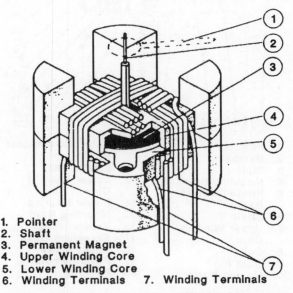

1. Pointer
2. Shaft
3. Permanent Magnet
4. Upper Winding Core
5. Lower Winding Core
6. Winding Terminals
7. Winding Terminals

Figure 14-17 The construction of the main section of a cross coil, analogue gauge.

- two coils of wire wound at right angles to each other, and a third coil connected by insulated cable to the sender unit.

Operation

The pointer attached to the moving magnet is held in its stop position by the retraction magnet. The opposing fields from these magnets create a rotational torque (towards stop position) on the moving magnet and shaft.

Two of the coils wound onto the gauge formers are supplied with a constant voltage. A third coil is supplied with a varying voltage from a sender unit. The magnetic force developed, at the third coil, will increase as sender voltage increases. This magnetic force will cause the moving magnet to rotate away from the zero

position. The pointer on the gauge will therefore vary its position as a result of the changes to the opposing magnetic forces.

SYSTEM OPERATION

Systems to be described within this section include:

- speedometer;
- tachometer;
- fuel level indicator;
- oil pressure indicator;
- engine coolant temperature indicator;
- trip computer.

SPEEDOMETER

The function of the speedometer is to:

- indicate the speed of the vehicle;
- show the total distance the vehicle has travelled (odometer);
- measure the distance of a specific trip and allow the display to be reset to zero (tripmeter).

Refer to Figure 14-19, the operation of the speedometer system is explained in three stages:

1 Input, speed transducer
2 Signal processor
3 Output, signal display.

1 **A speed transducer** driven by gearing from the transmission output shaft, provides the input. It is a 'Hall effect' type, consisting of a revolving multi-pole magnet assembly and a Hall effect magnetic switch.

The transducer produces a pulse the frequency of which is proportional to the speed of the vehicle. This pulse is sensed by the signal processing unit located in the speedometer assembly.

2 **The signal processor** contains electronic circuitry which produces two outputs from one input. The input signal is derived from the speed transducer (pulse generator). The input signal is filtered, pulse shaped and processed to produce regulated outputs. One

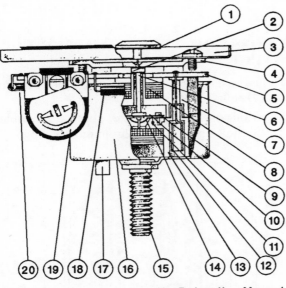

1.	Pointer	11.	Retraction Magnet Ring
2.	Damper	12.	Sleeve Plug
3.	Dial	13.	Lower Winding Core
4.	Mounting Bracket	14.	Lower Bearing
5.	Printed Circuit Board	15.	Mounting Stud
6.	Shaft	16.	Tubular Rivet
7.	Upper Winding Core	17.	Screening Case
8.	Coils	18.	Zener Diode
9.	Stop Pin	19.	Potentiometer
10.	Moving Magnet	20.	Resistor

Figure 14-18 The components of a cross coil type, analogue gauge.

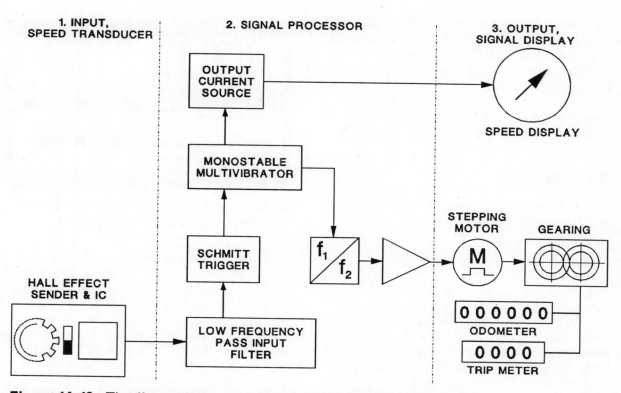

Figure 14-19 The three stages of an electronic speedometer system.

output provides a variable current flow to drive the analogue display unit. The other output provides a regulated pulse to drive a stepper motor for the odometer and trip meter display.

3 **Signal display** is a moving coil meter configured as a speedometer display unit. As the speed of the vehicle varies, the transmission mounted speed transducer sends a proportional varying voltage signal to the signal processor. A varying output voltage signal from the processor regulates the position of the speedometer's indicator needle. To obtain a steady needle reading, the rotation of the shaft is dampened by a silicone fluid.

Other types of displays register the vehicle speed and distance travelled by means of LCD and vacuum fluorescent assemblies. Their display patterns may be either a bar graph or numerical array (see Figure 14-20).

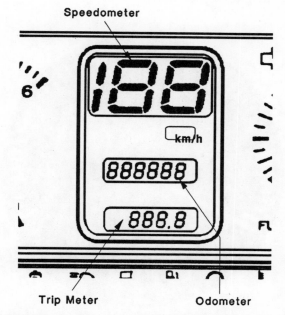

Figure 14-20 A signal display using several numerical arrays.

The odometer in these types of displays usually have a non-volatile memory, which means that the total distance/kilometre reading is not erased when the battery is disconnected.

The tripmeter has a volatile memory, which means that the display reading is erased when the battery is disconnected.

TACHOMETER

The **tachometer** display unit can be of the bar graph, analogue, or digital types.

OPERATION

When the engine is running, pulses from the negative side of the ignition coil are sensed by the ECU, which counts and processes these input pulses. The output from the ECU sends signals to the display unit. This display may be in the form of analogue, bar or numerical arrays.

Analogue display units (see Figure 14-21) are of the moving coil type and are similar in design and operation to the speedometer analogue display unit.

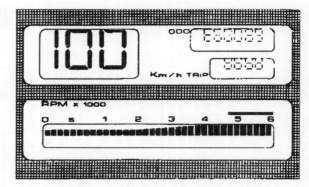

Figure 14-22 An electronic tachometer fitted with a bar graph display unit.

Bar graph types (see Figure 14-22) contain segments which, when illuminated, indicate the engine speed from zero to a designated value in steps of one hundred. When this value is reached, each illuminated segment represents steps of 200 r.p.m. until another designated value is reached. After this point the additional segments may display a different colour, to alert the driver of high engine r.p.m.

Numerical units indicate the engine r.p.m. in the form of numbers on a vacuum fluorescent display.

FUEL LEVEL INDICATOR

An electronic fuel level indicator system consists of a:

- sender unit located in the fuel tank;
- processor;
- display unit located in the instrument panel.

The sender must provide a reference voltage proportional to the level of fuel in the tank. The construction of this sender unit will vary, depending on whether the sender is based on resistive or capacitive sensing.

Resistive type senders will be linked to an ECU mounted in the dash panel area of the vehicle.

Capacitive type senders contain an inbuilt processor and conversion of a fuel level to a voltage, suitable for an input to a gauge or an ECU, which is performed within the sender unit assembly. This unit may be linked to a

Figure 14-21 An electronic tachometer fitted with an analogue display unit.

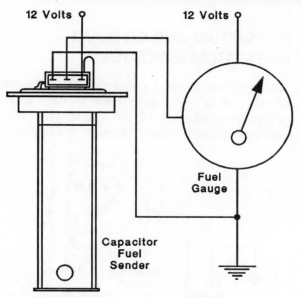

Figure 14-23 The circuit diagram for an electronic fuel level indicator.

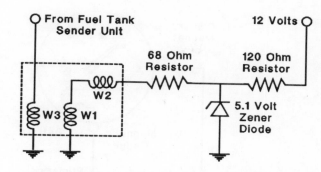

Figure 14-24 The component diagram of a cross coil fuel level indicator.

cross coil type gauge assembly mounted in the instrument panel or to the instrument panel ECU that controls the electronic display panels.

A typical example of one of these units is shown in Figure 14-23.

The IC unit contained within the sender has three connections: 12 volts supply, earth return, and a connection to the display unit (fuel gauge).

When the ignition is on and the fuel tank is empty, the voltage value on the supply wire to the gauge is zero.

While there is a zero voltage at the gauge sensor input, the indicator pointer on the dial of the cross coil type fuel gauge will move to the 'Empty' position. As the fuel level within the fuel tank is increased, the capacitance of the sender unit changes, thus varying the (gauge input) voltage signal. The voltage from the sender unit varies from zero volts (fuel tank empty) to 4.3 volts (fuel tank full).

Figure 14-24 shows a gauge circuit in a simplified form. Coils W1 and W2 are wound at right angles (90 degrees) to each other and operate at voltage controlled by a combination of the 120 ohm resistor, a 68 ohm resistor and a 5.1 volt zener diode.

Coil W3 is connected to the sender unit and its voltage signal varies proportionally to the fuel level in the fuel tank.

With the ignition switched on and the fuel tank empty, the voltage at coil W3 is zero. The combined magnetic effect of coils W1 and W2 will cause the needle to align with a 45 degree position between the two coils. The pointer on the gauge will be at Empty.

When the level in the fuel tank is half-full, the output voltage signal from the sender unit will be approximately 2.5 volts.

The increased sender voltage causes coil W3 magnetic field to strengthen and the needle moves towards W1. The indicating pointer will read half-full on the dial.

As more fuel is added to the supply tank, the increasing voltage signal from the sender unit, allows more current to flow through coil W3, progressively strengthening the magnetic fields around coils W3. The increased magnetic force, acting on the movable magnet, moves the indicating pointer towards the 'Full' position on the dial.

This type of cross-coil gauge movement (sometimes referred to as magnetic gauges) can be direct driven by a resistive sender unit (see Figure 14-25).

Another type of display that can be used to indicate the fuel level is a vacuum fluorescent bar graph (see Figure 14-26). The bar graph segments are placed on top of one another to form a column which changes colour as the fuel level drops. This type of unit may incorporate a flashing low fuel level indicator. The bar graph is driven by an ECU in response to voltage changes from a tank sensor.

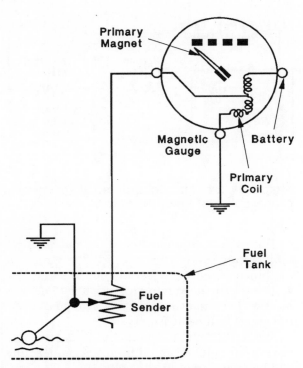

Figure 14-25 The component diagram of a cross coil meter and a resistive sender unit.

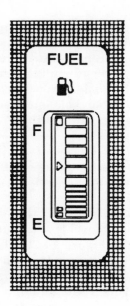

Figure 14-26 A vacuum fluorescent bar graph used to indicate the fuel level.

OIL PRESSURE AND COOLANT TEMPERATURE INDICATORS

The appearance and construction of these display units are comparable to fuel level indicators. Figure 14-27 illustrates the magnetic gauge controlled by resistive devices.

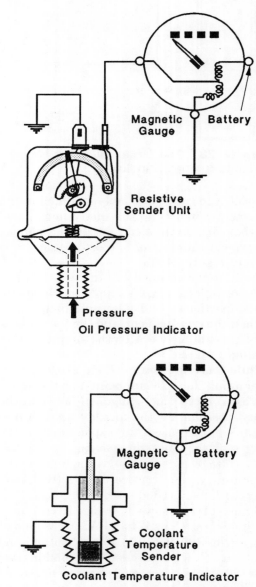

Figure 14-27 Various applications of the magnetic gauge with different resistive type sender units.

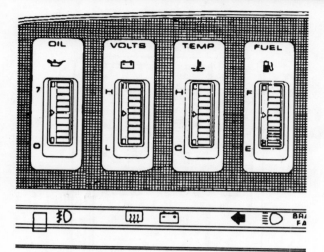

Figure 14-28 Bar graphs used in a combination display unit.

Figure 14-28 shows a combination display unit of the vacuum fluorescent type. This type of display requires an ECU to interface between the sensors and the display units.

A detailed explanation on the design and operation of bar graph display units has previously been described in this chapter.

TRIP COMPUTER

A trip computer (see Figure 14-29) provides the driver with information related to the operating conditions of the vehicle. This information includes:

- clock and calendar;
- fuel economy check—instant litres/100 km;
- trip economy check—average litres/100 km;
- average speed—average km/hour;
- range—distance to 'Empty' in km.

A trip computer can also have an audio 'beeper' warning system.

Manufacturers can program many different functions to the audio circuit, e.g. system malfunctions, low fuel warning, parklight reminder etc.

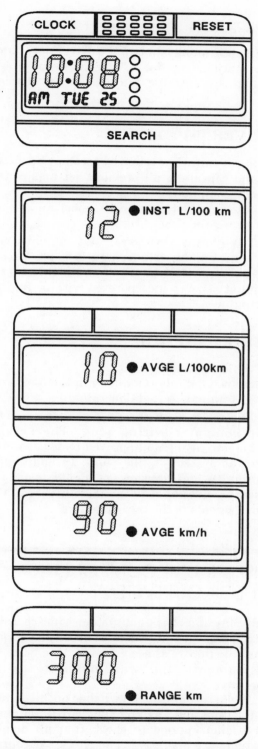

Figure 14-29 The layout and readouts of a typical trip computer.

OPERATION

At the beginning of a trip the driver may enter information (inputs) to the trip computer's system, by pressing buttons on the display panel.

DRIVER INPUTS

- initialising the system by zeroing trip meter and cancelling all previous settings in the trip computer;
- entering the total trip distance;
- setting the time clock.

The information and calculations which can be achieved by trip computers are varied depending on the type of system and the amount of input sensors incorporated within the trip computer.

SYSTEM INPUTS

- vehicle speed;
- quantity of fuel remaining in the fuel tank;
- instantaneous fuel flow rate.

Figure 14-30 shows a typical example of a trip computer wiring plug. The service technician, using an approved tester, can link into the system at this point. In addition to continuity and presence of signal tests, the tester provided by some manufacturers can also input signals to the trip computer.

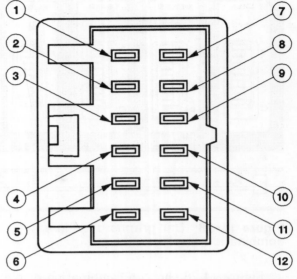

1. Light Dimmer Signal
2. EFI - Injector Signal (Fuel Consumption)
3. Spare
4. Battery Power
5. Fuel Gauge Signal
6. Ground (Earth)
7. Speedometer Signal
8. Ignition Power
9. Spare
10. Park Lamp Signal (Light Swith)
11. Buzzer Signal (Instrument Cluster)
12. Spare

Figure 14-30 The terminal layout of a trip computer wiring plug.

On completion of the check-out routine, the display units will revert to their normal operating condition.

DISPLAY UNIT COMPONENTS CHECK-OUT SYSTEM

A display unit check-out system is incorporated into most vacuum fluorescent display units. The function of this system is to ensure all instruments and warning systems are operative.

When the ignition is switched on, all display units, icons and warning lights are illuminated for a short period to check that all display systems are operative. The check-out routine may also place the warning display units into a flashing sequence for a few seconds.

SYSTEM WARNING DISPLAYS

Warning signals (display and/or audio) are an integral part of many operational systems. Their function is to alert the driver of any malfunction to specific systems within the vehicle.

Warning displays can be in the form of an illuminated symbol, a flashing light or an audible signal.

This chapter describes several system conditions in which warning display signals may be used:

- **fluid levels**, low fluid levels in the cooling, brake, fuel and wiper washer systems;

- **pressures**, low pressures from the engine oil, or brake systems;
- **position**, doors open, park brake in the 'on' position, seatbelts disconnected;
- **temperature**, high coolant temperature;
- **lighting system**, stop/tail lights, park lights, headlights.

Fluid level warning systems usually consists of a float assembly connected by wiring to a warning display icon.

When the fluid level drops to a predetermined level, the float makes a connection with an 'earth' terminal, completing the circuit and the warning system/display is activated.

Pressure warning systems can be activated by:

- the movement of a valve, due to unequal pressures within the brake hydraulic system. The movement of the valve allows a switch to complete the 'earth return' circuit within the warning system, causing a warning light to illuminate;
- a voltage signal from an engine oil pressure sender unit being outside the predetermined value range of the ECU program. When this occurs, the ECU switches on the low pressure warning circuit activating an audible alarm or a visual display.

Position warning systems include:

- doors open (ajar);
- park brake;
- seat belt.

The **door ajar** warning system consists of a spring-loaded switch fitted to each of the door pillars and a warning display. This system can be powered through the ignition switch or an ECU.

ECU type system: When a door is open, the spring-loaded switch completes the 'earth return' circuit. The ECU senses the change in the voltage signal within the circuit and the warning system/display is switched on. A separate warning display can be shown for each door.

Park brake: A switch at the park brake mechanism completes the 'earth return' circuit when the park brake is applied. This action causes the warning system/display to be switched on.

Seat belts use the belt buckle as a switch. When the belt is disconnected, the 'earth return' circuit is completed and the warning system/display is switched on. Most seat belt warning systems are switched off after a period of approximately 60 seconds: this is achieved by incorporating a capacitor circuit within the system.

Temperature: When the engine's coolant temperature increases to an unsafe level, a warning display, usually in the form of an icon, is illuminated. The high temperature acting on the coolant sensor reduces its resistance value. When the resistance decreases to a specific low value the warning display is switched on.

Lights warning system: Tail lights and stop light warning systems inform the driver when one or more lights are inoperative.

The ECU monitors the required operational voltage and amperage within the circuits. It activates the warning/system display unit when it senses any variance to the required voltage and/or amperage readings.

When the park lights and/or the headlights are left on and the ignition is switched off, a sensor unit detects this position and activates the warning/system display unit.

This system can be incorporated with the door ajar system to provide a warning to the driver. If the driver opens the door to leave the vehicle and the lights are on, an audible warning will sound while the door is ajar.

Audible warning signals can be in the form of a sounding horn, an intermittent beeper, a tape recorded message or a synthesised voice warning system.

Audible warning systems are incorporated in a number of vehicles and are proving to be an effective method of warning the driver of a potential danger or malfunction within an operating system.

MONITORING SYSTEMS

Monitoring systems constantly check on the operating conditions of the vehicle, and provide a warning signal when one or more of the systems malfunction. The engine management system has a number of monitoring systems; these systems are detailed in Chapter 11. Other types of monitoring systems include the engine diagnostic display and the trip computer.

The **engine diagnostic display** is usually in the form of an illuminated warning lamp, and/or a flashing icon symbol.

This system is designed to provide the driver and the service technician with information on the condition of specific engine operating systems. The warning lamp indicates to the driver that a fault is in a system.

The service technician can switch the system into the diagnostic mode. In this mode, the number and frequency of one or more flashing warning lights are interpreted by the service technician. The information gained from the diagnostic mode assists the service technician to quickly diagnose problems within the various systems. The ECU has the minimum and maximum voltage readings for each of the systems programmed into a memory IC. When the ECU senses voltage signals which are constantly outside the voltage settings in the memory program, it (ECU) switches on or pulses the warning display device.

FUTURE TRENDS

Each year more electronic control, display and monitoring systems are being fitted to motor vehicles. These systems improve the performance, reliability and the safety of the vehicles.

The monitoring of these operating systems and the information shown on the instrument display unit, can assist the driver in:

- making decisions on vehicle servicing;
- obtaining reliable information on vehicles' operating conditions;
- preventing vehicle breakdowns.

Systems being investigated by vehicle manufacturers are:

- navigation aids;
- 'heads up' displays;
- collision avoidance systems;
- CRT displays.

15

ACCESSORIES

This chapter describes the function and operation of accessories which are designed to improve the safety, and comfort of the driver and his/her passengers. The systems described in this chapter are:

- cruise control
- automatic climate control

CRUISE CONTROL

A vehicle, having a cruise control system, offers advantages to the driver. A cruise control:

- ensures the vehicle maintains a constant speed without having to regulate the accelerator pedal;
- maximises fuel economy by minimising the changes to engine speed during varying road and load conditions;
- reduces driver fatigue during long distances.

A cruise control actuator is fitted into the throttle system of the vehicle to maintain a pre-set speed under varying engine loads and conditions. Many types of sensors and switches are located on the vehicles ancillary systems to monitor and send information to the ECM where it is processed. The ECM decides on the action that is necessary to maintain the pre-set cruise speed and adjusts the throttle position. The cruise control system can be overridden by the conventional throttle. Safety features are included to ensure that the vehicle's operation is not impeded by a malfunction in the cruise control system.

COMPONENTS

A cruise control system like many other electronic systems contains:

- input devices;
- processor (ECM);
- output device;
- additional components.

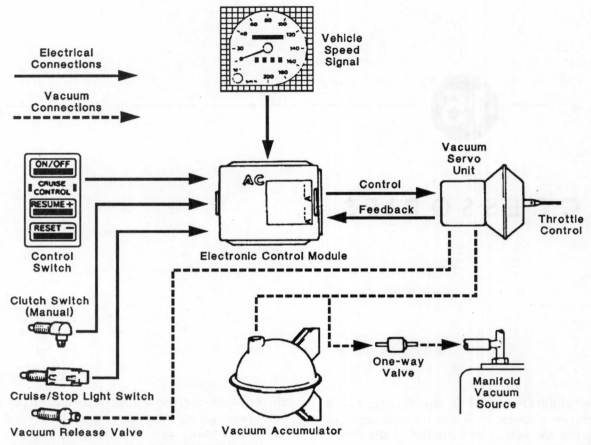

Figure 15-1 Cruise control system.

INPUT DEVICES

The input devices include:

- a control switch assembly unit;
- a vehicle speed sensing system;
- cancel switches fitted to the clutch and brake pedals.

LOCATION AND FUNCTION

The **cruise control switch** assembly normally consists of a cluster of push type or rocker type switches, fitted within a panel on the dash assembly. A variation to this is a 'stalk' mount switch assembly, which is usually fitted to the steering column.

A typical **switch panel** (see Figure 15-2) has the following switch arrangement:

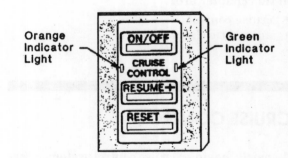

Figure 15-2 Cruise control switch assembly.

- on/off push type switch;
  —supplies power to the cruise control system;
  — a light indicates when the power is 'on';
- 'resume' + push type switch;

— provides a method of varying the 'set' cruise speed;

- 'reset'—push type switch;
  — engages the cruise control system;
  — a light indicates when the cruise system is engaged.

Vehicle speed sensor may be located on the tail shaft, in the speedometer or within the transmission. The types of sensors used for speed detection include the following:

1 Hall effect type sensor;
2 infra-red type sensor;
3 reed switch type sensor;
4 magnet and pulse generator type.

1 **Hall effect type** consists of a transducer (a Hall effect signal generator), fitted to the transmission (see Figure 15-3). The transducer shaft is rotated by a gear on the transmission's output shaft (similar to a speedo cable drive arrangement). The transducer generates and sends electrical pulses to the cruise control ECM. The electronics within the ECM accurately calculates the speed of the vehicle by processing the incoming pulses.

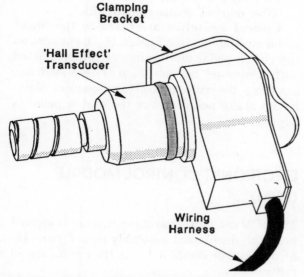

Figure 15-3 Transmission mounted (hall effect) vehicle speed sensor.

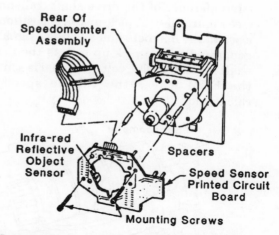

Figure 15-4 Instrument panel mounted (infra-red) vehicle speed sensor.

2 **Infra-red type** speed sensor consists of a printed circuit board and a stationary infra-red reflective object sensor attached to the speedometer assembly (see Figure 15-4).

The infra-red reflective object sensor transmits and receives light signals. Within the speedometer assembly a cup and an eight-lobed bright metal reflector strip rotate at the same speed as the speedometer cable.

The rotating lobes on the reflector strip reflect the infra-red light into the stationary sensor eight times per revolution. The cup is painted mat black and it absorbs the infra-red light.

The sensor sends these light generated pulses to the electronic control module. The electronics within the ECM calculates the speed of the vehicle by processing the incoming pulses.

3 **Reed switch type** sensors are fitted to cable driven speedometer assemblies. A rotating drum and reed switch assembly is contained within the speedo head (see Figure 15-5). As the speedometer drum rotates, the reed switch makes and breaks the control unit circuit four times per revolution. The electronics within the ECM calculates the speed of the vehicle by processing the incoming switching pulses.

4 The **magnet type** sensor has four permanent magnets placed around the

circumference of the drive shaft (tailshaft) (see Figure 15-6). A pick-up coil is positioned close to the magnets and a voltage signal is generated in the pick-up coil as the drive shaft is rotated. The voltage signal is sent to the ECM where the vehicle speed is calculated.

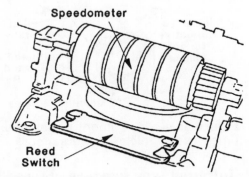

Figure 15-5 Instrument panel mounted (reed switch) vehicle speed sensor.

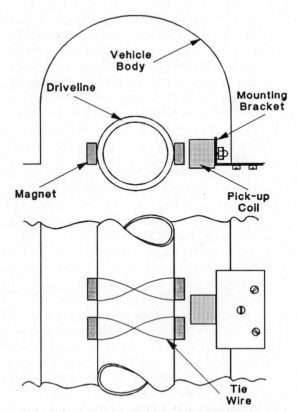

Figure 15-6 Driveline mounted (inductive pulse) vehicle speed sensor.

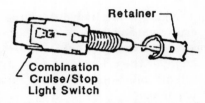

Figure 15-7 Brake light switch can be used as cancel switch.

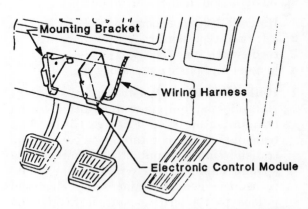

Figure 15-8 An electronic control module can be located under the instrument panel.

This type of sensor is usually supplied in 'after market' cruise control kits.

Cancel switches are fitted to the brake pedal mechanism (see Figure 15-7), and also on the clutch pedal mechanism of vehicles fitted with a manual transmission. These switches disengage the cruise control function immediately the brake pedal and/or the clutch pedal is depressed.

ELECTRONIC CONTROL MODULE (ECM)

The ECM can be a separate unit, usually located near the dashboard assembly (see Figure 15-8), or it can be combined with the engine speed control unit.

The function of the ECM includes:

- the interpretation of the position of the servo unit's variable inductance position sensor;
- receiving signals from the various switches

and sensors, processing and monitoring these input signals;

- the switching (on or off) of solenoids or circuits (outputs), in response to ECM input signals

OUTPUT DEVICE

The output device (engine speed control unit), located within the engine compartment, varies the throttle linkage position in response to signals from the ECM.

There are various engine speed control units, including:

1 vacuum servo type;
2 stepper motor type;
3 DC Motor and epicyclic gear type.

1 **Vacuum servo type** (see Figure 15-9), consisting of a unit, housing a spring-loaded diaphragm and movable central rod assembly. A variable inductance sensor locates and positions the moveable central rod.

Two electrically operated valves control the vacuum to and from the unit. A flexible cable connects between the unit's central rod and the throttle linkage.

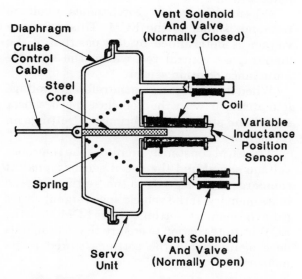

Figure 15-9 Vacuum servo throttle actuator.

2 **Stepper motor type** (see Figure 15-10), consisting of a combined stepper motor, gear train assembly, solenoid operated clutch and ECM. The stepper motor can operate in a clockwise and anti-clockwise direction. It provides torque through a mechanical gear train to the control cable. The control cable is connected between the stepper motor assembly and the throttle linkage.

3 **Electric DC motor and planetary gear type** (see Figure 15-11). This unit consists of an electric motor, a worm gear and wheel, planetary gear set, drive shaft, magnetic clutch and two limit switches.

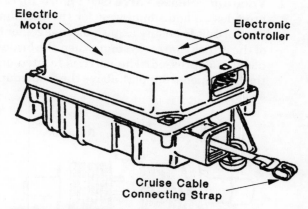

Figure 15-10 Stepper motor throttle actuator.

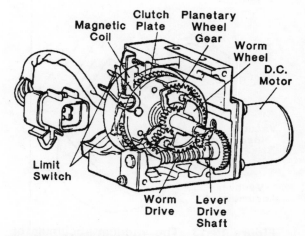

Figure 15-11 Electric motor, gear driven throttle actuator.

The ECM powers the DC motor and magnetic clutch assembly. The DC motor drives the planetary gear train and main shaft assembly, the limit switches control the degrees of rotation for the main shaft. A flexible cable is connected between a lever on the main shaft and the throttle linkage. The magnetic clutch assembly disengages the system when the clutch or brake pedal is applied.

ADDITIONAL COMPONENTS

1 **Vacuum release valve** (see Figure 15-12) provides an additional vent for the vacuum control unit to ensure a quick disengagement of the cruise control system, when the brake pedal is depressed. The valve is located on the brake pedal support above the stop lamp switch.

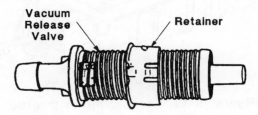

Figure 15-12 Brake pedal operated vacuum release valve.

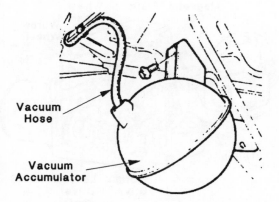

Figure 15-13 The vacuum accumulator acts as a vacuum reservoir.

2 **Vacuum accumulator** (see Figure 15-13) is connected to the engine intake manifold by a rubber flexible hose. Its function is to act as a storage vessel for the engine vacuum and to operate the cruise control system during periods of high engine loads (low vacuum).

OPERATION OF A CRUISE CONTROL SYSTEM (VACUUM SERVO UNIT TYPE)

Cruise control systems are designed to prevent engagement of the cruise control at vehicle speeds less than 40 km/h.

SYSTEM OPERATION (SEE FIGURE 15-14)

Assume the control switch is 'on' and the vehicle has been accelerated to a desirable speed. To engage the cruise control system, press and release the 'reset' button (switch). Both the 'power on' and the 'system engaged' indicator lights are 'on'.

Pressing the 'reset' switch sends a voltage reference signal to the ECM. This reference voltage is determined by the position of the diaphragm's central rod within the variable inductance position sensor.

When the ECM receives the reference voltage it matches it with the vehicle's speed sensor signal and stores this information 'settings' in its memory.

The ECM constantly monitors the reference voltage signal and the speed sensor signal; it compares these signals to the 'settings' stored in its memory. If the vehicle speed signal varies from the memory 'settings' in the ECM, then the ECM initiates action to ensure that the signals once again match the settings stored in its memory.

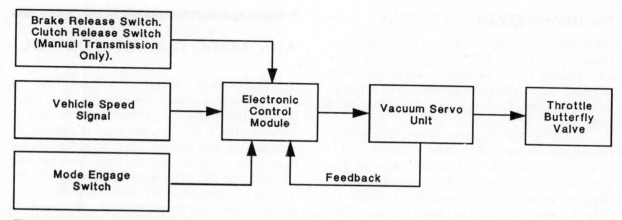

Figure 15-14 The block diagram illustrates the component relationship in the electronic cruise control system.

SPEED CHANGES

CRUISE SPEED

When the vehicle is cruising at a steady speed, and no load condition: The reference voltage and vehicle speed signal match the 'settings' in the memory of ECM. No action is required by the ECM.

Within the vacuum servo unit, both vacuum valves are closed. The servo unit has a constant vacuum acting on the diaphragm; the position of the central rod does not change within the variable inductance position sensor. The tensioned control cable will not alter the position of the throttle linkage.

DECREASING CRUISE SPEED

The ECM senses the decrease in the vehicle's speed by the change in the signal from the speed sensor.

When the vehicle road speed decreases by more than 1.5 km/h. the ECM senses this variance and pulses the vacuum inlet solenoid valve within the vacuum control unit. Each vacuum inlet solenoid valve pulse has an average duration of approximately 10 milliseconds. The pulsing of the vacuum solenoid valve increases the vacuum within the control unit.

The increased vacuum, acting on the

diaphragm, moves the central rod. The control cable opens the throttle valve, resetting the vehicle to the pre-selected road speed.

The vacuum solenoid pulses are repeated until the speed of the vehicle resumes its pre-set position. The ECM has the ability to lengthen the duration of the pulses, depending on the engine load conditions. When the vehicle encounters a steep gradient, the speed of the vehicle will quickly decrease. The ECM senses the sudden change in the vehicle's speed, and lengthens the duration of the pulses, to ensure the vehicle quickly, resumes its pre-set cruise speed.

HEAVY LOAD CONDITION

When the engine is under a prolonged load, e.g. driving up a long steep hill, the manifold vacuum gauge will show a low value. The one-way valve (see Figure 15-1) fitted between the inlet manifold and the vacuum accumulator, closes, blocking the vacuum supply from the manifold.

The vacuum in the accumulator controls the operation of the servo unit, until the manifold vacuum increases to a sufficient value to resume control of the servo unit.

INCREASING CRUISE SPEED

When the vehicle road speed increases by more than 1.5 km/h, the ECM senses this variance and pulses open (to atmosphere) the vent solenoid valve. The pulsing of the vent solenoid valve decreases the vacuum within the control unit.

The decreased vacuum causes the spring-loaded diaphragm and the central rod assembly to move in the direction of the control cable. The tensioned control cable, which is connected to the central rod, allows the throttle linkage and valve to move towards the closed position. This action decreases the engine speed and allows the vehicle to return to the pre-selected road speed.

The cruise system will disengage when any one of the following occurs:

- the brake and/or clutch pedal is depressed;
- a loss of electrical power to the cruise control system;
- the ignition is switched off.

When any of the above actions occur, the vacuum control servo unit goes into the open vent valve position, the vacuum release valve provides an additional vent to atmosphere for the vacuum control unit. The spring-loaded diaphragm, having no vacuum acting on it, moves the central rod partially out of the variable inductance position sensor. Tension on the control cable is released, allowing the throttle valve to close.

There are manual overrides of the system that the driver can select through the control switch assembly. These overrides feature:

- resumption of set cruise speed after system disengagement (this feature is not available when power has been removed from the system);
- increase to the pre-set speed;
- decrease to the pre-set speed;
- 'tap up' or 'tap down' for minor pre-set speed changes.

AUTOMATIC CLIMATE CONTROL

The functions of the automatic climate control system are to maintain a constant temperature within the passenger compartment: and to regulate the air flow from various outlets, as selected by the driver. These functions have been achieved by integrating the vehicle heating system and the air conditioning system. The system also provides for individual preferences, which allows the driver to manually override the fan speed settings and direct air flow to various outlets. When the vehicle is operating in hot sunny conditions, the system compensates for these conditions by providing additional cooling to maintain the desired comfort for the driver and passengers.

The **climate control system** (see Figure 15-15) will automatically provide heating or cooling as required. Sensors are positioned in locations which provide information relative to the operating conditions of the system. This is achieved by using voltage signals from these sensors. The voltage signals are 'inputs' to the ECU. The ECU processes the input signals, refers to the initial settings on the control panel and to a pre-set program in its memory. The ECU decides on a specific 'output' action after processing all the relevant information from its inputs and memory program. The 'outputs' are in the form of switching circuits on and off and energising solenoid valves to control the distribution of vacuum to servo motors. The servo motors control the air distribution and the heater's hot water valve.

SYSTEM COMPONENTS

The automatic climate control system consists of three sections (see Figure 15-16):

- inputs;
- processor (ECU);
- outputs.

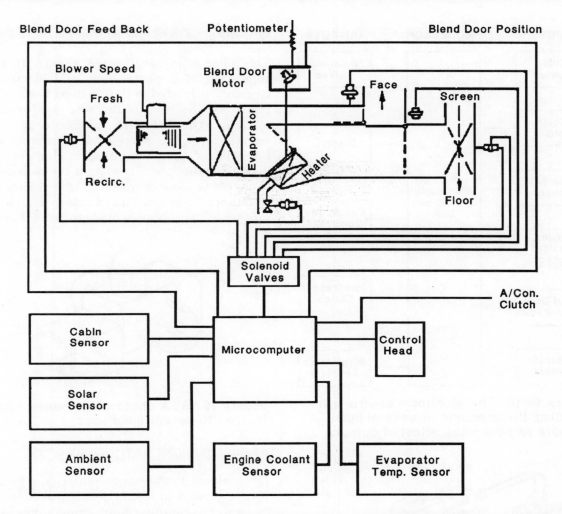

Figure 15-15 Relationship of the components in an automatic climate control system.

INPUTS

The system has many sensors throughout the vehicle to monitor and supply information to the ECU. These sensors are:

- cabin temperature;
- solar load;
- ambient temperature;
- engine temperature;
- blend door feedback;
- evaporator temperature.

CABIN TEMPERATURE SENSOR

The cabin temperature sensor unit (see Figure 15-17), located behind the instrument display panel, consists of a small 'sniffer' motor and an NTC type thermistor. Its function is to measure the air temperature within the passenger compartment. The 'sniffer' motor draws air over the thermistor, to ensure an accurate sampling of the inside air temperature. As the cabin air temperature increases, the resistance of the thermistor decreases.

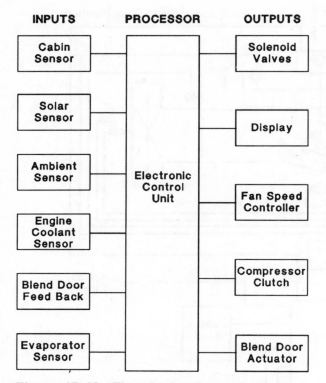

INPUTS	PROCESSOR	OUTPUTS
Cabin Sensor		Solenoid Valves
Solar Sensor		Display
Ambient Sensor	Electronic Control Unit	Fan Speed Controller
Engine Coolant Sensor		Compressor Clutch
Blend Door Feed Back		Blend Door Actuator
Evaporator Sensor		

Figure 15-16 The electronic control unit monitors the condition of several input sensors and provides selected outputs.

SOLAR LOAD SENSOR

The solar load sensor (see Figure 15-18), consisting of a photo-diode and electronic circuitry, is located in the crash pad extension panel. The function of the solar sensor is to measure the intensity of sunlight. The resistance of the photo-diode varies as the intensity of light acting on it changes. By increasing the intensity of light the resistance in the photo-diode decreases. These changes in resistance are measured and sent to the control unit where a decision is made to provide extra cooling.

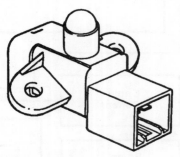

Figure 15-18 A photo-diode based sensor is used to detect solar load.

AMBIENT TEMPERATURE SENSOR

The ambient temperature sensor (see Figure 15-19), located in a safe position outside the passenger compartment, measures the temperature of the 'outside' air. The sensor contains a thermistor, which varies its resistance according to temperature changes. As the ambient air temperature increases, the resistance of the thermistor decreases.

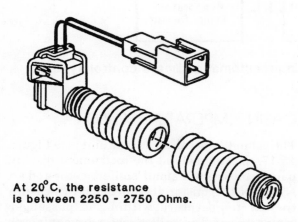

At 20°C, the resistance is between 2250 - 2750 Ohms.

Figure 15-17 Cabin temperature is sensed by an NTC thermistor.

At 20°C, the resistance is between 1900 - 2300 Ohms.

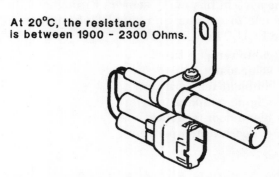

Figure 15-19 An NTC thermistor is used to sense ambient temperature.

ENGINE TEMPERATURE SENSOR

The engine temperature sensor (see Figure 15-20) is screwed into the engine cooling system's thermostat housing. It consists of a threaded housing: having an insulated electrical connector at one end and an NTC-type thermistor at its 'wet' end. As the temperature of the engine's coolant changes, the resistance of the thermistor varies accordingly.

Note: This sensor is specifically for the automatic climate control system and should not be confused with the engine management system's coolant sensor.

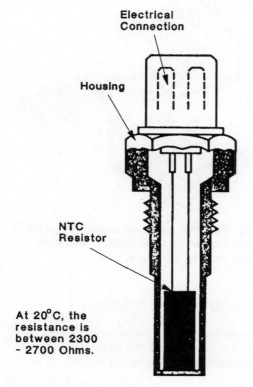

At 20°C, the resistance is between 2300 - 2700 Ohms.

Figure 15-20   A separate engine coolant temperature sensor is used with automatic climate control.

BLEND DOORS POSITION FEEDBACK

The blend doors feedback signal is derived from an integral potentiometer, located in the blend door actuator unit (see Figure 15-21). The potentiometer voltage signal, which corresponds to the blend doors opening position, is sent to the ECU. The ECU refers to this voltage signal when making adjustments to the climate control system.

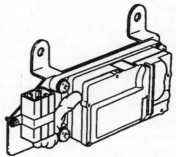

Figure 15-21   The blend door actuator contains a potentiometer to advise the ECU of the blend door position.

EVAPORATOR TEMPERATURE SENSOR

The evaporator temperature sensor (see Figure 15-22) is located at the air conditioner's evaporator housing. It consists of an NTC type thermistor and its function is to measure the temperature of the air at the outlet of the evaporator.

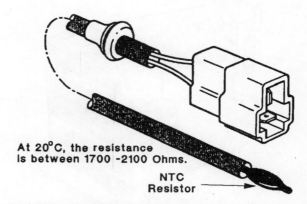

At 20°C, the resistance is between 1700 -2100 Ohms.

NTC Resistor

Figure 15-22   An NTC thermistor is used to measure evaporator temperature.

PROCESSOR

ELECTRONIC CONTROL UNIT

The ECU, which is an integral part of the automatic climate control's display panel (see Figure 15-23), consists of electronic circuitry, a pre-programmed IC (ROM) and a driver programmable IC (KAM). The function of the ECU is to:

- process data from the various input sensors, manual inputs, and blend door actuator sensor;
- store the system settings in memory after the ignition has been switched off;
- monitor the position of the blend door actuator;
- provide control of the air distribution through the various ducts;
- provide a variable fan motor signal;
- provide a voltage signal to the air conditioning system's compressor clutch unit circuit.

OUTPUTS

The outputs from the ECU consist of a:

- vacuum solenoid pack;
- fan motor speed control;
- air conditioning clutch switch;
- blend door actuator signal.

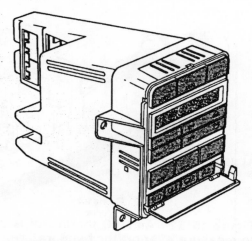

Figure 15-23 The ECU is contained within the climate control display panel.

VACUUM SOLENOID PACK

The vacuum solenoid pack (see Figure 15-24) consists of six electrical solenoids, which control the vacuum supply to the servo motors. Five of the servo motors control the air direction in the cabin. The remaining servo motor controls the heater's hot water valve. The output signals from the ECU turn solenoid valves on or off in response to the various input signals and its pre-set program.

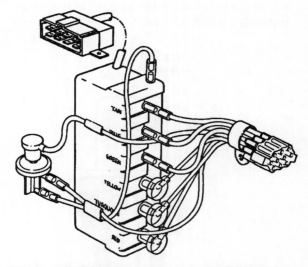

Figure 15-24 The vacuum solenoid pack uses electrical inputs to switch vacuum to various output devices.

FAN MOTOR SPEED CONTROL

The fan motor controller unit (see Figure 15-25) is usually located behind the dash panel.

The function of the fan speed controller unit is to:

- amplify the low current from the ECU to a level high enough to power the fan motor;
- vary the speed of the fan motor;
- provide a speed delay function (this ensures a gradual increase or decrease in speed);
- provide a manual override.

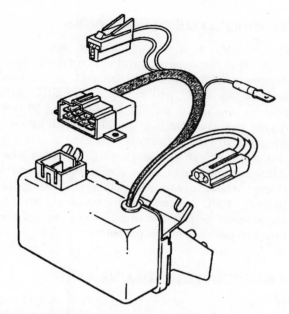

Figure 15-25 The fan is operated by the fan speed controller.

AIR CONDITIONER CLUTCH SWITCH.

The air conditioner clutch switch is an integral part of the ECU. The air conditioning compressor clutch is switched on or off by the ECU in response to the various inputs from the sensors and its pre-set program.

BLEND DOOR ACTUATOR

Three functions of the blend door device (see Figure 15-26) are to:

- mix hot air from the heater and cold air from the air conditioner in proportion to the temperature setting;
- provide cold air only;
- provide hot air only.

The blend door actuator unit consists of a stepper motor, an integral potentiometer and electronic circuitry. The ECU sends signals to the electronic circuitry, which in turn operates the stepper motor. The stepper motor controls the opening and closing positions of the blend door. The blend door position is measured by the integral potentiometer which sends a voltage signal to the ECU. This input signal is monitored

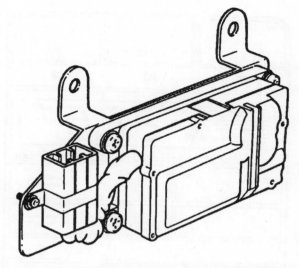

Figure 15-26 Movement of the blend door is provided by the blend door actuator.

by the ECU. If the blend door requires repositioning, an output signal is sent from the ECU to the stepper motor circuitry. The blend door will be moved until the potentiometer feedback signal indicates the door is correctly positioned.

SYSTEM OPERATION

There are four modes, set from the control panel, which determine the operation of the climate control system. These are the:

- automatic mode;
- economy mode;
- manual override mode;
- air discharge override mode.

CONTROL PANEL

The control panel is located within the vehicle's instrument display unit. It consists of control switches, LEDs and an LCD display panel (see Figure 15-27). The control panel enables the driver to:

- turn the system 'on' or 'off';
- vary the default setting for individual preferences;

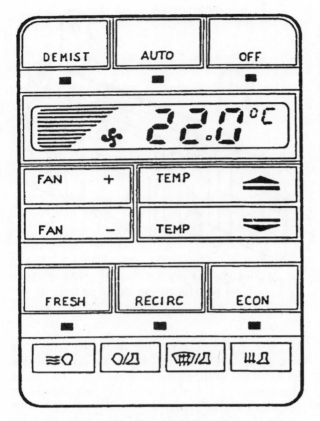

Figure 15-27 Automatic climate control display panel.

- control fan speed;
- direction of the air flow;
- select fresh or recirculated air;
- select auto, economy or manually override the system.

AUTOMATIC MODE

When the climate control system is operating in the fully automatic mode, the comfort settings are retained within the memory of the ECU, therefore they do not require resetting every time the engine is started. The operation in 'auto' mode can be explained by investigating the system's behaviour when the temperature increases above or decreases below the comfort setting.

TEMPERATURE INCREASES

If the cabin temperature sensor detects an increase in temperature and/or the solar sensor detects a high sun load, the voltage signals sent to the ECU will change. The ECU will process the input signals by referring to the driver initial settings in KAM and the pre-set program in its memory. It will then decide on the appropriate output action to lower the temperature, within the passenger compartment, to the driver's initial settings. The outputs from the ECU may switch on the air conditioning system or increase the flow of external air into the passenger compartment.

TEMPERATURE DECREASES

If the cabin temperature sensor detects a decrease in temperature and/or the solar sensor detects a low sun load, the voltage signals sent to the ECU will change. The ECU will process the input signals by referring to the driver's initial setting in the KAM and the pre-set program in its memory. It will then decide on the appropriate output action to raise the temperature, within the passenger compartment, to its initial settings. The outputs from

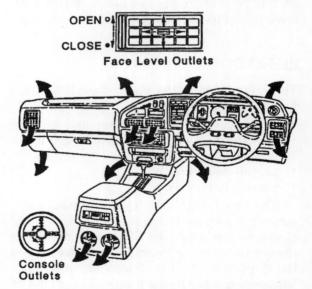

Figure 15-28 The diagram indicates the possible air flow paths.

the ECU may switch off the air conditioner, close the external air blend door or increase the opening of the hot water valve. The resultant heating of the passenger compartment air will restore the temperature to its initial setting. The ECU constantly monitors the system and makes adjustments to ensure the temperature and direction of air flow, within the passenger compartment, is the same as the initial settings.

ECONOMY MODE

The economy mode is selected by pressing the 'econ' button. When the ambient air temperature is above the cabin setting, the air conditioner will be switched on to cool the incoming air.

MANUAL OVERRIDE MODE

The automatic features of fan speed, temperature setting and air circulation can be overridden by the driver using manual selection of the required button. Most systems will revert to the default settings when the engine is stopped then restarted.

AIR DISCHARGE OVERRIDE MODE

In this mode the driver can control the outlets from which air will be discharged. The combination of vents available for selection are typically:

- face outlets and rear console outlets;
- face and floor outlets;
- screen and floor outlets;
- floor outlets only.

DIAGNOSIS AND TESTING

The automatic climate control system usually has a self-test system, which can diagnose electrical faults within the system. The on-board diagnostics described below is typical of the type of fault finding information supplied to the servicing technician. When the ECU detects a fault, in one or more of the sensors or circuits, it logs an error code in its memory and uses a default value for the faulty circuit/system. When the ignition is switched on, any faults detected and stored within the ECU will cause the display panel to flash for approximately 7 seconds. The system's diagnostic mode is selected by pressing the Floor button and the 'off' button at the same time. When in the diagnostic mode the control unit will check:

- the ECU's memory and processing circuits;
- each of the sensors and their circuits;
- the operation of the blend door servo through a full travel cycle.

The diagnostic test procedure will take 30 seconds and during this period all segments of the LCD display and LEDs are illuminated. At the completion of the system tests, the display will show an E (for Error) and a number between 1 and 7. Each error number (E1: E2...) refers to a specific component or system. If more than one fault is detected, the separate error codes can be displayed by pressing the 'auto' button. For more details on error codes consult the manufacturer's service instructions. When no fault is detected, the display will show 'off' and its LED will be illuminated.

16

TESTING AND DIAGNOSING

To test and diagnose faults with electronic components is a complex subject requiring a broad knowledge of individual component and system operation.

In the automotive application of electronics, it is the peripheral components and wiring attached to the microprocessors that cause the most problems. There are basic tests, at the input and output level, that can be carried out on a microprocessor by the servicing technician. Testing the components inside the 'black box' requires a sound knowledge of advanced electronics. Computer repair is a separate field of study and will not be discussed within this text.

A logical method is the best approach to diagnosing and then testing the components. It will save time and protect other components. It is not advisable to use the substitution method with electronic components because the replacement part may be seriously damaged when it is put into the circuit.

DIAGNOSING PROBLEMS

Most problems are indicated to the driver, by a warning light or a 'no operation' situation. When the vehicle is presented for service, the technician should discuss the circumstances related to the problem with the driver.

The technician will use two steps to locate the problem:

1 The fault codes stored in the computer's memory.
2 Examination of the work unit's or sensor's operation.

1 COMPUTER FAULT CODES

When the computer detects a fault, it stores a pre-set code into a special area of memory. This memory can be examined through a diagnostic link attached to computer. Each manufacturer provides the link through a set

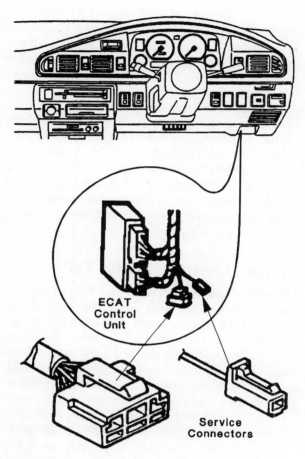

Figure 16-1 A link is made with the computer (ECU) through a set of plugs.

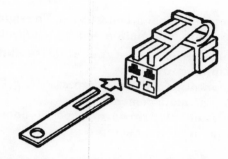

Figure 16-2 A bridging tool is used to 'earth' the computer.

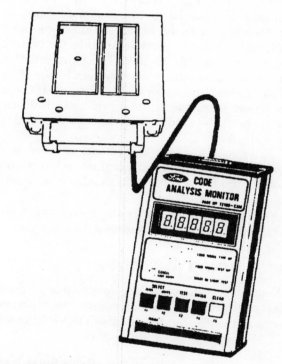

Figure 16-3 A digital display unit connected to the diagnostic plugs of the computer.

of plugs, located in a convenient place, in the wiring harness (see Figure 16-1). Some computers are equipped with red and green LEDs which flash the fault codes.

Most vehicle computers are placed in the diagnostic mode by connecting one of the service link wires to earth. The earth connection may be by a 'jumper' wire or a bridging tool (see Figure 16-2). When any doubts occur as to the wire that has to be connected to earth, consult the appropriate workshop manual.

The diagnostic plugs can be connected to a:

- diagnostic computer;
- special decoder;
- low wattage globe (LED);
- multimeter;
- digital display.

DIAGNOSTIC COMPUTER

A diagnostic computer can be connected to the diagnostic (service) link. When the diagnostic mode is selected, the diagnostic computer will display the code in figures and give the meaning of the code. It will also display the signature of the computer that is being tested. The signature can contain information which includes the:

- microprocessor type and its date of manufacture;
- application of the microprocessor (e.g. ECAT, engine management);
- version of firmware (type of operating system) in the microprocessor.

SPECIAL DECODER

Special decoders are produced by both the vehicle manufacturers and independent equipment suppliers. These devices can be unique to one model or can be produced as a relatively complex device suitable for testing several makes and models. The more complex units can 'drive' the entire system to simulate both normal and abnormal conditions. If these units are available to the servicing technician, they should be used in preference to alternative procedures.

The code appears on the test devices as a pulse. This causes a light to flash, a bar graph to rise and fall, a numerical display or a needle to flick back and forth.

The number of the pulses and the time between each group of pulses give the code.

FLASHING LIGHT TYPE

Assuming a manufacturer sets a code 21 for a low range reading from an engine temperature sensor, the test light would flash twice followed

by a 1.6 second pause, then it would flash once followed by a 4 second pause. To assist with the reading of the pulses, the tens (2 in this case) are held for three times the length of the units (1 in this case).

If only one fault exists, the code would be repeated until the diagnostic link is broken. When more than one code is logged into memory, each code will be 'flashed out' in sequence and then repeated until the links are broken. The code ('21') only indicates that there could be a problem in the engine temperature sensor. The next task is to test the circuit and the component to locate the fault.

TWIN FLASHING LIGHT TYPE

The previous method used an LED to flash the code. This decoder uses two LEDs. One LED is red and the other is green. The red LED flashes the tens and the green LED flashes the units. To flash a code 13, the red LED would flash once followed by a pause and the green LED would flash three times. After a slightly longer pause the code would be repeated. Some computers have LEDs, like these, built into their case (see Figure 16-5). They work in the same manner, when the diagnostic mode has been selected.

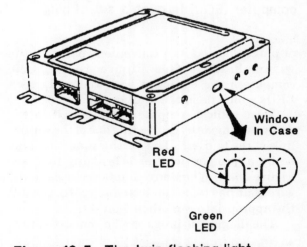

Figure 16-5 The twin flashing light decoder comprises two LEDs mounted on the vehicle's ECU board inside the case. A small window, in the case, allows the LEDs to be observed.

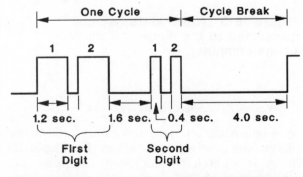

Figure 16-4 A graphic representation of the malfunction code 22.

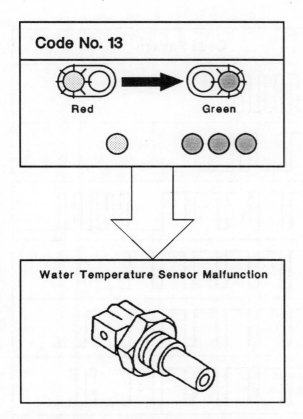

Figure 16-6 A graphic representation of a twin light decoder flashing a malfunction code 13.

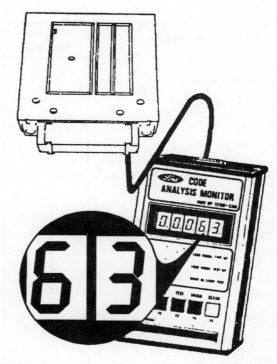

Figure 16-7 A direct fault code reading is shown on a digital display unit. In this instance, it is a code 63.

DIGITAL DISPLAY TYPE

The digital display (see Figure 16-7) shows the fault code directly on the display. A code 63 would appear as figures on the display.

MULTIMETER

The multimeter, when used to read codes, must be set to its voltage scale.

A dial type meter uses the needle to give the code (see Figure 16-8). The needle rises with each pulse. To indicate a code 12, the needle would rise, pause slightly, and fall. It would pause again and two short rises would follow the first but these pulses would be quick. A long pulse would be followed by a repeat of the code.

A digital multimeter uses a bar graph to flash the code in a similar manner to that described above.

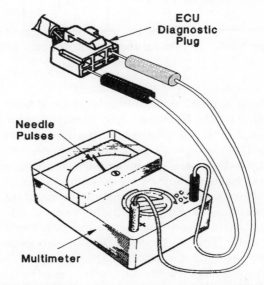

Figure 16-8 An analogue type multimeter can be used to read fault codes logged into the ECU.

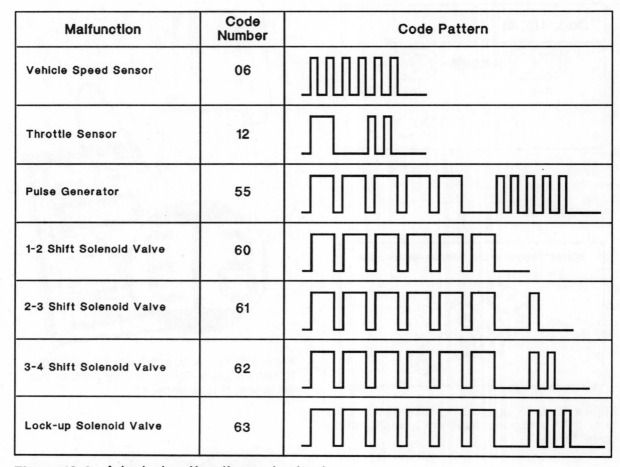

Figure 16-9 A typical malfunction code sheet.

The diagnostic information may have been supplied by the flashing LED, moving needle, bar graph or the digital display. This information can be translated to a number which represents a fault code. There are no uniform fault code sheets. Each manufacturer supplies a set of code sheets to cover the range of microcomputers in their vehicles. Figure 16-9 illustrates a code sheet for an electronic controlled automatic transaxle. This code sheet shows the code number and the malfunction description.

To successfully diagnose faults in electronic circuits, a code sheet is essential. Consult the appropriate manufacturer's workshop manual to obtain these code sheets.

2 WORK UNIT OR SENSOR OPERATION

This method requires the use of sight, hearing and feel to diagnose a problem in an electronic circuit. Although electric current cannot be seen, its effects can be detected by observation. Parts move, lights glow, horns and motors make sound, and heater elements give off heat. A knowledge of the work that a component does, when it is in good operating condition, is essential.

There are some components that do not display any detectable sign that they are working. An example of this is a water temperature sensor. The sensor changes resistance as the temperature changes. The sensor operation cannot be seen, heard or felt. However, the physical side of its circuit can be quickly checked by looking for loose or broken wires, wriggling the connectors and touching the sensor body to detect that temperature change has occurred.

Caution: Some components will be very hot, so care must be taken when feeling is being used to detect a problem.

When a work unit or sensor fails to operate correctly, the circuit will have to be tested to determine whether it is the component or the wiring at fault.

CIRCUIT TESTING

The list of equipment needed to test a circuit includes:

- voltmeter;
- ammeter;
- ohmmeter;
- set of small spanners to loosen the terminals;
- set of small 'jumper' leads fitted with alligator clips;
- wiring diagram;
- specifications on the work unit or sensor.

The voltmeter, ammeter and ohmmeter are generally combined into a multimeter. These meters must have a very high input impedance to protect the electronic components under test from excessive current flow.

TESTING PROCEDURE

The steps to determine a problem with a circuit are to:

- establish the nature of the circuit;
- connect the meters to the circuit;
- observe and record the meter readings;
- compare the meter readings with manufacturer's specifications;
- decide whether the work unit or sensor require removal for further testing or renewal;
- test the wiring in a circuit;
- check the voltages at the computer's inputs and outputs.

NATURE OF THE CIRCUIT

The information about the circuit is very important and must include:

- all connection points;

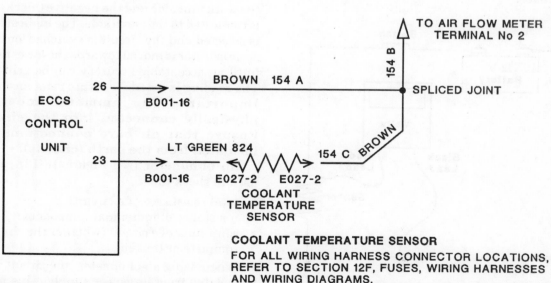

Figure 16-10 A manufacturer's wiring diagram provides important information about a particular circuit.

- applied voltage;
- current flow;
- work unit or sensor specifications.

This information is obtained from the various sections in the appropriate workshop manual.

The wiring diagram will supply the connection points and the colour coding of the wires. The colour code for the full length of the wire must be noted because the wire may change colour when it has passed through a connector or has been connected into another wire.

The specifications of the work unit or sensor will give the voltage, current rate, wattage, resistance or frequency values. It must be remembered that these specifications apply to a work unit or sensor that is operating correctly.

CONNECTING METERS

A **voltmeter** is the easiest of the meters to connect to a circuit (Figure 16-11). Its positive (red) lead is connected to a terminal on the work unit or sensor and the negative (black) lead is connected to earth (chassis). The highest scale on the meter is selected and the circuit is switched 'on'. The scale selector is moved toward the lowest scale until an acceptable accuracy can be achieved. The reading is noted for future reference.

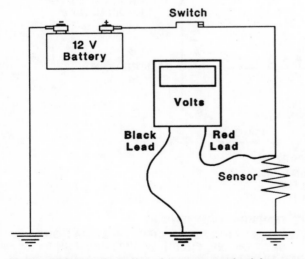

Figure 16-11 A voltmeter connected to an electrical circuit.

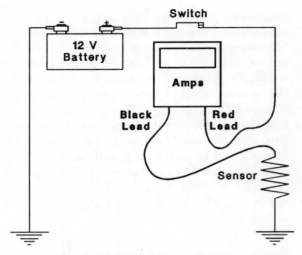

Figure 16-12 An ammeter connected into an electrical circuit.

An **ammeter** must be connected into a circuit (see Figure 16-12). A wire, in the circuit, can be disconnected at any convenient place to provide a connection point for the ammeter. If the wire between the battery and work unit has been disconnected, the positive (red) lead is connected to the battery side and the negative (black) lead is connected to the work unit side. If the wire between the work unit and earth has been disconnected, the positive (red) is connected to the work unit side and the negative (black) lead is connected to the earth side. The highest scale is selected and the circuit is switched 'on'. The scale selector is moved towards the lowest scale until an acceptable accuracy can be achieved. The reading is noted for future reference.

Important: The ammeter must be physically connected into the circuit. Ensure that all bare connections are insulated from the earth (chassis).

An **ohmmeter** (see Figure 16-13) can be used to check the:

- total resistance of a circuit;
- resistance of individual components;
- continuity of a circuit (whether the circuit is complete or broken).

When using an ohmmeter, the circuit must be isolated from its power supply. This meter has its own power supply which may be as high

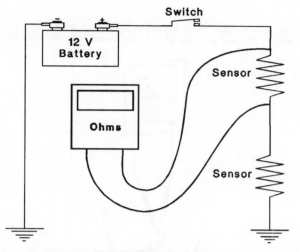

Figure 16-13 An ohmmeter connected across a component in an electrical circuit.

as 15 volts. Care must be taken to ensure that this meter is not connected across a voltage-sensitive component because it could cause permanent damage to the component.

The meter is prepared by selecting the highest scale, placing the two test leads together and adjusting the zero reading (see Figure 16-14). After the component has been disconnected from the circuit, the two test leads are placed on its terminals. The meter reading is noted. To improve the accuracy of the reading, a lower scale can be selected but remember that the zero must be adjusted for each scale.

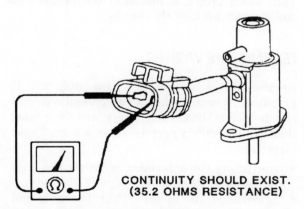

CONTINUITY SHOULD EXIST.
(35.2 OHMS RESISTANCE)

Figure 16-14 An ohmmeter used to check a solenoid for continuity and resistance.

READING METERS

The differences with reading meters are related to the types of display. An analogue meter has a needle and dial, and a digital meter has light emitting diode (LED) or liquid crystal display (LCD).

When an analogue meter is being used:

- check and adjust its zero;
    —the zero is adjusted by carefully turning the screw located at the base of the needle;
    —when the zero cannot be adjusted, note the error reading;
- take the reading while observing the dial from a position directly above the needle;
- subtract the zero error, if necessary, from the reading.

The numerical value can be read directly from the display of a digital meter but do not forget the scale factor. The scale factor shows as a small letter in one corner of the display. An 'M' indicates that the reading must be multiplied by one million. A 'k' indicates that the reading must be multiplied by one thousand.

COMPARING THE READINGS WITH SPECIFICATIONS

The manufacturer's specifications include a tolerance. Tolerance means there is a range, between upper and lower values, in which a reading is acceptable. When the reading is outside this range, the component is faulty. An example could be an engine coolant temperature sensor that has a resistance reading of 1500 ohms at 20 degrees Celsius. The manufacturer's specification for the sensor at 20 degrees Celsius is 2500 ohms plus or minus 400 ohms. This specification means that the upper value must not be more than 2900 ohms and the lower value must not be less than 2100 ohms. Assume the sensor is still installed in the engine when the test readings are taken at the computer plug (see Figure 16-15). The reading recorded is below the limit, therefore either the sensor or the circuit is faulty.

The readings that have been taken from the meters and noted are compared with the

Disconnect the 16 pin connector at the ECU and measure the resistance between terminals 23 and 26.

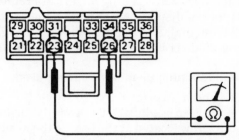

Cylinder Head Temperature	Resistance
Above 20° C	Below 2.9 kΩ
Below 20° C	Above 2.1 kΩ

Figure 16-15 The diagram shows an ohmmeter being used to measure the resistance of a coolant temperature sensor via its wiring harness.

manufacturer's specifications. Because the readings are outside the range, further testing should be carried out and a decision made on the serviceability of the component.

DECIDING TO RENEW COMPONENT

By testing the component in isolation, closely controlled conditions can be maintained throughout the tests.

Assuming that the same engine temperature sensor has been removed from the engine, the temperature can be properly monitored during the test. The sensor's resistance can be measured at several temperature values and compared with the specified range (see Figure 16-16). When the readings are outside the range, the component is unserviceable and it must be renewed.

When the component has passed the tests but the fault still remains, the problem could be associated with the circuit's wiring.

Dip the sensor into water maintained at a temperature shown on the chart and read its resistance.

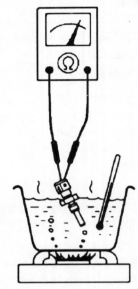

Water Temperature	Resistance
20° C	2.1 - 2.9 kΩ
50° C	0.68 - 1.0 kΩ
80° C	0.26 - 0.39 kΩ
100° C	0.18 - 0.20 kΩ

Figure 16-16 Controlled conditions allow a more accurate set of readings to be taken. By comparing these readings with a manufacturer's chart, a decision on its serviceability can be quickly made.

TESTING THE WIRING

As previously stated, a wiring diagram is essential to ensure that a component's circuit can be correctly tested. There are four tests that can be quickly performed on a wire. These tests are:

- continuity (to determine whether the wire is broken);
- resistance (to ensure that there is not excessive resistance in the wire);
- short to earth (to ensure that the wire's insulation has not rubbed through and its core is not touching a metal part);

- short to another wire (to establish whether the core of one wire is touching the core of another wire).

An ohmmeter, set on a low scale, is used for the four tests. If ohmmeter testing is to be effective, the wire must be disconnected from all components in the circuit. The wiring circuit diagram should also be checked to ensure that a second circuit is not linked to the first wire within the wiring loom. Either condition, the wire not disconnected or a second circuit, will make the meter reading invalid.

The testing steps are:

1. Place one test probe into one end of the wire and the other test probe into the other end of the wire.

   Note: This may require the use of jumper leads, but ensure that the jumper leads have been tested first.

2. Observe and record the reading.

   When there is an infinity reading (on some meters—overload or full scale deflection), the wire lacks continuity (open circuit) and must be repaired or renewed.

   When the reading is greater than one ohm, there is too much resistance in a join or terminal connection.

3. Remove one test probe and place it on a metal part of the chassis.

   When the reading is low, the wire has a short to earth and it must be repaired.

4. Remove the test probe from the chassis and place it into a terminal in the same socket of the wire being tested.

   When there is a reading on the meter, a short has occurred between the cores of two wires and a repair to the wires must be performed.

   Repeat this for other wires in the harness.

CHECK VOLTAGES AT INPUTS AND OUTPUTS OF THE COMPUTER

The reason for checking voltages at the various input and output pins of the computer is to establish whether the computer is causing the problem (see Figure 16-17). It must be remembered that the computer will make its decisions on the information received from the inputs. When the inputs are incorrect, the computer may send wrong values to the output. One method of determining whether inputs are correct is to measure their voltages at the computer socket. If the input voltages are correct and the output voltages are incorrect then it may be assumed that the computer is faulty.

The equipment needed to check the voltage at the inputs and outputs of a vehicle's computer is a:

- diagram of the pin configuration of the input and output plugs and sockets;
- chart of the voltage levels and the condition to which they apply;
- multimeter—must be high impedance type;
- set of small jumper leads.

After selecting the voltage scale on the multimeter, the voltages are checked systematically at each pin on the input plug (see Figure 16-18).

The leads of the meter must be connected as shown on the manufacturer's chart and these pins must be probed from the back of the socket (see Figure 16-19). Back probing a pin in the socket will prevent damage to the connection area of the pin.

As each reading for the condition indicated on the chart is observed, the voltage value must be compared with the stated value on the chart and it must be within the limits. When the voltage falls outside the limits, the circuit being tested must be carefully examined to locate the fault.

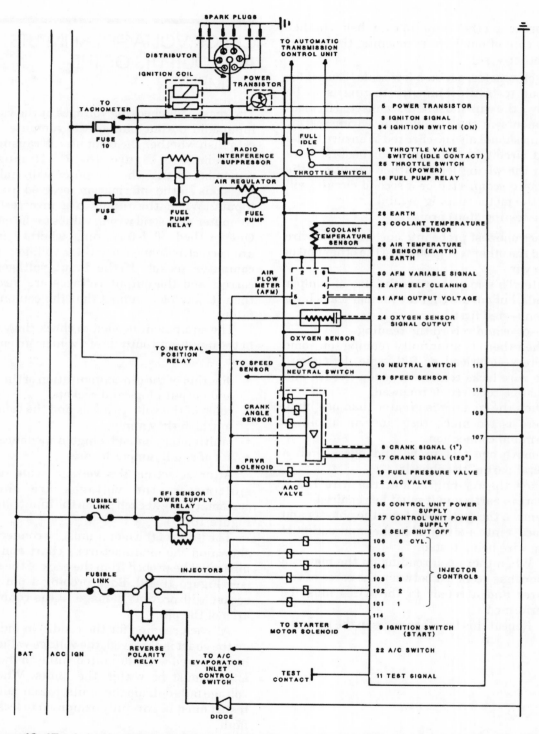

Figure 16-17 A typical manufacturer's diagram of the inputs and outputs to and from the electronic control unit.

CONTROL UNIT TERMINAL NUMBERS

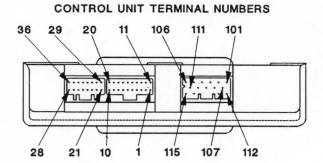

Figure 16-18 The diagram shows the configuration of the input and output sockets of the ECU. The designation of these pin numbers is found on the manufacturer's chart.

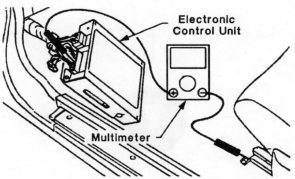

Figure 16-19 Back probing a pin of an ECU plug or socket will prevent pin damage and allow a reading to be taken while the system is operating.

TEST NO.	TEST	TERM-INAL NO.	IGN. SWITCH	COMMENTS	CONNECTION POINTS AND TESTER RANGE	CORRECT TEST VALUE
17	AIR REGULATOR	16	ON			Battery Voltage (After 5 Seconds)
18	CRANK ANGLE SENSOR	35, 36	ON	Voltage Supply	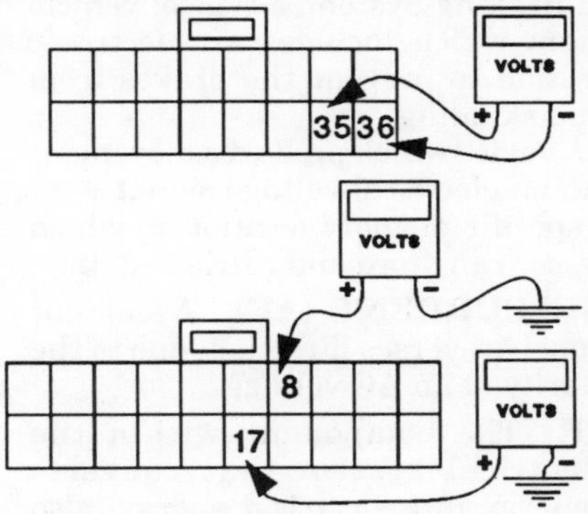	Battery Voltage
		8	Start Engine	1° Slot		IGN On - 5 Volts. At Idle - 2-3 Volts.
		17		120° Slot		IGN On - 0.5 Volts. At Idle - 0.2 Volts.
19	POWER SUPPLY TO CONTROL UNIT	35	ON		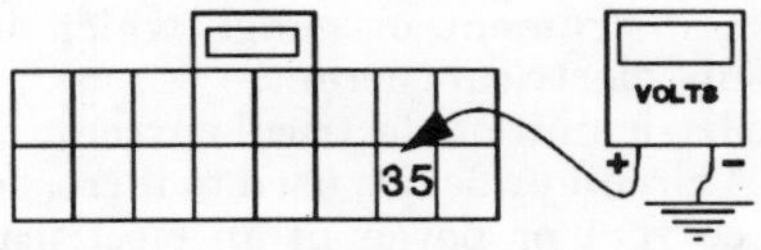	Battery Voltage
20	ECU OUTPUT TO POWER TRANSISTOR	5	Start Engine		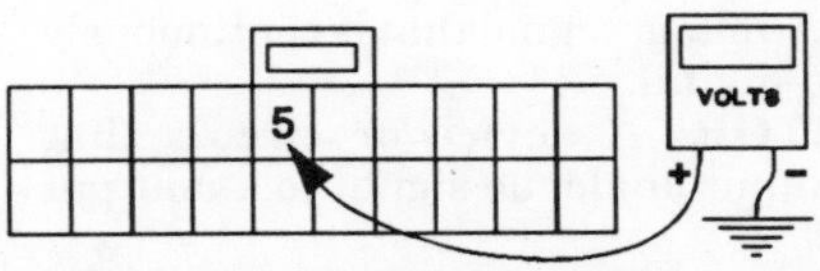	Approx. 0.5 Volts

Figure 16-20 A typical manufacturer's chart showing the pin numbers in the sockets, setting the conditions for the test and providing the correct values for each test.

GLOSSARY

ABS Anti-skid Braking System, a type of vehicle braking system which includes an electronic controlled section to prevent the brakes from locking up and skidding.

ACTUATOR A device which produces movement in response to an electrical voltage signal.

ADDRESS A specific memory location to which a microprocessor can store and retrieve data.

ALTERNATING CURRENT (AC) A current that continuously reverses direction, due to the changing polarity of an AC voltage.

ALTERNATOR The component within the vehicle's electrical system that converts mechanical energy into electrical energy, also known as an AC generator.

AMMETER An instrument or gauge which is used to measure electrical current.

AMPERE The basic unit of electrical current.

AMPLIFIER A circuit or device used to increase the voltage, current or power of an electrical signal.

ANALOGUE A voltage signal that is continuously variable, non-digital.

A/D CONVERTOR A device or circuit that converts an input analogue signal to an output digital signal.

ANODEM The negative electrode/terminal of a component.

ARCM A spark caused by an electrical current flowing across an air gap.

ARMATURE A part moved by or through a magnetic field, to produce motion or to complete a circuit. The rotating component of an electric motor.

BAC Bypass Air Control valves are fitted to electronic fuel injection systems to control the idle speed of an engine.

BALLAST RESISTOR A resistance in the ignition primary circuit to reduce battery voltage to the coil during engine operation.

BATTERY An electrochemical device that is used for storing energy in a chemical form so that it can be released as electricity. A group of electric cells interconnected within a housing.

BIMETALLIC STRIP A strip consisting of two dissimilar metals, each having a different coefficient of expansion. Changes in temperature causes the bimetallic strip to bend.

BDC Refers to the position of a piston in its bore when the piston is close to the crankshaft.

BRUSH A carbon conductor which makes contact with a moving surface to make a good electrical connection, such as a slip ring on an alternator rotor.

BTDC Before Top Dead Centre.

CALIBRATE To adjust a meter to a set position. To adjust or restore the setting of a component to its original value.

CANDLEPOWER The standard unit of light intensity.

CANDELA The metric unit of light bright-ness.

CAPACITANCE The ability to store elec-trons.

CAPACITOR A component used to temporar-ily store an electrical charge.

CATHODE The positive electrode/terminal of a component.

CHASSIS The main frame of a micro-computer used for housing and attaching components.

CHIP One or more miniaturised electronic integrated circuits combined in a small package.

CHOKE An induction coil used to 'choke' or 'block' high frequencies.

CIRCUIT BREAKER A thermal or electro-mechanical switch that causes an 'interrupt' or 'break' in a circuit when the current flow exceeds a predetermined value.

CIRCUIT PROTECTION A device within an electrical circuit to protect against 'circuit overloading', e.g. fuses, circuit breakers and voltage regulators.

CLOSED LOOP Refers to an electronic controlled system in which the output of the system is monitored and the inputs to the system are adjusted to achieve the required results from the system's output.

COIL A spiral of wire—ignition coil a 'step up' transformer.

COMPUTER A device which receives information, processes it, makes decisions and outputs those decisions.

CONDENSERM A common name within the automotive industry for a capacitor. A condenser is located within the primary circuit of the ignition system; it is connected across the points to reduce arcing.

CONDUCTORM Any material or substance that allows current to flow easily.

CONNECTOR A device or housing that allows wires/cables to be connected and disconnected.

CONTACT BREAKER POINTS The stationary and movable points within the primary circuit of the ignition circuit; their function is to make and break the current flow through the primary circuit.

CPI Centre Point Injection systems are electronic controlled systems that have one or two petrol fuel injectors located in a body attached to the inlet manifold in the place of the carburettor.

CPU Central Processing Unit, the main processing unit within a computer which decides the output action as a result of the input data.

CRANKSHAFT POSITION SENSOR A device (sensor) which relays the actual position of the crankshaft to a micro-processor.

CRUISE CONTROL A vehicle speed control system which can be set to a speed determined by the driver; the system incorporates safety override features.

CYCLE A series of events that repeat themselves.

CRYSTAL SENSOR A type of sensor that uses a crystal to vary the resistance, commonly used in MAP sensors and engine knock sensors.

CURRENT A flow of electrons measured in amps.

CUTOUT A device that switches off the current flow within a circuit; it can be voltage or current controlled.

DETONATION An uncontrolled second explos-ion within the engine's combustion chamber, commonly known as 'pinging'.

DFI Direct Fire Ignition—a type of ignition system that does not use a distributor.

DIALECTIC The insulating material between the plates of a capacitor.

DIGITAL A signal that has two states: 'On' or 'Off'.

DIODE A solid state device that allows current to flow in only one direction; an electronic one-way valve.

DIRECT CURRENT (DC) Current which flows in one direction and does not alternate.

DISTRIBUTOR A component of the ignition system; its function is to direct the high voltage surge from the coil to the spark plugs at the correct time. It also houses the rotary switch for the primary circuit.

DUTY CYCLE Refers to a percentage of one complete cycle, e.g. a complete cycle for an injector is when it is closed (Off), open (On), and closes again. The percentage time period when the solenoid is open (On) is referred to as the duty cycle. If an injector was open for half of a complete cycle it would have a 50 per cent duty cycle.

DWELL ANGLE The number of degrees in which the contact breaker points are closed.

DWELL PERIOD The time period in which the contact breaker points are closed.

ECA Electronic Control Assemble is the name for the ignition module located on the distributor body.

ECM Electronic Control Module—the system computer that controls the input, processing and output functions.

ECUM Electronic Control Unit. *See* ECM.

EFI Electronic Fuel Injection—a petrol fuel system that has been introduced so that vehicles can meet the emission standards set down by the EPA.

EGR An Exhaust Gas Recirculation device is fitted between the inlet and exhaust manifolds of an engine to permit a metered amount of exhaust gas to enter the combustion chamber to cool the combustion process.

ELECTRODE An electrode can be positive or negative. In a spark plug there is a positive central electrode and the negative electrode is the section welded to the spark plug casing.

ELECTROLYTE The mixture of sulphuric acid and water used within a battery.

ELECTROMAGNET Magnetism is formed by a current flowing through a coil of wire wound around an iron core.

ELECTROMECHANICAL Refers to a device that uses electronic and mechanical principles in its operation.

ELECTRONIC Refers to any component that uses solid state semi-conductive devices and circuits.

ELECTRONS The negatively charged particle of an atom.

EMITTER One of the three terminals of a transistor.

ENERGY The ability or capacity to do work.

EPA Environment Protection Authority.

EST Electronic Spark Timing is an electronic controlled ignition system, in which sensors provide information on engine load and crank position to ensure correct ignition timing.

EXHAUST GAS OXYGEN SENSOR A component that provides information to the ECM on the oxygen content of the exhaust gases.

FARAD The unit of quantity in the measure-ment of electricity, commonly used in determining the 'capacity' of capacitors.

FUEL/AIR RATIO *See* Stoichiometric.

FUSE A thermal protective device which opens when current becomes excessive for the rated value of the fuse.

FUSIBLE LINK A type of circuit protector in which a special wire melts to open when the current becomes excessive.

GENERATOR A component which converts mechanical energy to electrical energy.

GERMANIUM A semiconductor material used in the manufacture of diodes and transistors.

HALL EFFECT The voltage produced across the edges of a semiconductor strip which is carrying a current and located in a magnetic field.

HEAT SINK A device for absorbing and trans-ferring heat, usually made from an alloy, diodes and transistors are mounted to the heat sink.

HERTZ (Hz) The unit of frequency. One hertz equals one cycle per second.

HIGH TENSION VOLTAGE The voltage produced by the transformer action of the coil in an ignition system, it is transferred through 'high tension' leads to the spark plugs.

HYBRID ELECTRONIC UNIT A type of unit which contains a combination of analogue and digital systems.

IGNITION COIL A step up transformer used to increase the voltage in an ignition system.

IGNITION MODULE The electronic unit used in ignition systems to control the dwell and switching of the primary circuit.

IMPEDANCE The total opposition to the flow of alternating current in a circuit containing resistance and reactance.

INDUCTANCE Inducing an electromotive force in a conductor.

INSULATOR A material that does not conduct electrical current.

INTEGRATED CIRCUIT (IC) A small sem-iconductor chip that contains com-ponents and circuitry, and can be used for many applications.

INTERNAL RESISTANCE The resistance contained within a battery, amplifier or power supply.

JOULE The unit of work and energy.

KAM Keep Alive Memory is powered by the vehicle battery and is part of the micro-computer; it stores information on the input sensors. It can identify intermittent 'Input' failures and assists in system diagnostics. KAM adopts some calibration parameters to compensate for changes within the vehicle's operating system.

KAPWR Keep Alive Power refers to connect-ion between the battery and the computer that is not interrupted by a switch, thus ensures a constant voltage is present at all times.

KNOCK SENSOR A component fitted to the engine crankcase; its function is to sense abnormal engine vibrations and provide information on engine vibrations to the engine control system.

LAMINATED CORE A stack of soft iron pieces used as the core of an armature or electromagnet.

LCD Liquid Crystal Display used in digital instrument displays and in some test equipment displays.

LED Light Emitting Diode—a type of diode that emits light when current is passing through it.

LIMP HOME This is a mode that a computer controlled management system will initiate, to allow the vehicle to be driven, should a malfunction occur in one or more of the vehicle's electronic controlled systems.

MAGNETIC Having the ability to attract iron, it can be by a current flow, known as electro-magnetic, or it can be permanent.

MAP Manifold Absolute Pressure—the pressure within the inlet manifold of an engine.

MEGA A prefix designating a million (M).

MICA A non-magnetic, non-conductive insulating material.

MICROCOMPUTER A miniature computer which receives information, usually from sensors, processes this information, makes decisions and outputs these decisions.

MPI Multi-point Injection systems have one petrol fuel injector for each cylinder of the engine, which is electronically controlled.

MULTIPLEXER A logic circuit which accepts several data inputs and allows only one input at a time to get through to the output.

MULTIMETER A test instrument designed for measuring two or more electrical quantities.

MUTUAL INDUCTANCE The inductance between two separate coils that share a common magnetic field.

NEGATIVE One terminal of any electrical device, the negative terminal of a battery is of smaller diameter and is indicated by a minus sign.

NOVRAM Non Volatile Random Access Memory—a memory integrated circuit which constantly stores and updates data. It can be used to constantly update the reading of an odometer and retains this information, without the use of a battery, for more than ten years.

NTC Negative Temperature Coefficient, the ability of a material to decrease its resistance as its temperature increases.

OHM'S LAW The law of electricity determin-ing the relationship between volts, amps and ohms.

OPEN LOOP An electronic control system in which there is not any feedback from the output, to the control system.

OXYGEN SENSOR A device fitted to the exhaust system, which measures the amount of oxygen in the exhaust gas. It sends this information as a varying voltage signal to the microprocessor.

OSCILLOSCOPE A high speed voltmeter, which displays on its screen, the intensity of voltage variations in the shape of wave forms.

PARALLEL CIRCUIT Two or more circuits formed in a way that their positive and negative terminals share a common source and can operate independently of each other.

PERMEABILITY A measure of the ability of a material to establish magnetic lines of force.

PHOTO DIODE A diode specifically designed for light detection.

PIEZOELECTRIC A method of producing a voltage by applying a pressure to a crystal.

PIEZORESISTIVE The inbuilt resistance within a piezoelectric device.

PIP Profile Ignition Pick-Up is identified in a type of Ford electronic ignition system. It is part of the Hall Effect switch.

POLARITY Refers to the positive and negative terminals in an electric circuit and to the north and south poles of a magnet.

POTENTIOMETER A variable resistor having three connections; its function is to vary the resistance applied to a circuit. A meter used to make accurate voltage measurements.

POWER AMPLIFIER A device which increases the power of an electrical signal by using an active component, such as a transistor.

PRIMARY WINDING The outer winding of an ignition coil, consisting of many turns of a relatively heavy copper wire.

PRINTED CIRCUIT BOARD A process of applying a conductive material to an insulating board in a pattern which provides for circuits between components mounted to the board.

PROCESSOR A component which houses a microcomputer and other electronic devices; its function is to control an electronic system.

PTC Positive Temperature Coefficient—the ability of a material to increase its resistance as its temperature increases.

PULSE GENERATOR A electromagnetic device which can be used as a type of sensor in an electronic control system. It consists of a coil and a permanent magnet. It generates a voltage signal, used in electronic ignition systems, as speed sensors and crank position sensors.

PULSE TRANSFORMER *See* pulse generator.

RAM Random Access Memory—a type of memory in a computer, which allows temporary storage of data. The data can be written to or read from RAM.

REACTANCE The opposition to the flow of alternating current by capacitance and/or inductance.

RECTANGULAR WAVE FORM A digital on/off type signal which produces a wave form which takes the shape of a rectangle or square.

RECTIFIER An electronic device or circuit which converts AC to DC.

RELAY A switching device which uses electro-magnetism to open or close one or more circuits.

RELUCTANCE Is the ability of a material to oppose the establishing of magnetic flux.

RESISTOR An electrical component which has an ability to resist current flow.

RESISTANCE The ability of a material to resist current flow.

RHEOSTAT A two terminal variable resistor used to vary current flow within a circuit.

ROM Read Only Memory—a type of memory in a computer, which is used to store data permanently. Data can be read from, but not written to ROM.

ROTARY SWITCH A type of switch having a number of terminals. A central rotary control knob/lever can switch on and off the various circuits connected to the terminals.

ROTOR A revolving part of a machine, alternator rotor, distributor rotor.

SCR *See* Thyristor.

SEMI CONDUCTOR A material which acts like a conductor under some conditions and as a resistor under other conditions.

SECONDARY WINDING The inner winding of an ignition coil, consisting of many turns of a relatively light copper wire.

SENSOR A component which measures an operating condition and provides an electrical input signal to a microcomputer.

SERIES CIRCUIT A type of circuit having two or more components, through which the current has only one path to flow.

SHORT CIRCUIT A defect in an electrical circuit which allows the current flow to take a 'shortcut' instead of following the prescribed path.

SHUNT A parallel circuit, used to divert current flow around a meter.

SILICON A semiconductor material used in the manufacture of electronic components.

SINE WAVE An alternating waveform whose value is related to the trigonometric sine function of time or angle.

SOLENOID COIL A coil having a movable plunger which operates by electromagnetism.

SOLID STATE Having no moving parts, transistors and diodes are examples of solid state devices.

SPOUT The spark out signal sent by the electronic control assemble to the thick film integrated (TFI) ignition module.

SPST Single Pole Single Throw—identification of a type of switch.

SQUARE WAVE SIGNAL A digital on/off signal having a rapid rise and fall, which takes the shape of a square or rectangle.

STEP DOWN TRANSFORMER A type of transformer in which the output voltage is less than the input voltage.

STEP UP TRANSFORMER A type of transformer in which the output voltage is more than the input voltage.

STEPPER MOTOR A special type of electric motor, fitted with an movable linkage and normally ECU controlled. The function of a stepper motor is to provide accurate positioning of the moveable linkage.

STOICHIOMETRIC Chemically balanced air/fuel mixture to achieve the best fuel economy (15 parts air to 1 part petrol, by mass).

SWITCH A device which opens and closes an electric circuit.

TBI Throttle Body Injection systems are electronic controlled systems that have one or two petrol fuel injectors located in a body attached to the inlet manifold in the place of the carburettor.

TDC Refers to the position of a piston; its bore when the piston is closest to the cylinder head.

TERMINALS Electrical connecting points of components.

TEST LIGHT A test instrument which allows a lamp to be illuminated when the circuit is complete.

TFI Thick Film Integrated ignition, a type of electronic ignition system commonly used by Ford.

THERMAL Pertaining to heat.

THERMISTOR A type of resistor having a NTC, commonly used as an engine temperat-ure sensor.

THROTTLE POSITION SENSOR (TPS) An electronic sensing device used in an engine control system; its function is to provide information to the ECU of the position of the throttle plate.

THYRISTOR A semiconductor device commonly used as a solid state switch; the two main types of thyristors are SCR and TRIAC.

TRIAC *See* Thyristor.

TRANSDUCER A device that converts a variable, such as speed, pressure or temperature, to an electrical signal.

TRANSFORMER An electrical device which by using electromagnetic induction, transfers electrical energy from a primary winding to a secondary winding. This transference can result in an increase or decrease to the AC voltage.

TRANSIENT A sudden temporary change in circuit conditions from one steady state to another.

TRANSISTOR A semiconductor component, usually having three terminals, used for electronic switching or amplification.

VLSI Very Large Scale Integration, is a type of digital IC containing a very large collection of resistors, diodes and transistors in the one chip.

VOLTAGE Electrical pressure.

VOLT DROP The reduction of voltage across a resistance.

VOLT REGULATOR A device used to regulate the voltage in a circuit, commonly used in alternator charging circuits and some instrument circuits.

VOLTMETER An instrument or gauge which is used to measure electrical pressure (voltage).

WATT A unit of measurement for electrical power. Volts multiplied by Amps equals Watts.

WAVE FORM The shape of a wave as shown on an oscilloscope screen when instantan-eous values of a varying quantity are plotted against time.

WHEEL SENSOR A type of sensor used in an ABS system. The function of this sensor is to determine the speed of a road wheel, convert the speed to an electrical value and send this value to the ECU.

WINDING The turns of coiled wire wound around a transformer or inductor.

WIPER The movable contact in a potentio-meter.

ZENER DIODE A type of diode which will reverse bias when it goes higher than the design voltage.

INDEX